Next-Generation Hydrogen Economy

Production, Storage, and Fuel Cell Technologies

Dr. Peeyush Phogat

Dr. Shreya Sharma

Satyam Rawat

WILEY VCH

Authors

Dr. Peeyush Phogat
NIScPR
H.No-2176, Sector-13, HUDA
Bhiwani, Haryana
India

Dr. Shreya Sharma
NIScPR
H.No-696, Sector-23, HUDA
Bhiwani, Haryana
India

Satyam Rawat
Netaji Subhas University of Technology
Sec-3, Dwarka
New Delhi
India

Cover Design: Wiley
Cover Image: © Just_Super/Getty Images

Typesetting: (Lumina Datamatics, Madurai, India)
Printing and Binding: CPI Group (UK) Ltd, Croydon, CR0 4YY

Library of Congress Card No.: applied for

British Library Cataloguing-in-Publication Data A catalogue record for this book is available from the British Library.

Bibliographic information published by the Deutsche Nationalbibliothek
The Deutsche Nationalbibliothek lists this publication in the Deutsche Nationalbibliografie; detailed bibliographic data are available on the Internet at <http://dnb.d-nb.de>.

Print ISBN: 9783527356010
ePDF ISBN: 9783527855100
epub ISBN: 9783527855094
oBook ISBN: 9783527855117

C9783527356010_261025

ॐ नमः पार्वती पतये, हर-हर महादेवः

कर्मण्येवाधिकारस्ते मा फलेषु कदाचन।

मा कर्मफलहेतुर्भूर्मा ते सङ्गोऽस्त्वकर्मणि॥

To our beloved parents and loved ones – thank you for your unwavering support, encouragement, and the sacrifices you made that allowed us to pursue our passions. This work would not have been possible without your constant belief in us.

We also dedicate this book to all present and future seekers of knowledge, whose curiosity, perseverance, and spirit of discovery continue to shape a better world.

> *"The time will come when diligent research over long periods will bring to light things which now lie hidden. A single lifetime, even though entirely devoted to the sky, would not be enough for the investigation of so vast a subject ... And so this knowledge will be unfolded only through long successive ages. There will come a time when our descendants will be amazed that we did not know things that are so plain to them ... Many discoveries are reserved for ages still to come, when memory of us will have been effaced."*
>
> —*Seneca*

May this book be a small contribution to the unfolding of that timeless journey.

Contents

Preface

The global energy landscape is undergoing a profound transformation as nations, industries, and societies confront the urgent challenges of climate change, fossil fuel depletion, and the pursuit of energy independence. In this context, hydrogen has emerged as a central pillar of the clean energy transition – promising a sustainable, versatile, and zero-emission alternative to conventional fuels across multiple sectors. From powering vehicles and storing renewable energy to decarbonizing hard-to-abate industries like steel, cement, and ammonia production, hydrogen technologies are paving the way toward a carbon-neutral future.

This book, *Next-Generation Hydrogen Economy: Production, Storage, and Fuel Cell Technologies*, is designed to serve as a comprehensive guide for academics, industry professionals, policymakers, and students who are actively engaged in or entering the hydrogen energy field. It brings together a detailed exploration of hydrogen's scientific foundations, technological innovations, infrastructure development, and future opportunities. Drawing on the latest research and real-world case studies, this book provides a clear and practical understanding of the key components necessary to scale-up and commercialize hydrogen systems globally.

The content spans 10 chapters, beginning with the historical evolution and current status of the hydrogen economy, followed by in-depth analyses of advanced production techniques, material innovations, storage strategies, and fuel cell mechanisms. The latter half of the book focuses on transportation and industrial applications, policy and economic frameworks, safety standards, and the challenges of industrial scale-up. Importantly, it also looks ahead to the future – exploring AI-driven research, smart hydrogen cities, and the role of hydrogen in achieving net-zero carbon targets.

Our motivation in writing this book stems from the need for an interdisciplinary resource that bridges chemistry, physics, materials science, engineering, and energy policy. The hydrogen transition is not only a technological endeavor but a societal one, requiring global cooperation and informed decision-making. We hope this book contributes to that mission by equipping readers with the knowledge needed to innovate, collaborate, and lead in the hydrogen era.

We express our gratitude to the researchers, engineers, institutions, and organizations whose work continues to push the boundaries of hydrogen technology. We also thank the publishing team and reviewers whose insights helped refine and shape this book.

We invite you, the reader, to explore the fascinating journey of hydrogen – one molecule with the power to fuel a more sustainable world.

Dr. Peeyush Phogat
Dr. Shreya Sharma
Mr. Satyam Rawat

Acknowledgments

We extend our heartfelt gratitude to **Prof. Ranjana Jha**, Vice Chancellor, Indira Gandhi Delhi Technical University for Women, for her inspiring leadership, guidance, and continued encouragement throughout our academic journey. We are deeply thankful to **Prof. Sukhvir Singh** and **Prof. Nandlal** of the Delhi Skill and Entrepreneurship University for their invaluable insights, mentorship, and unwavering support at every step of this endeavor.

Our sincere appreciation goes to the **Wiley publishing team** for recognizing the value of this work and providing us with the opportunity to share our vision with the global research community. Your professionalism and support have made this project a reality.

We also thank our **research team**, whose dedication, resilience, and collaborative spirit have been a constant source of strength and motivation. Your contributions, both seen and unseen, have helped shape the foundation of this book.

Lastly, to our families and loved ones – thank you for your endless patience, sacrifices, and unconditional love. This work is a reflection of all the support you have given us.

—Dr. Peeyush Phogat, Dr. Shreya Sharma, and Mr. Satyam Rawat

About the Authors

Dr. Peeyush Phogat completed his PhD in the Department of Physics at Netaji Subhas University of Technology. His research emphasizes the synthesis and characterization of advanced materials for energy applications, focusing on innovations in energy storage, photodetection, and sustainability. A prolific researcher, Peeyush has authored 36 research papers in reputed journals, contributing to advancements in electrochemical applications and renewable energy materials. He has also published many review articles providing in-depth insights into emerging trends in material science and energy technologies. In addition, Peeyush has contributed 18 book chapters across diverse domains in material science and co-authored three books: *Next-Generation Electrochemical Devices: From Principles to Applications*, *The Future of Energy: How Innovative Devices Shape Our Environment*, and *AI-Driven Innovations in Electrochemical Technologies for Sustainable Energy Solutions*. Driven by a commitment to scientific progress, Peeyush aims to advance sustainable energy solutions through material science innovations and interdisciplinary research.

Dr. Shreya Sharma is a physicist who earned her PhD in Physics from Netaji Subhas University of Technology, where her research was centered on the exploration of exceptional properties exhibited by nanomaterials and their potential contributions to advancing renewable energy technologies. Her primary focus lies in investigating the unique properties of two-dimensional (2D) materials and their electrochemical study, specifically transition metal dichalcogenides, with the aim of applying these findings to enhance solar cells, supercapacitors, photodetectors, hydrogen production, and electrochemical sensing. Her dedication to leveraging advanced materials for sustainable energy solutions reflects her vision of contributing to societal progress through innovative technology. Currently, she is working to identify sustainable energy resources, with a specific focus on blue energy foresight. She is also the author and co-editor of six scientific books and more than 40 research papers in esteemed journals. Driven by a deep commitment to environmental sustainability, Dr. Sharma's work embodies her passion for scientific advancement in service to future energy needs.

Mr. Satyam Rawat is currently pursuing an MSc in Physics, specializing in the synthesis and characterization of 2D materials like reduced graphene oxide. His research focuses on developing nanocomposites of these materials for advanced applications, including supercapacitors and photodetectors. His work aims to enhance the performance and versatility of energy storage and optoelectronic devices through innovative material design and optimization.

Disclosure of AI Use

In accordance with publishing guidelines and ethical authorship practices, the authors of this book affirm the following regarding the use of artificial intelligence (AI) tools in the preparation of this manuscript:

1. AI tools (e.g., OpenAI's ChatGPT) were utilized under human supervision to assist in:
   - Structuring content outlines and chapter flow,
   - Drafting preliminary versions of technical explanations,
   - Polishing of language and checking consistency,
   - Generating illustrative figure captions and table summaries. All AI-generated content was **thoroughly reviewed, edited, and verified** by the authors to ensure **accuracy, originality, and contextual relevance**.
2. All final content decisions, interpretations of scientific data, and conceptual frameworks presented in this book were **formulated, validated, and approved by the human authors**. AI was **not used for ideation, conclusion formulation, or original research interpretation**. The authors take **full responsibility for the integrity and reliability** of the content.
3. All figures, tables, and illustrations included in this book are either:
   - Created by the authors using professional scientific software,
   - Derived from publicly available and cited data,
   - Commissioned via professional graphic design based on author-provided input. No standalone generative AI image tools were used without human modification or verification.
4. No personal data, confidential research material, or copyrighted content was used in AI prompts or inputs. All data complies with **academic integrity standards, open-access policies, and third-party rights**.
5. This declaration is made to ensure transparency for readers, editors, and peer reviewers. The authors affirm that the use of AI did not replace critical thinking, domain expertise, or the ethical responsibilities of authorship.

1

The Hydrogen Paradigm – Past, Present, and Future

1.1 Introduction to Hydrogen as an Energy Carrier

1.1.1 Definition and Fundamental Properties of Hydrogen

Hydrogen, the lightest and most abundant element in the universe, plays a crucial role in both natural and engineered energy systems. It is represented by the chemical symbol H and has an atomic number of 1, meaning it contains only one proton in its nucleus. As a fundamental building block of the universe, hydrogen is a primary component of stars and gas giants and is widely found in water, hydrocarbons, and biological systems. Its significance in energy applications arises from its high-energy density and clean combustion properties, making it an ideal candidate for sustainable fuel alternatives [1].

The atomic structure of hydrogen is the simplest among all elements, consisting of a single proton and an electron. However, hydrogen exhibits multiple **isotopic forms**, which are distinguished by the number of neutrons in the nucleus. The three naturally occurring isotopes of hydrogen are:

- **Protium (^1H)** – The most abundant isotope, accounting for over 99.98% of naturally occurring hydrogen. It consists of a single proton and no neutrons.
- **Deuterium (^2H or D)** – This isotope contains one proton and one neutron, making it twice as heavy as protium. Deuterium is stable and occurs naturally in small quantities (about 0.015% of Earth's hydrogen). It is commonly used in nuclear fusion research and heavy water applications in nuclear reactors.
- **Tritium (^3H or T)** – A radioactive isotope containing one proton and two neutrons. Tritium has a half-life of 12.3 years and is primarily produced in nuclear reactions. It is used in specialized applications such as nuclear fusion research and self-luminous materials.

Hydrogen's placement in the periodic table highlights its unique properties and chemical versatility (as illustrated in Figure 1.1). Despite being positioned in Group 1 due to its single valence electron, hydrogen exhibits behavior distinct from alkali metals, and in some respects, resembles halogens. It readily forms bonds with other

Next-Generation Hydrogen Economy: Production, Storage, and Fuel Cell Technologies, First Edition.
Peeyush Phogat, Shreya Sharma and Satyam Rawat.

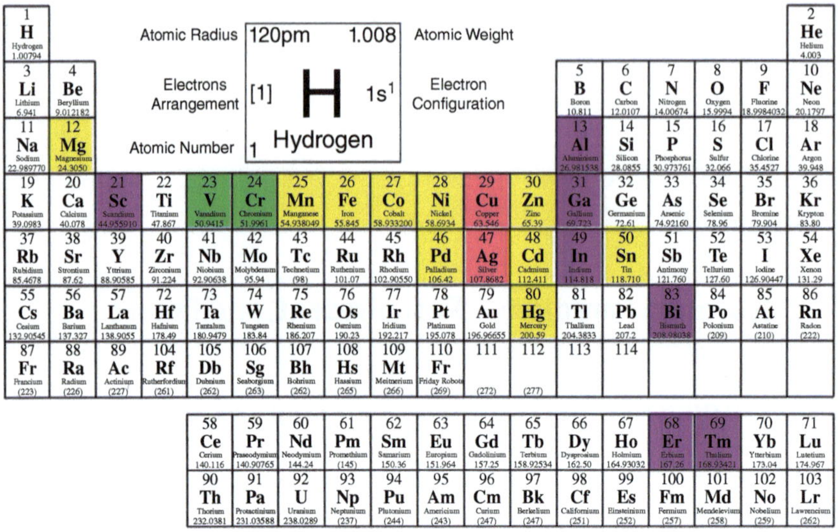

Figure 1.1 Hydrogen's position in the periodic table and its key properties. Reproduced with permission of [5]/John Wiley & Sons.

elements, contributing to a vast range of compounds, including water (H_2O), hydrocarbons (CH_4, C_2H_6), and acids (HCl, H_2SO_4) [2].

The properties of hydrogen that make it particularly attractive as an energy carrier are its lightweight nature, high-energy content, and environmental benefits [3]. Some key physical and chemical characteristics of hydrogen relevant to energy applications are as follows:

Low Atomic and Molecular Mass – Hydrogen is the lightest element, with an atomic mass of 1.008 u. Molecular hydrogen (H_2) has a molecular weight of 2.016 g/mol, making it significantly lighter than other fuels. Its low density (0.0899 g/L at standard conditions) presents challenges for storage and transportation, necessitating advanced containment techniques such as compression, liquefaction, or chemical storage. These storage challenges are particularly important in applications where large quantities of hydrogen need to be transported or stored over extended periods.

High-Energy Density – Despite its low volumetric density, hydrogen has an exceptionally high gravimetric energy density of 120 MJ/kg, nearly three times that of gasoline (44 MJ/kg). This property makes hydrogen a promising alternative for applications requiring lightweight, high-energy fuels, such as aerospace, fuel cell vehicles, and portable power systems. Its high-energy density per unit mass also positions hydrogen as an ideal candidate for energy applications in sectors that demand significant power with minimal weight, such as aviation and heavy-duty transport.

Chemical Reactivity and Combustion – Hydrogen exhibits strong reducing properties, allowing it to participate in various chemical reactions, including redox processes and catalysis. When burned in the presence of oxygen, hydrogen reacts exothermically:

$$2 + O_2 \rightarrow 2H_2O + \text{Energy} \tag{1.1}$$

This reaction releases a significant amount of energy and produces only water vapor as a byproduct, making hydrogen a zero-emission fuel when used in combustion engines or fuel cells (as illustrated in Figure 1.1). The clean nature of hydrogen combustion positions it as an attractive alternative to fossil fuels, which emit harmful greenhouse gases and pollutants.

Flammability and Explosion Limits – Hydrogen has a wide flammability range (4–75% in air) and a low ignition energy, making it highly reactive. This characteristic necessitates strict safety measures in hydrogen storage and handling. However, due to its low molecular weight, hydrogen dissipates quickly in the atmosphere, reducing the risk of prolonged exposure to explosive concentrations. These safety considerations are essential in ensuring the safe adoption of hydrogen in various energy systems.

Electrochemical Potential in Fuel Cells – Hydrogen's ability to release electrons in an electrochemical reaction makes it an ideal fuel for proton exchange membrane (PEM) fuel cells, where hydrogen is split into protons and electrons to generate electricity. This application is central to hydrogen-powered vehicles, backup power systems, and stationary fuel cell applications. The efficiency of hydrogen fuel cells, combined with their zero-emission nature, offers a promising solution for a wide range of applications, from transportation to stationary power generation.

Interaction with Metals and Storage Challenges – Hydrogen can diffuse through metals, leading to hydrogen embrittlement, which weakens structural materials over time. This property impacts pipeline transport and metal storage solutions, requiring specialized materials to contain hydrogen safely. Engineers are working to develop advanced materials that can withstand the effects of hydrogen embrittlement, ensuring the safe transport and storage of hydrogen.

Cryogenic Liquefaction – Hydrogen has a boiling point of −252.87 °C, necessitating cryogenic technology for liquid hydrogen storage. Liquefied hydrogen (LH_2) is often used in rocket propulsion and energy storage applications where high-density storage is required. The cryogenic liquefaction of hydrogen enables its storage in compact, dense forms, making it suitable for high-energy applications such as space exploration.

Hydrogen's distinct physical and chemical properties underscore its potential as a clean and sustainable energy source, while also presenting technical challenges that require innovative solutions for widespread adoption. As the world shifts toward decarbonization and renewable energy integration, hydrogen will continue to play a pivotal role in shaping the future energy landscape [4].

1.1.2 Why Hydrogen Is Considered a Clean Energy Carrier

Hydrogen has gained global recognition as a clean energy carrier due to its potential to provide a sustainable, zero-emission alternative to conventional fossil fuels. Its unique properties, including high-energy density, efficiency in energy conversion, and versatility across multiple sectors, make it a promising candidate for the future of energy. Hydrogen can be used in fuel cells, internal combustion engines, and industrial applications while producing minimal environmental impact. The key reasons hydrogen is considered a clean energy carrier include its zero-emission potential, high efficiency in energy conversion, and favorable comparison with fossil fuels and renewables [6].

One of the most compelling advantages of hydrogen as an energy carrier is its zero-emission potential when combusted or used in fuel cells. Unlike fossil fuels, which release carbon dioxide (CO_2), nitrogen oxides (NO_x), sulfur oxides (SO_x), and particulate matter, hydrogen produces only water vapor (H_2O) when reacted with oxygen. This clean combustion process is represented by the reaction (1.1). This reaction is highly exothermic, releasing a significant amount of energy while generating no carbon-based emissions. Due to this characteristic, hydrogen is considered a viable alternative to coal, oil, and natural gas in sectors where emissions are a major concern, such as transportation, power generation, and heavy industry. Additionally, hydrogen can be produced from renewable energy sources, such as solar, wind, and hydropower, through electrolysis, making it a truly sustainable energy carrier. Green hydrogen, produced via water electrolysis powered by renewables, has a net-zero carbon footprint, making it an ideal candidate for long-term energy storage and grid stabilization [7].

Hydrogen can also be blended with natural gas in existing pipeline infrastructure to reduce carbon emissions in heating applications. A blend of 20% hydrogen with natural gas can lower overall emissions without requiring major modifications to current gas infrastructure. In contrast, using pure hydrogen in fuel cells ensures completely carbon-free power generation, making it particularly suitable for electric vehicles, stationary power systems, and portable energy applications [8].

Hydrogen's efficiency in energy conversion is another reason for its growing popularity as a clean energy carrier. When used in fuel cells, hydrogen undergoes an electrochemical reaction rather than combustion, significantly improving efficiency and reducing energy losses. Fuel cells can achieve an efficiency of 50–60%, much higher than internal combustion engines, which typically operate at 20–30% efficiency due to heat losses. When combined with waste heat recovery systems, fuel cells can reach an overall efficiency of 80–90%, making them one of the most efficient energy conversion technologies available. In contrast, traditional coal-fired power plants have efficiencies of only 35–40%, with the remainder of the energy lost as heat. The sustainability of hydrogen is further enhanced by its ability to be produced domestically using locally available resources. Unlike fossil fuels, which are finite and concentrated in specific regions, hydrogen can be derived from water (H_2O), biomass, or industrial byproducts. This enhances energy security by reducing dependence on oil and gas imports, creating a more resilient and distributed energy system. Furthermore, hydrogen can act as an energy storage medium for intermittent renewables like solar and wind. Surplus electricity generated during high renewable output periods can be used to produce hydrogen via electrolysis, which can then be stored and converted back to electricity when demand is high (as depicted in Table 1.1). This capability makes hydrogen a key enabler of grid stability and energy balancing, addressing one of the major challenges of renewable energy integration [9].

A comprehensive comparison of hydrogen with other energy carriers, including electricity, natural gas, and gasoline, reveals its potential advantages in energy density, emissions, and efficiency (as shown in Table 1.1). Hydrogen has a higher energy content per unit mass (120 MJ/kg) than gasoline (44 MJ/kg) and natural gas (53 MJ/kg). This makes hydrogen particularly attractive for aerospace and heavy-duty transport, where weight considerations are critical. However, hydrogen's low volumetric energy density presents storage and transportation challenges, requiring compression, liquefaction, or solid-state storage solutions [10].

Emissions Profile:

- Hydrogen emits zero greenhouse gases when used in fuel cells.
- Natural gas, while cleaner than coal, still emits 50–60% of the CO_2 per unit energy compared to gasoline.
- Gasoline and diesel combustion produce CO_2, NO_x, SO_x, and particulate matter, contributing to air pollution and climate change.

Efficiency:

- Hydrogen fuel cells operate at 50–60% efficiency and can reach 80–90% with heat recovery.
- Internal combustion engines (ICEs) powered by gasoline or diesel typically have efficiencies of 20–30%, with significant losses due to heat dissipation.
- Battery electric vehicles (BEVs) achieve efficiencies of 85–90%, but their energy storage capacity is limited compared to hydrogen fuel cells.

Hydrogen, with its zero-emission potential, efficiency in energy conversion, and favorable emissions profile, stands out as a clean energy carrier capable of playing a pivotal role in the global transition toward a more sustainable and decarbonized energy future [11].

Table 1.1 Comparison of hydrogen with other energy carriers (electricity, natural gas, gasoline) in terms of energy density, emissions, and efficiency. (Placed at the end of this section to highlight advantages and limitations.)

Energy carrier	Gravimetric energy density (MJ/kg)	Emissions (CO_2 per MJ energy)	Efficiency (%)	Storage challenges
Hydrogen (H_2)	120	0 (Fuel cells)	50–60 (Fuel cells)	Requires compression/ liquefaction [12]
Gasoline	44	73 g CO_2/MJ	20–30 (ICEs)	Liquid fuel, stable storage [13]
Natural gas (CH_4)	53	50 g CO_2/MJ	30–40 (Gas turbines)	Requires pipelines or liquefied natural gas (LNG) storage [14]
Electricity (battery storage)	0.3 (Li-ion)	Depends on grid mix	85–90 (BEVs)	Limited by battery capacity and lifespan [15]

Hydrogen stands out as a clean, efficient, and sustainable energy carrier with the potential to replace fossil fuels in multiple sectors. Its zero-emission nature, high-energy conversion efficiency, and ability to store and transport renewable energy make it a key enabler of the global energy transition. While challenges such as storage, infrastructure, and cost remain, ongoing advancements in hydrogen production, distribution, and utilization will drive its widespread adoption. As global policies shift toward carbon neutrality, hydrogen is poised to play a pivotal role in achieving sustainable and emission-free energy systems [16].

1.1.3 Role of Hydrogen in Decarbonization Efforts

The transition to a low-carbon economy is crucial for mitigating climate change, and hydrogen has emerged as a key enabler of global decarbonization efforts. As a clean and versatile energy carrier, hydrogen can significantly reduce greenhouse gas (GHG) emissions by replacing fossil fuels in transportation, industry, and power generation. Moreover, its ability to integrate with renewable energy sources makes it an essential component of a sustainable energy system. By leveraging green hydrogen production, advancements in fuel cell technology, and hydrogen infrastructure development, the world can accelerate toward carbon neutrality and meet climate goals set by international agreements such as the Paris Agreement [17].

Carbon neutrality – achieving net-zero carbon dioxide (CO_2) emissions – requires a fundamental shift away from fossil fuels toward clean energy solutions. Hydrogen offers a zero-emission alternative when produced using renewable energy sources such as solar and wind power. Unlike coal, oil, and natural gas, which emit CO_2, methane (CH_4), and other pollutants, hydrogen can be used in fuel cells and combustion processes while producing only water (H_2O) as a byproduct. To achieve deep decarbonization, hydrogen must be produced, stored, and utilized in a way that minimizes emissions. This includes transitioning from gray hydrogen (derived from fossil fuels) to blue hydrogen (produced with carbon capture and storage) and ultimately to green hydrogen (generated via electrolysis powered by renewables). Large-scale deployment of green hydrogen is expected to reduce CO_2 emissions in hard-to-abate sectors and contribute significantly to global carbon neutrality goals [18].

Sectors Benefiting from Hydrogen-Based Decarbonization

a. **Transportation:** The transportation sector is one of the largest contributors to global CO_2 emissions, primarily due to the reliance on gasoline and diesel-powered vehicles. Hydrogen-based solutions, such as fuel cell electric vehicles (FCEVs), offer a zero-emission alternative to ICEs. Unlike BEVs, which require long charging times, FCEVs can be refueled quickly and have longer driving ranges, making them ideal for heavy-duty transport, buses, trains, and maritime shipping. Hydrogen-powered aviation is also under development, with companies exploring liquid hydrogen fuel for long-haul flights [19].
b. **Industry:** Heavy industries, such as steel, cement, and chemical manufacturing, account for a significant share of global carbon emissions. These industries rely on high-temperature processes that are difficult to electrify. Hydrogen,

especially in the form of green hydrogen, can serve as a clean alternative to coal and natural gas in steel production (via hydrogen direct reduction), ammonia synthesis, and chemical refining. By replacing fossil-based hydrogen (gray hydrogen) with green hydrogen, industries can achieve substantial CO_2 emission reductions while maintaining efficiency [20].

c. **Power Generation:** Hydrogen can play a vital role in stabilizing electricity grids and supporting renewable energy expansion. While solar and wind energy are crucial for decarbonization, their intermittent nature presents challenges for grid stability. Hydrogen can be used to store excess renewable energy during periods of low demand and later converted back to electricity via fuel cells or hydrogen turbines when needed. This ability to act as a long-term energy storage solution makes hydrogen an essential component of future power systems. Countries are already testing hydrogen-fired gas turbines to replace traditional natural gas power plants, further reducing carbon emissions from the power sector [21].

Hydrogen's compatibility with renewable energy is a major advantage in global decarbonization efforts. When solar and wind farms generate excess power, this surplus energy can be used for water electrolysis, producing green hydrogen. This hydrogen can then be stored and transported to be used in fuel cells, industrial applications, or power plants. The concept of "power-to-gas" (P2G) allows renewable energy to be converted into hydrogen, effectively addressing energy storage and grid-balancing challenges. Furthermore, hydrogen pipelines can integrate with existing gas networks, allowing for blending hydrogen with natural gas to progressively reduce carbon footprints. Some countries are already experimenting with 20–30% hydrogen blending, with the goal of transitioning to 100% hydrogen-based heating and electricity generation [22].

Hydrogen is a critical enabler of global decarbonization efforts, offering solutions for carbon neutrality in transportation, industry, and power generation. By leveraging green hydrogen production and integrating it with renewable energy systems, nations can significantly reduce GHG emissions, enhance energy security, and create a sustainable energy future. While challenges remain, including infrastructure development and production costs, ongoing technological advancements and policy support will accelerate the global hydrogen economy, making it a cornerstone of climate change mitigation strategies [23].

1.2 Evolution of Hydrogen as an Energy Carrier

1.2.1 Early Uses of Hydrogen

Hydrogen has a long history of use, beginning with its discovery in the eighteenth century and gradually expanding into various industrial and scientific applications. The first documented production of hydrogen dates back to 1671, when the English scientist Robert Boyle observed the gas while experimenting with metals and acids. However, it was not until 1766 that the British chemist Henry Cavendish identified hydrogen as a distinct element, isolating it and recognizing its flammability.

His experiments revealed that hydrogen, when burned, produced water, a crucial discovery that laid the foundation for understanding its chemical properties [24].

One of the earliest and most iconic applications of hydrogen was in balloons. In the late eighteenth and early nineteenth centuries, hydrogen gas became widely used to fill balloons, offering a practical means of achieving flight due to its low molecular weight and high buoyancy. The French brothers Joseph-Michel and Jacques-Étienne Montgolfier are often credited with the development of the first hot air balloons, but hydrogen's ability to lift heavier loads made it an attractive alternative. In 1783, the first hydrogen-filled balloon flight took place, and soon after, hydrogen was used for a series of daring aerial adventures. It was particularly instrumental in early ballooning expeditions, including those by famous figures such as Jean-Pierre Blanchard and Henry Coxwell. Despite its potential for flight, the flammability of hydrogen eventually led to its replacement by safer, less volatile gases like helium. Beyond ballooning, hydrogen found early use in various industrial processes. In the nineteenth century, it became an essential component in the production of ammonia through the Haber process, a critical advancement for fertilizer production that revolutionized agriculture. Hydrogen's role in hydrogenation processes also began to take shape during this period, facilitating the conversion of unsaturated fats and oils into solid forms, such as margarine. This process, known as catalytic hydrogenation, is still widely used in the food industry today. Overall, the early use of hydrogen paved the way for its later applications in energy and chemical processes. Its discovery and the subsequent exploration of its potential led to significant technological and industrial developments, establishing hydrogen as a key element in the scientific and industrial revolutions of the eighteenth and nineteenth centuries [25].

1.2.2 Development of Hydrogen Production Technologies

The development of hydrogen production technologies has evolved significantly from early methods like electrolysis to the sophisticated reforming processes used today. Hydrogen's potential as an energy carrier has driven scientists and engineers to explore various methods for its generation, resulting in numerous breakthroughs over the centuries. In the late eighteenth century, the first recorded instance of hydrogen production occurred via electrolysis, a process that remains relevant to this day. In 1800, William Nicholson and Anthony Carlisle demonstrated the electrolysis of water, breaking it down into hydrogen and oxygen using an electric current. This was the first time hydrogen had been intentionally produced from water, marking a major milestone in hydrogen energy development. Electrolysis became a foundational technology for hydrogen production, as it provided a means of generating hydrogen from abundant and renewable water sources [26].

The next significant breakthrough came in the nineteenth century, with the development of steam reforming, which remains one of the most widely used methods for hydrogen production today. In the early 1800s, chemists discovered that high temperatures and the presence of a catalyst could be used to extract hydrogen from hydrocarbons, such as methane. Steam reforming, which involves reacting methane with steam at high temperatures (700–1,000 °C) to produce hydrogen, carbon monoxide, and a small amount of carbon dioxide, became a major industrial method. This process, still widely used in refineries and chemical plants, is efficient but produces carbon-based byproducts, making it a less environmentally friendly option compared to newer technologies like water electrolysis powered by renewable energy sources [27]. In the mid-twentieth century, advances in hydrogen production continued with the introduction of more efficient methods, including partial oxidation and auto-thermal reforming. These processes combined steam reforming with other chemical reactions to increase hydrogen yield and reduce energy consumption. Additionally, during this time, the industrial-scale production of hydrogen saw rapid growth, especially with its increasing use in the petroleum industry and for the production of ammonia [28]. The late twentieth and early twenty-first centuries saw a surge in research aimed at improving the sustainability of hydrogen production. The introduction of "green hydrogen" production methods, particularly through electrolysis powered by renewable energy sources such as wind, solar, and hydropower, represented a breakthrough in decarbonizing hydrogen production. These methods have gained momentum in recent years, driven by the need for cleaner energy alternatives to fossil fuels. Green hydrogen is produced by splitting water molecules into hydrogen and oxygen using electricity derived from renewable sources, resulting in a zero-emission production process that holds great promise for reducing global carbon footprints [29].

Alongside green hydrogen, developments in biomass gasification, which uses organic materials to produce hydrogen, have also emerged as another potential source of hydrogen. This method converts biomass into hydrogen-rich gases, making it an attractive option for regions with abundant agricultural or forest waste. As shown in Figure 1.2, the timeline of hydrogen energy development highlights the key milestones in hydrogen production, from the early electrolysis experiments to the modern reforming processes and the ongoing push for sustainable hydrogen generation technologies. This progression reflects the growing understanding of hydrogen's potential

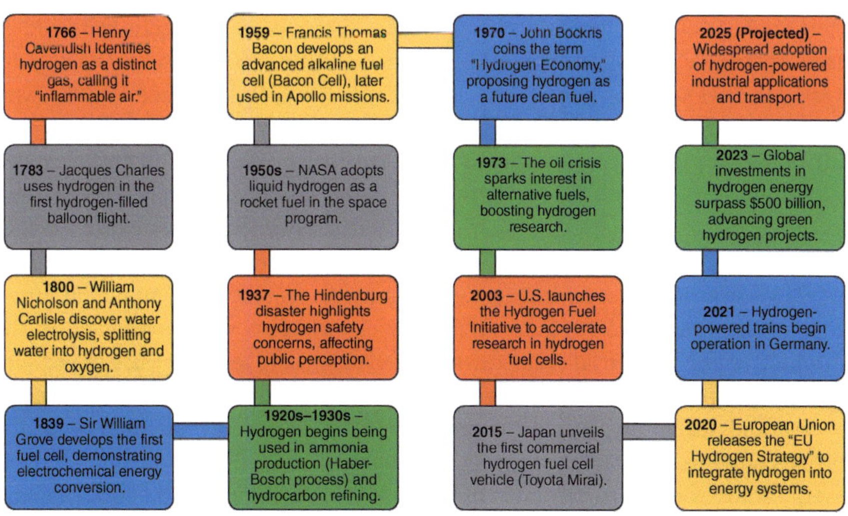

Figure 1.2 Timeline showing key milestones in hydrogen energy development.

as a clean energy carrier and the increasing demand for efficient, environmentally friendly hydrogen production methods to meet the energy needs of the future. The future of hydrogen production lies in further innovations, such as advanced electrolyzers and more cost-effective renewable energy integration, ensuring its role in the global transition to a cleaner energy system [30, 31].

1.2.3 The Rise of Fuel Cells and Space Applications

The rise of hydrogen fuel cells in space applications marked a significant turning point in both the development of fuel cell technology and the broader use of hydrogen as an energy carrier. In the mid-twentieth century, NASA's space programs became one of the most prominent arenas for demonstrating the potential of hydrogen fuel cells, a technology that would later find its way into various commercial applications. The development of hydrogen fuel cells for space missions began in the 1950s, with the realization that traditional power sources would not meet the unique needs of space exploration. Fuel cells, which generate electricity through the electrochemical reaction between hydrogen and oxygen, provided a lightweight, reliable, and long-lasting energy source. This made them ideal for powering spacecraft, where weight and efficiency are critical. NASA's interest in fuel cells led to significant advancements in both the technology and its application in space missions [32, 33].

In the early 1960s, NASA began using hydrogen fuel cells for its Gemini and Apollo space programs. The Apollo spacecraft, which carried astronauts to the Moon, relied on fuel cells to generate electricity and supply potable water for the astronauts. The fuel cell system was highly efficient, producing both electricity and water as a byproduct, which was particularly important in the confined environment of the spacecraft. The Apollo missions demonstrated that hydrogen fuel cells could operate efficiently in space, providing a continuous power source for critical systems while also producing water for consumption by the astronauts. The success of hydrogen fuel cells in the Apollo missions catalyzed their further development and adoption [34]. By the 1970s, fuel cells were widely used in space exploration, and NASA's space shuttle program relied heavily on hydrogen fuel cells for power. The space shuttle, which launched in 1981, was equipped with fuel cells that generated electricity from hydrogen and oxygen during flight. These fuel cells provided electrical power for the shuttle's systems and were essential for supporting the crew's needs during both launch and re-entry. The use of hydrogen fuel cells in space applications was a testament to their reliability, long life, and efficiency under extreme conditions. Parallel to NASA's work in space, hydrogen fuel cells began to receive commercial attention. In the 1960s and 1970s, several early experiments were conducted in the automotive and industrial sectors, where fuel cells were tested for use in vehicles and backup power systems. Although the technology faced challenges in terms of cost and scalability, it laid the groundwork for future developments. In the 1990s, commercial interest in fuel cells increased as the need for clean, efficient energy sources grew, particularly in the context of environmental concerns and the development of alternative energy solutions [35].

The early experiments and commercial developments of hydrogen fuel cells, although limited in scope at the time, demonstrated their potential to provide clean energy for various applications. Today, hydrogen fuel cells have moved from the realm of space exploration to terrestrial applications, including transportation, stationary power systems, and backup generators. The development of fuel cells for NASA's space programs played a pivotal role in proving the viability of hydrogen as an energy carrier, and it continues to be a driving force in the development of fuel cell technologies today. As the world moves toward cleaner, more sustainable energy solutions, hydrogen fuel cells are positioned to play a key role in decarbonizing sectors such as transportation and power generation, building upon the legacy of their success in space applications [36, 37].

1.2.4 Hydrogen's Transition to an Energy Vector in the Twenty-first Century

Hydrogen has undergone a remarkable transformation over the last few decades, evolving from a simple industrial gas to a cornerstone of clean energy solutions in the twenty-first century. Historically, hydrogen was primarily used in industries such as refining, ammonia production, and chemical synthesis. It was seen as a useful but niche industrial product, and its applications were largely confined to these sectors. However, as the world faces mounting challenges associated with climate change, air pollution, and the depletion of fossil fuels, hydrogen's potential as a clean energy carrier has gained global attention, setting the stage for its significant role in the future energy landscape. The shift of hydrogen from an industrial gas to a clean energy solution is largely driven by the increasing demand for decarbonization in various sectors, particularly in transportation, power generation, and heavy industry. One of the most compelling aspects of hydrogen is its versatility: it can be used in fuel cells for vehicles, stored as an energy carrier, and even integrated into industrial processes to replace carbon-intensive fuels. As concerns about GHG emissions have intensified, hydrogen has emerged as a key solution to reduce emissions in hard-to-abate sectors like steel production and long-distance transportation [38].

In the early twenty-first century, hydrogen began to be seen not just as a chemical feedstock but as a vector for storing and transporting renewable energy. The rise of renewable energy sources, such as solar and wind, has posed significant challenges to grid stability due to their intermittent nature. Hydrogen offers a solution to this challenge by acting as an energy carrier. Surplus renewable energy can be used to produce hydrogen through electrolysis, and this hydrogen can then be stored for use when renewable energy generation is low. This ability to store and transport energy over long distances makes hydrogen a critical player in the transition to a low-carbon energy system [39].

The global policy landscape has also shifted significantly to support hydrogen adoption. Countries around the world are recognizing hydrogen as an essential part of their energy transition strategies. In Europe, for example, the EU has set ambitious goals for hydrogen development, focusing on the production of green hydrogen and its integration into the energy grid. The EU Hydrogen Strategy, unveiled in 2020, aims to scale up the production of renewable hydrogen, with a target of producing 10 million tons of green hydrogen annually by 2030 [40]. In Japan, hydrogen has been central to the country's energy strategy, with policies supporting hydrogen fuel cell vehicles, infrastructure development, and international hydrogen trade. The Japanese government's commitment to hydrogen is exemplified by the country's early leadership in hydrogen fuel cell vehicles and its establishment of a national hydrogen roadmap [41]. Similarly, governments in South Korea, China, and the United States have been prioritizing hydrogen in their energy policies. South Korea has launched a hydrogen roadmap aimed at boosting hydrogen production and adoption in various sectors, including transportation, while China's commitment to hydrogen has been reinforced by large-scale pilot projects and funding for hydrogen infrastructure. In the United States, hydrogen is increasingly seen as a key component of decarbonizing the industrial sector, with both state and federal initiatives focused on developing hydrogen technologies and expanding infrastructure [42].

At the same time, there has been an explosion of research and development in hydrogen production technologies. Early methods of hydrogen production, such as steam methane reforming (SMR) and electrolysis, have been refined to improve efficiency and reduce costs. SMR, which remains the most common method of hydrogen production, has been traditionally criticized for its high carbon emissions [43]. However, with advancements in carbon capture and storage (CCS) technologies [44], low-carbon versions of SMR are now being explored. Meanwhile, electrolysis, which splits water into hydrogen and oxygen using electricity, is gaining momentum, particularly when powered by renewable sources of energy. Innovations in electrolyzer technologies are rapidly increasing the efficiency of electrolysis, driving down costs and making green hydrogen more competitive in the market. The continuous evolution of hydrogen production methods is depicted in Table 1.2, which shows key milestones in hydrogen generation technologies, along with trends in efficiency and cost. The table highlights the progression from early industrial hydrogen production methods to modern techniques, such as green hydrogen production via electrolysis, and illustrates the declining costs and increasing efficiencies over time. As these technologies mature and scale, hydrogen is becoming an increasingly viable option for large-scale, clean energy systems.

As hydrogen becomes more integrated into the global energy landscape, the role it plays in decarbonization efforts is expected to expand dramatically. The shift from industrial gas to clean energy vector is already underway, with research, innovation, and policy support driving hydrogen's role in addressing climate change. This transition is critical not only for reducing emissions but also for ensuring energy security and creating sustainable, low-carbon economies in the future. As the world accelerates its push toward a greener, more resilient energy system, hydrogen's role as an energy vector will continue to grow, unlocking new opportunities for innovation and collaboration across the globe.

Table 1.2 Evolution of hydrogen production methods with efficiency and cost trends.

Hydrogen production method	Description	Efficiency (%)	Cost trends (per kg)	Key milestones
Electrolysis (early methods)	Electrolysis of water using direct current (DC)	~50–60	High ($10–$12)	Early nineteenth century: discovery and initial use [45]
SMR	Hydrogen produced by reacting natural gas with steam	~65–75	$1–$3	Industrial scale-up in the twentieth century [46]
Coal gasification	Hydrogen produced from coal through a gasification process.	~65–70	$1.5–$4	Twentieth century: used for industrial purposes [47]
Electrolysis (modern)	Modern electrolyzers powered by renewable electricity	70–85	~$3–$6	Twenty-first century: adoption of green hydrogen [48]
Biomass gasification	Hydrogen produced from organic materials (e.g., biomass)	~60–70	~$2–$6	Increasing interest with bioenergy [49]
Blue hydrogen (SMR + CCS)	SMR with CCS to reduce emissions	~75–80	~$2–$4	2020s: focus on decarbonizing hydrogen [50]
Green hydrogen (electrolysis)	Electrolysis powered by renewable sources (solar, wind, etc.)	70–85	~$3–$6	2020s: scaling up with renewable energy [51]

1.3 Why Hydrogen? Comparison with Fossil Fuels and Renewables

1.3.1 Hydrogen vs. Fossil Fuels

Table 1.3 compares the characteristics of hydrogen with fossil fuels (gasoline and natural gas) and lithium-ion batteries, focusing on key aspects such as emissions, energy efficiency, energy density, and transport/storage capabilities. This comparison provides valuable insights into how hydrogen fares against traditional and emerging energy carriers in terms of sustainability, cost-effectiveness, and performance. The table below highlights these differences:

1.3.2 Hydrogen vs. Batteries

Table 1.4 compares hydrogen with batteries in terms of key performance metrics such as storage duration, charging time, efficiency, scalability, long-term performance, and emissions. The table also provides a detailed comparison of emissions

Table 1.3 Comparison of hydrogen with other energy carriers (electricity, natural gas, gasoline) in terms of energy density, emissions, and efficiency.

Characteristic	Hydrogen	Gasoline	Natural gas	Lithium-ion batteries
Energy density (per kg)	120 MJ/kg	44 MJ/kg	53 MJ/kg	150 – 200 Wh/kg (0.54 – 0.72 MJ/kg)
Energy density (per L)	0.0108 MJ/L (at 700 bar)	33.6 MJ/L	22.8 MJ/L	~0.25 – 0.5 MJ/L
Emissions (CO_2 equivalent)	Zero emissions (when used in fuel cells)	2.31 kg CO_2/L (per liter of gasoline)	1.85 kg CO_2/m^3 (natural gas)	Zero emissions at point of use (electricity production emissions vary)
Efficiency	50–60% (fuel cells), up to 90% with heat recovery	20–30% (internal combustion engine)	30–35% (fossil fuel power plants)	85–90% (for electric vehicles and stationary applications)
Storage	Requires high pressure (700 bar), cryogenic storage (−253 °C), or metal hydride storage	Stored as liquid or vapor at standard temperature and pressure	Stored in pressurized containers, or liquefied for transport	Stored as chemical energy in cells (solid state)
Transport	High-pressure cylinders, pipelines, or liquid transport	Trucks, pipelines, tanks	Pipelines, trucks, LNG	Transported in individual cells or packs
Cost (per kWh)	~\$4–\$6 (green hydrogen)	~\$0.10–\$0.25 (depending on region)	~\$0.04–\$0.10 (depending on region)	~\$0.12–\$0.18 per kWh
Availability	Requires infrastructure development (pipelines, fueling stations)	Widely available at fuel stations	Widely available at gas stations and through pipelines	Widely available for electric vehicle charging stations

Table 1.4 Comparison of hydrogen and battery energy storage systems in terms of storage duration, charging time, efficiency, scalability, and long-term performance.

Characteristic	Hydrogen	Batteries (lithium-ion)	Coal	Natural gas	Renewables (solar, wind, hydro)
Energy storage duration	Can store energy for weeks to months when in large-scale storage	Typically hours to days, depending on battery capacity and demand	N/A	N/A	N/A
Charging/ refueling time	Few minutes (fuel cell refueling)	Hours (charging time for electric vehicles and stationary storage)	N/A	N/A	N/A
Energy efficiency	50–60% (fuel cells), up to 90% with heat recovery	85–90% (for electric vehicles and stationary applications)	~35–40% (in coal-fired power plants)	~30–35% (fossil fuel power plants)	85–90% (for electricity generation)
Scalability	Highly scalable for large-scale storage and transport	Less scalable for long-term storage (limited by battery capacity)	Limited scalability; dependent on available coal reserves	Limited scalability in areas with pipeline infrastructure	Highly scalable, dependent on local resources
Long-term performance	High durability, especially in stationary applications and for grid balancing	Limited by number of charge-discharge cycles (~1,000–5,000 cycles)	High environmental impact over time (pollution and resource depletion)	Lower long-term performance due to depletion of fossil reserves	Long-term performance depends on system maintenance (e.g., wind turbine or solar panel lifespan)
Emissions (CO_2 equivalent)	Zero emissions at point of use (green hydrogen production via electrolysis has low emissions)	Zero emissions at point of use (electricity production emissions vary)	0.9 – 1.1 kg CO_2/kWh (varies by coal type and efficiency)	0.4 – 0.5 kg CO_2/kWh (natural gas power generation)	0 – 0.2 kg CO_2/kWh (depending on energy mix and system efficiency)
Cost (per kWh)	~$4–$6 (for green hydrogen)	~$0.12–$0.18 (for lithium-ion batteries)	~$0.04–$0.06 (coal-fired power)	~$0.04–$0.10 (natural gas)	~$0.05–$0.10 (depending on region and resources)
Storage capacity	Can store large amounts of energy, especially when liquefied or stored as metal hydrides	Limited by battery capacity and current technological constraints	N/A	N/A	N/A

associated with different energy sources, including hydrogen, coal, natural gas, and renewables. This comparison allows for a better understanding of how hydrogen and batteries perform as energy carriers in different sectors, including transportation, stationary power, and industrial applications.

1.3.3 Hydrogen's Role in the Renewable Energy Ecosystem

Hydrogen is increasingly being recognized as a critical component of the renewable energy ecosystem, providing an effective solution to some of the challenges posed by intermittent renewable energy sources like solar, wind, and hydropower. As the global energy landscape shifts toward cleaner, more sustainable energy systems, hydrogen's role as an energy carrier, storage medium, and means of balancing supply and demand becomes more vital. One of the primary advantages of hydrogen lies in its ability to store excess energy produced by renewable sources during periods of high generation and release it when demand is high or renewable generation is low. This capability is crucial for addressing the intermittency of renewable energy generation, which often fluctuates due to weather conditions, time of day, or seasonality [52].

One of the key challenges with renewable energy sources such as solar and wind is their variable output. Solar power generation, for instance, depends on sunlight, which can be erratic depending on time of day, season, or weather conditions. Similarly, wind power is highly dependent on wind speeds, which fluctuate. This variability makes it difficult to ensure a stable and continuous energy supply from these sources. Hydrogen provides a solution to this problem by acting as an energy carrier, storing surplus renewable energy that can be converted back into electricity when needed. The process of producing hydrogen from renewable energy, often referred to as "green hydrogen," involves using excess electricity generated by solar, wind, or hydropower plants to split water into hydrogen and oxygen through a process called electrolysis. This hydrogen can then be stored and transported for later use, either in fuel cells to generate electricity or in combustion systems for heat or power. In essence, hydrogen acts as a storage medium for excess renewable energy, providing a means to stabilize the grid and store energy for times when generation from renewables is low (such as at night for solar or during wind lulls) [53].

The integration of hydrogen into renewable energy systems is a promising pathway for achieving energy resilience and sustainability. When combined with solar and wind energy, hydrogen can enable a more reliable and flexible energy system. For example, when renewable generation exceeds demand, surplus electricity can be diverted to electrolyzers to produce hydrogen, which can be stored for later use. This stored hydrogen can be converted back into electricity using fuel cells or turbines when renewable generation falls short, such as on cloudy days for solar or during periods of low wind. Hydropower, which provides a stable and continuous source of electricity, can also be used in combination with hydrogen to enhance grid stability. In times of high hydropower generation, excess electricity can be used to produce hydrogen, providing a buffer that can be drawn upon during periods of low renewable energy output. In addition, hydrogen can be used to balance the grid by integrating with other low-carbon technologies such as bioenergy and geothermal power, enhancing the overall flexibility and reliability of renewable energy systems [54].

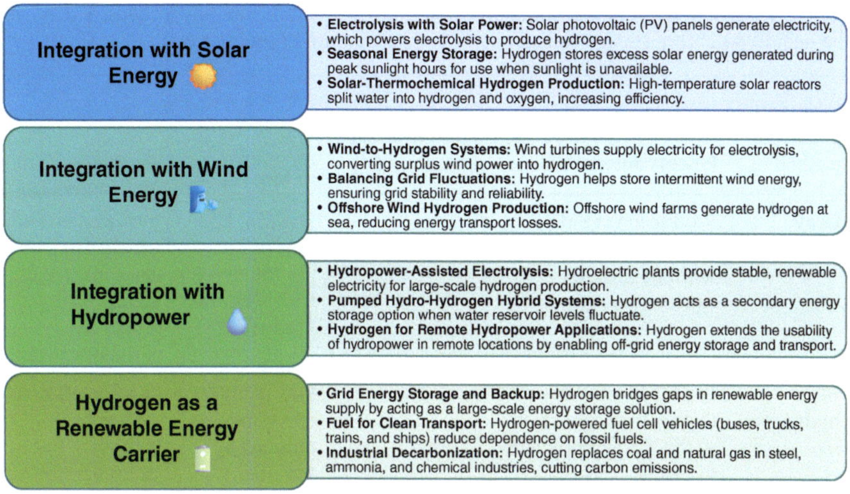

Figure 1.3 Hydrogen's role in the renewable energy ecosystem (integrating solar, wind, and hydropower).

Hydrogen also offers significant potential in heavy-duty applications that are difficult to decarbonize with electricity alone, such as long-haul transport, heavy industry, and large-scale storage. In these sectors, hydrogen can serve as a clean alternative to fossil fuels, facilitating the decarbonization of industries like steelmaking, shipping, and aviation. By utilizing renewable hydrogen for these hard-to-abate sectors, the overall carbon footprint of industrial processes can be significantly reduced, contributing to global decarbonization goals. As shown in Figure 1.3, hydrogen's role in the renewable energy ecosystem is central to enabling the large-scale integration of solar, wind, and hydro-based power systems. The diagram illustrates how hydrogen production, storage, and conversion fit within the broader renewable energy landscape, helping to manage fluctuations in renewable generation and ensure a stable energy supply. By connecting hydrogen production with renewable energy generation, we can create a more resilient and sustainable energy system, capable of meeting future energy demands while reducing carbon emissions [55, 56].

1.4 Current Global Hydrogen Economy and Future Roadmaps

1.4.1 Global Hydrogen Production and Consumption

Hydrogen has gained significant attention as a key component of the global energy transition, with its demand and supply experiencing rapid growth as countries move toward carbon-neutral economies. The hydrogen economy is evolving with various production methods and consumption applications, contributing to its rising importance in sectors like energy, industry, and transportation. In recent years, a global shift toward cleaner hydrogen production methods has intensified as part of a broader push to decarbonize key industries and reduce carbon emissions [57]. As outlined in Table 1.5,

Table 1.5 Global hydrogen production by method (gray, blue, green hydrogen). (Placed at the beginning of this section to establish the current state of production.)

Hydrogen production method	Process description	Current share of global hydrogen production (%)	CO_2 emissions	Key features
Gray hydrogen	Produced via SMR from natural gas, without carbon capture	95	High (8–12 tons of CO_2 per ton of hydrogen produced)	Inexpensive and widely used, but environmentally damaging due to CO_2 emissions [61]
Blue hydrogen	Produced similarly to gray hydrogen, but with CCS technologies to capture CO_2 emissions	4	Low (CO_2 emissions reduced by 50–90%)	Lower emissions than gray hydrogen, but still reliant on fossil fuels [62]
Green hydrogen	Produced via water electrolysis powered by renewable energy sources (wind, solar, hydropower)	1	Zero (net-zero emissions)	Sustainable and clean, with potential for large-scale production in the future [63]
Turquoise hydrogen	Produced through methane pyrolysis, where methane is split into hydrogen and solid carbon	Emerging	Zero or low emissions (depending on carbon management)	A potential new method for producing hydrogen with lower CO_2 emissions [64]
Pink hydrogen	Produced via electrolysis powered by nuclear energy	Emerging	Zero (no direct emissions)	Can provide a low-carbon hydrogen alternative, but dependent on nuclear energy [65]

the global hydrogen production is largely divided into three primary categories: gray hydrogen, blue hydrogen, and green hydrogen. These categories are distinguished by the methods used for production and their associated environmental impacts.

Gray hydrogen is currently the dominant form of hydrogen production, accounting for about 75–80% of global hydrogen supply. It is primarily produced through natural gas reforming, a process known as SMR. In this method, methane (CH_4) reacts with steam (H_2O) to produce hydrogen and carbon dioxide (CO_2). However, gray hydrogen is considered environmentally harmful because it results in the release of significant amounts of CO_2, contributing to GHG emissions. Despite its low production costs, gray hydrogen does not align with global decarbonization goals, making its reliance unsustainable in the long term. In response to the need for cleaner hydrogen production, blue hydrogen has emerged as a more sustainable alternative.

Blue hydrogen is produced using the same SMR process as gray hydrogen, but it incorporates carbon capture, utilization, and storage (CCUS) technologies to prevent CO_2 emissions from being released into the atmosphere. By capturing and storing CO_2, blue hydrogen significantly reduces its environmental impact, although it still depends on fossil fuels for its production. While blue hydrogen presents a transitional solution, it still carries some challenges related to CCUS infrastructure and the need for continued reliance on natural gas [58].

The green hydrogen revolution, however, represents the most promising future for hydrogen production. Green hydrogen is produced via the electrolysis of water, using renewable energy sources such as solar, wind, or hydropower to split water molecules into hydrogen and oxygen. The process does not involve any fossil fuels and results in zero carbon emissions, making green hydrogen the most environmentally friendly option. Although green hydrogen is currently more expensive to produce than gray or blue hydrogen, falling renewable energy costs and improvements in electrolyzer efficiency are expected to bring down its costs significantly over the coming decades. Green hydrogen is seen as the cornerstone of a sustainable hydrogen economy, with its potential to decarbonize heavy industries, provide long-term energy storage solutions, and contribute to the broader energy transition. As the global hydrogen economy continues to grow, demand for hydrogen is expanding across a range of sectors. In particular, hydrogen plays a vital role in decarbonizing hard-to-abate industries such as steel production, cement manufacturing, and heavy-duty transportation. Hydrogen is also being used for energy storage and balancing in grid systems, where its ability to store excess renewable energy and discharge it when needed makes it a valuable asset in the renewable energy ecosystem [59, 60].

The geopolitical landscape is also shifting in favor of hydrogen adoption, with countries like Japan, Germany, and the United States investing heavily in hydrogen infrastructure and research. Several nations have already established hydrogen roadmaps and policy frameworks to facilitate the scaling up of green hydrogen production and integrate it into existing energy systems. By 2030, it is expected that hydrogen will play a critical role in achieving carbon neutrality and facilitating the transition to cleaner energy worldwide. Overall, while hydrogen production methods are currently dominated by gray hydrogen, the future of hydrogen lies in blue and green production. Table 1.5 provides a comprehensive breakdown of global hydrogen production by method, demonstrating the progress in cleaner hydrogen production and highlighting the necessary transition toward more sustainable hydrogen solutions. The demand for hydrogen is set to grow exponentially, and as production technologies evolve, hydrogen will continue to play a pivotal role in the global energy transition.

1.4.2 Investments and Policies Driving Hydrogen Adoption

The global push toward hydrogen adoption is supported by a combination of government policies and private sector investments aimed at fostering the development of a hydrogen economy. These initiatives are designed to promote research, infrastructure development, and scaling up hydrogen production while addressing the key challenges related to cost, storage, and distribution. Both public and private sectors have recognized hydrogen's potential as a critical solution for achieving carbon neutrality and energy security, prompting a surge in investments across the hydrogen value chain [66].

Government initiatives play a significant role in hydrogen adoption, as many countries have recognized the importance of hydrogen in their energy strategies. Governments worldwide are introducing policies, incentives, and funding programs to accelerate the development and deployment of hydrogen technologies. The EU, for example, has outlined a comprehensive hydrogen strategy as part of its Green Deal, with a target to produce 10 million tons of green hydrogen by 2030. The European Commission has committed substantial financial support to green hydrogen projects, with billions allocated for research and development, as well as for funding infrastructure and market development. In Germany, the government has earmarked €9 billion as part of its National Hydrogen Strategy, aiming to establish the country as a leader in hydrogen production and technologies [67].

In the United States, hydrogen is a key component of the Biden administration's climate agenda. The US Department of Energy (DOE) has launched various programs to support hydrogen innovation, such as the Hydrogen and Fuel Cell Technologies Office, which funds research on hydrogen production, storage, and fuel cell technologies. The DOE has also pledged substantial funding to establish hydrogen hubs in different regions, particularly focusing on integrating hydrogen into industrial applications, transportation, and power sectors. The United States is aiming to produce 50 million metric tons of clean hydrogen annually by 2050, positioning hydrogen as an essential element of its energy transition strategy [68].

Japan has also been at the forefront of hydrogen adoption, positioning itself as a leader in the global hydrogen market. Japan's government has made significant investments in hydrogen research and development, with a long-term goal to establish a hydrogen-based society. The country has invested heavily in hydrogen fuel cell technologies for vehicles, power generation, and industrial applications. In addition to these governmental efforts, Japan has entered into international collaborations with other nations, such as Australia, to secure a reliable hydrogen supply chain for its energy needs [69].

Private sector investments are equally crucial in driving hydrogen adoption. Leading companies in the energy, automotive, and industrial sectors are investing billions in hydrogen technologies. Energy giants like Shell, BP, and TotalEnergies have committed to significant investments in hydrogen projects, particularly green hydrogen. These companies are focusing on expanding their portfolios of renewable energy sources while simultaneously developing hydrogen production capabilities to meet future demand. In the automotive industry, companies such as Toyota, Hyundai, and Honda have made substantial advancements in hydrogen fuel cell vehicles (FCVs) and have partnered with governments to promote hydrogen infrastructure development, such as hydrogen refueling stations. Startups and technology firms are also playing a vital role in the hydrogen ecosystem by driving innovations in hydrogen production, storage, and transportation. Companies specializing in electrolyzers, which are critical for green hydrogen production, are attracting substantial venture capital investment, helping to accelerate the commercialization of renewable hydrogen technologies. Both government policies and private sector investments are critical drivers of hydrogen adoption. Through strategic funding, technological advancements, and international collaborations, these initiatives are laying the foundation for a global hydrogen economy. With strong financial support and a clear focus on overcoming the current technological barriers, hydrogen is poised to play an increasingly central role in the transition to a sustainable and decarbonized energy future [70].

1.4.3 Key Players in Hydrogen Development

Key players in hydrogen development are strategically shaping the global hydrogen landscape, with major initiatives underway in Japan, the EU, the United States, China, and India. These regions are making substantial investments in research, technology, and infrastructure to accelerate the adoption of hydrogen as a clean energy solution. Japan has been a pioneer in hydrogen technologies and continues to lead the way in hydrogen adoption. The country has set ambitious targets, aiming to establish a hydrogen-based society by 2050. Japan's government has invested heavily in hydrogen production, fuel cell technologies, and infrastructure development. Key initiatives include the development of hydrogen FCVs and the establishment of a nationwide network of hydrogen refueling stations. Japan is also focusing on international partnerships to secure reliable hydrogen supply chains, with projects underway in Australia and other regions to import clean hydrogen.

In the EU, hydrogen is central to the Green Deal and the EU's strategy for decarbonization. The EU has committed to producing 10 million tons of green hydrogen by 2030, supported by policies and funding for hydrogen research and infrastructure. Key players in the EU hydrogen ecosystem include Germany, which has launched its National Hydrogen Strategy with a €9 billion investment, and France, which is focusing on hydrogen for heavy industries and transportation. The EU's hydrogen strategy also emphasizes the importance of cross-border collaboration to create a connected hydrogen market across member states [71].

The United States is making significant strides in hydrogen development under the Biden administration, which has prioritized clean energy and hydrogen in its climate agenda. The US DOE has funded various hydrogen projects, with a focus on hydrogen production, storage, and fuel cell technologies. The United States is also investing in hydrogen hubs to promote regional hydrogen economies, particularly in areas like transportation, power generation, and industry. California, a major hydrogen hub in the United States, is leading the charge with a network of hydrogen refueling stations and initiatives to reduce emissions in the transportation sector [72].

China, the world's largest emitter, is increasingly focusing on hydrogen as part of its efforts to decarbonize. The country has set ambitious goals for hydrogen production and infrastructure, with a particular focus on green hydrogen. China is investing in hydrogen FCVs and infrastructure, with plans to build a large network of hydrogen refueling stations in major cities [73].

India, with its growing energy demand, is also exploring hydrogen as a key part of its energy strategy. The Indian government is focusing on green hydrogen production using renewable energy sources like solar and wind. India has set a target to become a global hub for green hydrogen production and has introduced policies to encourage private sector investment and development of hydrogen infrastructure [74].

1.4.4 Future Hydrogen Economy Trends and Projections

a. **Hydrogen Market Outlook**
   - **Increasing Market Demand**: The global hydrogen market is projected to experience exponential growth in the coming decades. By 2050, the hydrogen economy could be worth over $2.5 trillion, driven by demand in sectors such as transportation, industry, and power generation. Analysts predict that hydrogen could account for up to 18% of global energy consumption by mid-century, making it a key player in decarbonization strategies [75].
   - **Rising Green Hydrogen Demand**: As nations set more ambitious decarbonization targets, the demand for green hydrogen, produced using renewable energy sources like wind, solar, and hydroelectric power, will increase significantly. Green hydrogen is expected to dominate the hydrogen market, with countries and corporations investing heavily in electrolysis technologies to generate it sustainably [76].
   - **Hydrogen as a Long-term Energy Solution**: With growing concerns about energy security and the environmental impact of fossil fuels, hydrogen is seen as a long-term solution to mitigate carbon emissions and diversify energy sources. As industries transition toward sustainable practices, hydrogen's role as a versatile and scalable energy carrier is set to expand [77].

b. **Emerging Hydrogen Technologies**
   - **Electrolysis Advancements**: The efficiency of electrolyzers (the devices used to produce hydrogen from water) will continue to improve with the development of new materials and technologies. Solid oxide electrolyzers (SOEC) and PEM electrolyzers are expected to become more efficient, reducing the cost of green hydrogen production. Advances in electrolyzer design and performance will help drive down the price of hydrogen and make it more competitive with fossil fuels [78].
   - **Storage and Transport Solutions**: Hydrogen storage technologies, including high-pressure tanks, liquefied hydrogen, and chemical hydrogen carriers, are evolving rapidly. Innovations in solid-state hydrogen storage and cryogenic storage will make hydrogen transport more cost-effective, especially for long-distance shipments. Storage solutions will also focus on improving energy density and minimizing hydrogen loss during storage [79].
   - **Fuel Cell Advancements**: Fuel cell technologies are expected to continue progressing, offering higher efficiency, durability, and reduced cost. PEM fuel cells and solid oxide fuel cells will see significant improvements, enabling hydrogen-powered vehicles, stationary power systems, and portable devices to become more mainstream [80].
   - **Hydrogen Blending**: In regions with existing natural gas infrastructure, hydrogen blending (mixing hydrogen with natural gas in existing pipelines) will become a key transitional strategy. This method can significantly reduce carbon emissions from natural gas without requiring extensive new infrastructure [81].

c. **Infrastructure Development**
   - **Expansion of Hydrogen Refueling Stations**: Infrastructure for hydrogen refueling stations is expected to expand, especially in countries with strong hydrogen strategies like Japan, Germany, and the United States. Investments in refueling infrastructure will be critical to support hydrogen-powered vehicles, including trucks, buses, and cars, making hydrogen more accessible to consumers [82].
   - **Hydrogen Pipelines**: To support the large-scale distribution of hydrogen, a global network of hydrogen pipelines will be developed, linking hydrogen production areas with demand centers. The expansion of hydrogen pipeline networks will allow for a more efficient and cost-effective distribution of hydrogen, making it a viable alternative to fossil fuels [83].
   - **Hydrogen Storage Facilities**: Large-scale hydrogen storage facilities will be built to facilitate the integration of hydrogen with renewable energy sources. These facilities will store hydrogen produced during periods of high renewable energy generation, providing a flexible and reliable energy supply when demand peaks or renewable output decreases [84].

d. **Cost Reductions**
   - **Declining Production Costs**: Hydrogen production costs, especially for green hydrogen, are expected to fall significantly. As renewable energy generation costs continue to decrease and electrolyzer technologies improve, the cost of green hydrogen will become more competitive with fossil fuels. By 2030, it is expected that the cost of green hydrogen will decrease by up to 50%, making it more accessible for industrial and commercial applications.
   - **Cost-Effective Storage and Transport**: As hydrogen storage and transportation technologies evolve, their costs will also decrease. Innovations in storage materials, pipeline construction, and transportation methods (such as liquid organic hydrogen carriers (LOHCs) or ammonia) will reduce the overall cost of delivering hydrogen to end-users, making it more economically feasible for widespread adoption.
   - **Economies of Scale**: As demand for hydrogen increases, economies of scale will drive down costs across the hydrogen value chain, from production to storage, transportation, and distribution. This will help lower the cost of hydrogen and improve its competitiveness compared to fossil fuel alternatives [85].

e. **Hydrogen Adoption in Various Sectors**
   - **Heavy Industry and Manufacturing**: Hydrogen is poised to play a critical role in decarbonizing industries like steel, cement, and chemicals, where electrification and direct renewable energy use are difficult. Hydrogen can replace fossil fuels in high-temperature industrial processes, significantly reducing emissions in these hard-to-abate sectors.
   - **Transportation and Mobility**: Hydrogen is expected to become a leading fuel for sectors such as heavy-duty transport, shipping, and aviation. Long-range, zero-emission vehicles powered by hydrogen fuel cells will be deployed at scale, alongside hydrogen-powered ships and planes, contributing to the decarbonization of the transport sector.

- **Power Generation**: Hydrogen is seen as a key solution for integrating intermittent renewable energy sources like wind and solar into the grid. Hydrogen can be stored and converted back into electricity when needed, providing a stable and reliable source of power for the grid. This will help solve the intermittency issues associated with renewable energy generation [86].

The hydrogen economy is poised for a transformative growth phase in the coming decades. Technological advancements, infrastructure expansion, and cost reductions will unlock hydrogen's potential as a clean and sustainable energy carrier. With ongoing global investments and the rise of hydrogen-driven industries, hydrogen will be central to achieving net-zero emissions and a resilient energy future.

1.5 Key Challenges in Hydrogen Adoption and Infrastructure Development

1.5.1 Scalability and Cost Barriers

The widespread adoption of hydrogen as a clean energy carrier faces significant scalability and cost challenges. While hydrogen is recognized for its potential to decarbonize industries and transportation, its large-scale deployment remains constrained by high production costs, infrastructure limitations, and economic feasibility concerns. Overcoming these barriers requires technological advancements, cost-reduction strategies, and strong policy frameworks to accelerate adoption [87].

One of the most pressing challenges in hydrogen deployment is its high production cost, particularly for green hydrogen, which is produced through electrolysis using renewable energy. The cost of green hydrogen currently ranges from $4 to $6 per kilogram, significantly higher than gray hydrogen (produced from fossil fuels) at $1 to $2 per kilogram. This cost disparity is largely due to expensive electrolyzer technology, high-energy consumption, and limited production capacity [88].

To reduce costs, ongoing research focuses on:

- Electrolyzer efficiency improvements through advanced catalysts and membrane materials.
- Economies of scale in hydrogen production plants, which can lower per unit costs.
- Integration of renewable energy sources with electrolysis to reduce electricity expenses.
- Development of modular and distributed hydrogen production systems to cut transportation costs [89].

Significant reductions in capital expenditures and operational expenditures are necessary for hydrogen to be cost-competitive with fossil fuels. Experts project that with technological advancements and increased production, green hydrogen costs could fall below $2 per kilogram by 2030, making it more viable for widespread use.

Large-scale hydrogen deployment faces logistical and economic hurdles related to storage, transportation, and refueling infrastructure. Unlike fossil fuels, hydrogen has a low volumetric energy density, requiring compression, liquefaction, or conversion into carriers like ammonia or LOHCs. These processes add significant costs and energy losses, reducing overall system efficiency. Additionally, the lack of a global hydrogen supply chain makes distribution expensive. Many regions lack dedicated pipelines or refueling stations, making long-distance transportation unfeasible. Addressing these challenges requires massive infrastructure investment and strategic planning to integrate hydrogen into existing energy networks [90]. Government policies play a crucial role in driving down costs and scaling hydrogen adoption. Many countries have introduced:

- Subsidies and tax incentives for hydrogen production and infrastructure.
- Carbon pricing mechanisms to make fossil-based hydrogen less competitive.
- Public-private partnerships to accelerate research, development, and deployment.

With strong policy support, technological advancements, and investment in infrastructure, hydrogen can become a cost-effective and scalable clean energy solution.

1.5.2 Infrastructure Gaps

The current hydrogen refueling infrastructure is limited and geographically dispersed, making hydrogen-powered transportation less practical compared to battery electric alternatives. Expanding hydrogen fueling stations requires significant investment in high-pressure storage, compressors, and distribution networks. Similarly, pipeline infrastructure for hydrogen transport is underdeveloped. Existing natural gas pipelines can be retrofitted to transport hydrogen, but concerns over material degradation (hydrogen embrittlement) and leakage risks present engineering challenges. Developing dedicated hydrogen pipelines is essential for large-scale industrial and mobility applications. Table 1.6 provides a comparative analysis of the costs associated with hydrogen infrastructure versus conventional energy infrastructure, highlighting the investment needs for hydrogen adoption.

1.5.3 Technological Hurdles

There are several technological hurdles to overcome in hydrogen adoption. One of the primary challenges lies in catalyst development for efficient electrolysis and fuel cells. Current catalysts, such as platinum, are expensive and scarce, limiting the scalability and affordability of hydrogen production. Research is ongoing to develop more cost-effective and abundant alternatives, such as nickel-based catalysts, to make hydrogen production more sustainable. Another significant hurdle is hydrogen storage. Storing hydrogen safely and efficiently remains a challenge due to its low volumetric energy density. Advanced materials such as metal hydrides, porous carbons, and LOHCs are being explored, but none have achieved widespread commercial success yet. Efficient hydrogen storage solutions will be critical for practical hydrogen utilization, particularly in transport and mobile applications. As these innovations mature, the cost-effectiveness and scalability of hydrogen technologies will improve, paving the way for broader adoption [101].

Table 1.6 Comparison of hydrogen infrastructure costs vs. conventional energy infrastructure.

Infrastructure type	Hydrogen (estimated cost)	Conventional energy (natural gas/ petroleum)	Challenges and considerations
Storage tanks (high pressure)	\$800–\$1,200 per kg H_2 capacity [91]	\$50–\$150 per barrel [92]	High-cost due to material strength and safety requirements
Cryogenic storage facilities	\$3–\$5 per kg H_2 [93]	N/A (fossil fuels stored at ambient temperatures) [94]	Requires extreme cooling, leading to high-energy losses
Pipeline infrastructure	\$1–\$2 million per mile [95]	\$0.2–\$0.5 million per mile [96]	Hydrogen embrittlement requires pipeline material modifications
Refueling stations	\$1–\$2 million per station [97]	\$200K–\$500K per gasoline station [98]	Requires high-pressure compression and safety measures
Transport (trucks, ships, rail)	\$5–\$15 per kg for delivery [99]	\$1–\$3 per barrel of oil [100]	Hydrogen's low energy density increases cost per unit of energy delivered

1.5.4 Public Perception and Market Readiness

Public perception of hydrogen is crucial for its acceptance and widespread adoption. Consumer awareness about the benefits and safety of hydrogen is still relatively low compared to other energy technologies, such as BEVs. Hydrogen is often perceived as a risky or untested technology, partly due to its flammability and the historical association with hydrogen explosions (e.g., the Hindenburg disaster). To address these concerns, consumer education campaigns are essential to highlight hydrogen's safety features and its potential as a clean energy solution. In terms of industry readiness, while there has been significant investment in hydrogen technologies, the pace of adoption varies by sector. Some industries, such as heavy-duty transport and aerospace, are more prepared for hydrogen integration due to their higher energy demands. However, the transition to hydrogen in sectors like residential heating or light-duty transport is slower. Industry readiness depends on the resolution of cost, efficiency, and infrastructure barriers, as well as clear government policies and incentives. As these challenges are addressed, market readiness for hydrogen adoption will increase, driving its widespread use [86].

1.5.5 Environmental and Safety Concerns

Hydrogen is widely regarded as a clean energy carrier with the potential to reduce global carbon emissions significantly. However, the large-scale deployment of hydrogen requires addressing environmental and safety concerns, particularly in production, storage, transport, and utilization. Hydrogen's flammability, low density, and wide explosive range present unique risks that necessitate strict safety protocols and robust regulatory frameworks to ensure safe handling. Additionally, hydrogen leakage and its indirect environmental effects must be carefully managed to maintain the sustainability benefits of hydrogen energy [102].

One of the primary safety concerns in hydrogen production arises from the high-pressure and high-temperature processes involved in its generation. Electrolysis, although a cleaner method, involves high-voltage electrical systems that pose shock and explosion hazards if not properly managed. SMR and coal gasification require high-temperature reactors, increasing the risk of industrial accidents. Additionally, in blue hydrogen production, CCS systems must be monitored carefully to prevent CO_2 leakage, which could undermine hydrogen's environmental advantages [103].

Storage and transport of hydrogen present further challenges due to its low volumetric energy density and tendency to leak. Hydrogen can be stored as compressed gas (350–700 bar), cryogenic liquid (−253 °C), or in chemical carriers such as ammonia and LOHCs. Each method has safety trade-offs. High-pressure storage requires strong containment materials to prevent ruptures, while cryogenic hydrogen storage faces challenges related to boil-off losses and insulation efficiency. Chemical hydrogen carriers mitigate some of these risks but introduce new challenges, such as toxicity (ammonia) and complex conversion processes (LOHCs and metal hydrides). Transportation through pipelines, trucks, and tankers also requires specialized infrastructure to prevent hydrogen embrittlement, leaks, and ignition risks. Hydrogen refueling stations, critical for transportation applications, must be equipped with high-pressure compressors, flame detectors, and emergency venting systems to ensure operational safety [104].

Hydrogen leakage is a major concern due to its small molecular size, making it prone to escaping through tiny gaps in pipes, tanks, and valves. Unlike natural gas, hydrogen burns with an almost invisible flame, making fire detection difficult. In industrial and refueling applications, advanced leak detection systems, such as infrared sensors and optical gas imaging technologies, must be implemented to ensure safety. Additionally, hydrogen embrittlement-resistant materials should be used in pipelines and containment systems to prevent structural failures. Beyond direct safety risks, hydrogen leakage also has environmental consequences. While hydrogen itself is not a GHG, excessive leakage into the atmosphere can contribute to climate change indirectly. Hydrogen can react with atmospheric hydroxyl radicals (OH•), which are responsible for breaking down methane (CH_4), a potent greenhouse gas. If hydrogen leakage is not controlled, it could prolong the atmospheric lifetime of methane, reducing the overall climate benefits of hydrogen energy. Thus, minimizing hydrogen leaks across the supply chain is crucial to ensure its net-zero carbon potential [105].

Regulatory frameworks play a vital role in establishing global hydrogen safety standards. Organizations such as the International Organization for Standardization (ISO) and the Occupational Safety and Health Administration have set guidelines for hydrogen handling, refueling stations, and industrial applications. The EU Hydrogen Strategy and Japan's Ministry of Economy, Trade, and Industry Hydrogen Safety Guidelines also outline stringent safety regulations for hydrogen infrastructure and transport. Ensuring strict compliance with these standards, along with ongoing advancements in leak detection, fire suppression, and infrastructure resilience, will be key to mitigating risks and making hydrogen a truly safe and scalable energy solution [106]. As hydrogen adoption continues to grow, addressing safety concerns, improving infrastructure, and minimizing leakage risks will be critical for its long-term sustainability. Innovations in materials science, AI-driven predictive safety systems, and enhanced regulatory frameworks will play a major role in ensuring that hydrogen is not only clean but also safe and reliable for widespread use in industrial, transportation, and grid applications. By prioritizing both environmental and safety measures, hydrogen can fulfill its potential as a key enabler of the global energy transition [107].

Figure 1.4 illustrates the key challenges in the adoption and development of hydrogen infrastructure through a circular flow diagram. At the center, "Challenges in Hydrogen Adoption and Infrastructure Development" serves as the focal point, highlighting the interconnected nature of the issues. The diagram identifies seven major challenges: scalability, cost barriers, infrastructure gaps, technological hurdles, public perception, market readiness, and environmental and safety concerns. Scalability remains a critical challenge due to the difficulty in expanding hydrogen production and infrastructure on a large scale. Cost barriers, including the high costs of production, transportation, and storage, further hinder adoption. Infrastructure gaps, such as the limited number of refueling stations and inadequate distribution networks, restrict hydrogen's widespread use. Technological hurdles involve the need for advancements in fuel cell efficiency and hydrogen production methods. Public perception remains a significant challenge, as skepticism and safety concerns

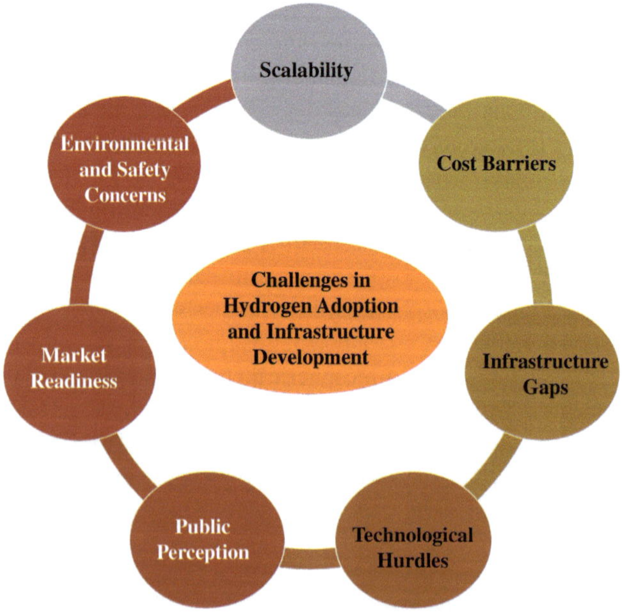

Figure 1.4 Key challenges in hydrogen adoption and infrastructure development.

affect consumer confidence. Market readiness is another barrier, with insufficient market demand and commercial viability slowing down progress. Finally, environmental and safety concerns, including the risks associated with hydrogen leakage and production emissions, further complicate adoption. The circular design of the diagram emphasizes the interdependence of these challenges, suggesting that addressing one aspect will likely influence others, highlighting the need for a comprehensive and integrated approach to overcoming these barriers.

References

1 Lubitz W, Tumas W. Hydrogen: An overview. *Chem Rev* [Internet]. 2007 Oct [cited 2025 Feb 1];107(10):3900–3. Available from: https://pubs.acs.org/doi/full/10.1021/cr050200z.

2 Rigden JS. *Hydrogen: The Essential Element*. New York: Harvard University Press; 2003.

3 Keçebaş A, Kayfeci M. Hydrogen properties. In: *Solar Hydrogen Production: Processes, Systems and Technologies*. Massachusetts: Academic Press; 2019. 3–29.

4 Idriss H, Scott M, Subramani V. Introduction to hydrogen and its properties. In: *Compendium of Hydrogen Energy: Hydrogen Production and Purification*: Cambridge, UK: Woodhead Publishing Series in Energy; 2015. 3–19.

5 Wang F, Shifa TA, Yu P, He P, Liu Y, Wang F, et al. New Frontiers on van der Waals layered metal phosphorous trichalcogenides. *Adv Funct Mater* [Internet]. 2018 Sep 1 [cited 2025 Apr 6];28(37):1802151. Available from: https://onlinelibrary.wiley.com/doi/full/10.1002/adfm.201802151.

6 Hosseini SE, Wahid MA. Hydrogen from solar energy, a clean energy carrier from a sustainable source of energy. *Int J Energy Res* [Internet]. 2020 May 1 [cited 2025 Feb 1];44(6):4110–31. Available from: https://onlinelibrary.wiley.com/doi/full/10.1002/er.4930.

7 Hosseini SE, Wahid MA. Hydrogen production from renewable and sustainable energy resources: Promising green energy carrier for clean development. *Renew and Sustain Energy Rev*. 2016 May 1; 57:850–66.

8 Hassan Q, Algburi S, Sameen AZ, Jaszczur M, Salman HM. Hydrogen as an energy carrier: Properties, storage methods, challenges, and future implications. *Environ Syst Decis* [Internet]. 2024 Jun 1 [cited 2025 Feb 1];44(2):327–50. Available from: https://link.springer.com/article/10.1007/s10669-023-09932-z.

9 Zhang L, Jia C, Bai F, Wang W, An S, Zhao K, et al. A comprehensive review of the promising clean energy carrier: Hydrogen production, transportation, storage, and utilization (HPTSU) technologies. *Fuel*. 2024 Jan 1;355:129455.

10 Akpasi SO, Michael I, Anekwe S, Kweinor Tetteh E, Amune UO, Ishola Mustapha S, et al. Hydrogen as a clean energy carrier: Advancements, challenges, and its role in a sustainable energy future. *Clean Energy* [Internet]. 2025 Jan 22 [cited 2025 Feb 1];9(1):52–88. Available from: https://dx.doi.org/10.1093/ce/zkae112.

11 Mazloomi K, Gomes C. Hydrogen as an energy carrier: Prospects and challenges. *Renew and Sustain Energy Rev*. 2012 Jun 1; 16(5):3024–33.

12 Yue M, Lambert H, Pahon E, Roche R, Jemei S, Hissel D. Hydrogen energy systems: A critical review of technologies, applications, trends and challenges. *Renew and Sustain Energy Rev*. 2021 Aug 1;146:111180.

13 Fraidl G, Kapus P, Vidmar K. The gasoline engine and RDE challenges and prospects. In: Bargende, M, Reuss, HC, Wiedemann, J. (eds) 16. *Internationales Stuttgarter Symposium*. 2016 [cited 2025 Feb 1];257–83. Available from: https://link.springer.com/chapter/10.1007/978-3-658-13255-2_20.

14 Rathnayaka RIA, Ranjith PG. Hydrogen storage in unlined rock caverns: An insight on opportunities and challenges. *Int J Hydrogen Energy*. 2024 Jun 27; 72:1023–37.

15 Saldarini A, Longo M, Brenna M, Zaninelli D. Battery electric storage systems: Advances, challenges, and market trends. *Energies* 2023, Vol 16, Page 7566 [Internet]. 2023 Nov 14 [cited 2025 Feb 1];16(22):7566. Available from: https://www.mdpi.com/1996-1073/16/22/7566/htm.

16 Tarhan C, Çil MA. A study on hydrogen, the clean energy of the future: Hydrogen storage methods. *J Energy Storage*. 2021 Aug 1;40:102676.

17 Evro S, Oni BA, Tomomewo OS. Carbon neutrality and hydrogen energy systems. *Int J Hydrogen Energy*. 2024 Aug 12; 78:1449–67.

18 Chen L, Msigwa G, Yang M, Osman AI, Fawzy S, Rooney DW, et al. Strategies to achieve a carbon neutral society: A review. *Environ Chem Lett* [Internet]. 2022 Aug 1 [cited 2025 Feb 2];20(4):2277–310. Available from: https://link.springer.com/article/10.1007/s10311-022-01435-8.

19 Maestre VM, Ortiz A, Ortiz I. Challenges and prospects of renewable hydrogen-based strategies for full decarbonization of stationary power applications. *Renew and Sustain Energy Rev*. 2021 Dec 1; 152:111628.

20 Bartlett J, Krupnick A. Decarbonized hydrogen in the US power and industrial sectors: Identifying and incentivizing opportunities to lower emissions. [Internet]. 2020, Dec 1. Available from: https://www.rff.org/publications/reports/decarbonizing-hydrogen-us-power-and-industrial-sectors/.

21 Ameli H, Strbac G, Pudjianto D, Ameli MT. A review of the role of hydrogen in the heat decarbonization of future energy systems: Insights and perspectives. *Energies* 2024, Vol 17, Page 1688 [Internet]. 2024 Apr 2 [cited 2025 Feb 2]; 17(7):1688. Available from: https://www.mdpi.com/1996-1073/17/7/1688/htm.

22 Gondal IA. Hydrogen integration in power-to-gas networks. *Int J Hydrogen Energy*. 2019 Jan 15; 44(3):1803–15.

23 Shah JK, Sharma M, Gunasekaran A, Joshi S. Modeling critical enablers of hydrogen supply chains for decarbonization: Insights from emerging economies. [Internet] 2024 [cited 2025 Feb 2]. Available from: https://papers.ssrn.com/abstract=5013821.

24 Züttel A, Remhof A, Borgschulte A, Friedrichs O. Hydrogen: The future energy carrier. *Philosophical Transactions of the Royal Society A: Mathematical, Physical and Engineering Sciences* [Internet]. 2010 Jul 28 [cited 2025 Feb 2]; 368(1923):3329–42. Available from: https://royalsocietypublishing.org/doi/10.1098/rsta.2010.0113.

25 Rosen MA, Koohi-Fayegh S. The prospects for hydrogen as an energy carrier: An overview of hydrogen energy and hydrogen energy systems. *Energy Ecol Environ* [Internet]. 2016 Feb 1 [cited 2025 Feb 2];1(1):10–29. Available from: https://link.springer.com/article/10.1007/s40974-016-0005-z.

26 Kalamaras CM, Efstathiou AM, Al-Assaf Y, Poullikkas A. Hydrogen production technologies: Current state and future developments. *Conference Papers in Science* [Internet]. 2013 Jan 1 [cited 2025 Feb 2];2013(1):690627. Available from: https://onlinelibrary.wiley.com/doi/full/10.1155/2013/690627.

27 Melaina MW. Market Transformation Lessons for Hydrogen from the Early History of the Manufactured Gas Industry. *Hydrogen Energy and Vehicle Systems* [Internet]. 2016 Apr 19 [cited 2025 Feb 2];150–85. Available from: https://www.taylorfrancis.com/chapters/edit/10.1201/b13046-12/market-transformation-lessons-hydrogen-early-history-manufactured-gas-industry-marc-melaina.

28 Yusaf T, Faisal Mahamude AS, Kadirgama K, Ramasamy D, Farhana K, Dhahad HA, et al. Sustainable hydrogen energy in aviation – A narrative review. *Int J Hydrogen Energy*. 2024 Jan 2; 52:1026–45.

29 Johnston B, Mayo MC, Khare A. Hydrogen: The energy source for the 21st century. *Technovation*. 2005 Jun 1; 25(6):569–85.

30 Kovač A, Paranos M, Marciuš D. Hydrogen in energy transition: A review. *Int J Hydrogen Energy*. 2021 Mar 3; 46(16):10016–35.

31 Bockris JOM. The hydrogen economy: Its history. *Int J Hydrogen Energy*. 2013 Feb 27; 38(6):2579–88.

32 Spazzafumo G, Sørensen B. *Hydrogen and Fuel Cells: Emerging Technologies and Applications*. Ed. 3. Academic Press; 2018.

33 Okolie JA, Patra BR, Mukherjee A, Nanda S, Dalai AK, Kozinski JA. Futuristic applications of hydrogen in energy, biorefining, aerospace, pharmaceuticals and metallurgy. *Int J Hydrogen Energy*. 2021 Feb 19; 46(13):8885–905.

34 Baroutaji A, Wilberforce T, Ramadan M, Olabi AG. Comprehensive investigation on hydrogen and fuel cell technology in the aviation and aerospace sectors. *Renew and Sustain Energy Rev*. 2019 May 1; 106:31–40.

35 Luo N, Miley GH, Gimlin R, Burton R, Rusek J, Holcomb F. Hydrogen-peroxide-based fuel cells for space power systems. [Internet]. 2012 May 23 [cited 2025 Feb 2];24(3):583–9. Available from: https://arc.aiaa.org/doi/10.2514/1.31522.

36 Giacoppo G, Hovland S, Barbera O. 2 kW Modular PEM fuel cell stack for space applications: Development and test for operation under relevant conditions. *Appl Energy*. 2019 May 15; 242:1683–96.

37 Guo Q, Ye F, Guo H, Ma CF. Gas/water and heat management of PEM-based fuel cell and electrolyzer systems for space applications. *Microgravity Sci Technol* [Internet]. 2017 Feb 1 [cited 2025 Feb 2];29(1–2):49–63. Available from: https://link.springer.com/article/10.1007/s12217-016-9525-6.

38 Capurso T, Stefanizzi M, Torresi M, Camporeale SM. Perspective of the role of hydrogen in the 21st century energy transition. *Energy Convers Manag*. 2022 Jan 1; 251:114898.

39 Farias CBB, Barreiros RCS, da Silva MF, Casazza AA, Converti A, Sarubbo LA. Use of hydrogen as fuel: A trend of the 21st century. *Energies* [Internet]. 2022 Jan 3 [cited 2025 Feb 2];15(1):311. Available from: https://www.mdpi.com/1996-1073/15/1/311/htm.

40 Majewski S, Zhao X, Vivanco-Martín B, Iranzo A. Analysis of the European strategy for hydrogen: A comprehensive review. *Energies* 2023, Vol 16, Page 3866 [Internet]. 2023 May 2 [cited 2025 Feb 2];16(9):3866. Available from: https://www.mdpi.com/1996-1073/16/9/3866/htm.

41 Noussan M, Raimondi PP, Scita R, Hafner M. The role of Green and Blue hydrogen in the energy transition—A technological and geopolitical perspective. *Sustainability* 2021, Vol 13, Page 298 [Internet]. 2020 Dec 31 [cited 2025 Feb 2];13(1):298. Available from: https://www.mdpi.com/2071-1050/13/1/298/htm.

42 Andrews J, Shabani B. Re-envisioning the role of hydrogen in a sustainable energy economy. *Int J Hydrogen Energy*. 2012 Jan 1; 37(2):1184–203.

43 Simpson AP, Lutz AE. Exergy analysis of hydrogen production via steam methane reforming. *Int J Hydrogen Energy*. 2007 Dec 1; 32(18):4811–20.

44 Rosa L, Mazzotti M. Potential for hydrogen production from sustainable biomass with carbon capture and storage. *Renewable and Sustainable Energy Reviews*. 2022 Apr 1;157:112123.

45 Godula-Jopek A. *Hydrogen production: by electrolysis*. Wiley; 2015.

46 Barelli L, Bidini G, Gallorini F, Servili S. Hydrogen production through sorption-enhanced steam methane reforming and membrane technology: A review. *Energy*. 2008 Apr 1; 33(4):554–70.

47 Midilli A, Kucuk H, Topal ME, Akbulut U, Dincer I. A comprehensive review on hydrogen production from coal gasification: Challenges and opportunities. *Int J Hydrogen Energy*. 2021 Jul 21; 46(50):25385–412.

48 Stenina I, Yaroslavtsev A. Modern technologies of hydrogen production. *Processes*. 2023, Vol 11, Page 56 [Internet]. 2022 Dec 26 [cited 2025 Feb 2];11(1):56. Available from: https://www.mdpi.com/2227-9717/11/1/56/htm.

49 Chang ACC, Chang HF, Lin FJ, Lin KH, Chen CH. Biomass gasification for hydrogen production. *Int J Hydrogen Energy*. 2011 Oct 1; 36(21):14252–60.

50 Ishaq H, Crawford C. Towards less carbon-intensive blue hydrogen: Integrated natural gas reforming and CO_2 capture approach. *J Environ Chem Eng*. 2025 Apr 1; 13(2):115043.

51 Shiva Kumar S, Lim H. An overview of water electrolysis technologies for green hydrogen production. *Energy Reports*. 2022 Nov 1; 8:13793–813.

52 Bhandari R, Adhikari N. A comprehensive review on the role of hydrogen in renewable energy systems. *Int J Hydrogen Energy*. 2024 Sep 11; 82:923–51.

53 Bampaou M, Panopoulos KD. An overview of hydrogen valleys: Current status, challenges and their role in increased renewable energy penetration. *Renew and Sustain Energy Rev*. 2025 Jan 1;207:114923.

54 Noyan OF, Hasan MM, Pala N. A global review of the hydrogen energy ecosystem. *Energies*. 2023, Vol 16, Page 1484 [Internet]. 2023 Feb 2 [cited 2025 Feb 2];16(3):1484. Available from: https://www.mdpi.com/1996-1073/16/3/1484/htm.

55 Chapman A, Itaoka K, Hirose K, Davidson FT, Nagasawa K, Lloyd AC, et al. A review of four case studies assessing the potential for hydrogen penetration of the future energy system. *Int J Hydrogen Energy*. 2019 Mar 8; 44(13):6371–82.

56 Hassan Q, Nassar AK, Al-Jiboory AK, Viktor P, Telba AA, Awwad EM, et al. Mapping Europe renewable energy landscape: Insights into solar, wind, hydro, and green hydrogen production. *Technol Soc*. 2024 Jun1; 77:102535.

57 Ishaq H, Dincer I, Crawford C. A review on hydrogen production and utilization: Challenges and opportunities. *Int J Hydrogen Energy*. 2022 Jul 22; 47(62):26238–64.

58 Panchenko VA, Daus YV, Kovalev AA, Yudaev IV, Litti YV. Prospects for the production of green hydrogen: Review of countries with high potential. *Int J Hydrogen Energy*. 2023 Feb 8; 48(12):4551–71.

59 Squadrito G, Maggio G, Nicita A. The green hydrogen revolution. *Renew Energy*. 2023 Nov 1;216:119041.

60 Sebbagh T, Şahin ME, Beldjaatit C. Green hydrogen revolution for a sustainable energy future. *Clean Technol Environ Policy* [Internet]. 2024 Dec 1 [cited 2025 Feb 2];26(12):4017–40. Available from: https://link.springer.com/article/10.1007/s10098-024-02995-9.

61 Dumančić A, Vlahinić N, Skok M. Replacing gray hydrogen with renewable hydrogen at the consumption location using the example of the existing fertilizer plant. *Sustainability (Switzerland)* [Internet]. 2024 Aug 1 [cited 2025 Feb 2]; 16(15):6437. Available from: https://openurl.ebsco.com/contentitem/doi:10.3390%2Fsu16156437?sid=ebsco:plink:crawler&id=ebsco:doi:10.3390%2Fsu16156437.

62 Bauer C, Treyer K, Antonini C, Bergerson J, Gazzani M, Gencer E, et al. On the climate impacts of blue hydrogen production. *Sustain Energy Fuels* [Internet]. 2021 Dec 21 [cited 2025 Feb 2];6(1):66–75. Available from: https://pubs.rsc.org/en/content/articlehtml/2022/se/d1se01508g.

63 Moritz M, Schönfisch M, Schulte S. Estimating global production and supply costs for green hydrogen and hydrogen-based green energy commodities. *Int J Hydrogen Energy*. 2023 Mar 22; 48(25):9139–54.

64 Diab J, Fulcheri L, Hessel V, Rohani V, Frenklach M. Why turquoise hydrogen will be a game changer for the energy transition. *Int J Hydrogen Energy*. 2022 Jul 19; 47(61):25831–48.

65 Fernández-Arias P, Antón-Sancho Á, Lampropoulos G, Vergara D. Emerging trends and challenges in pink hydrogen research. *Energies*. 2024, Vol 17, Page 2291 [Internet]. 2024 May 10 [cited 2025 Feb 2];17(10):2291. Available from: https://www.mdpi.com/1996-1073/17/10/2291/htm.

66 Cheekatamarla P. Hydrogen and the global energy transition—Path to sustainability and adoption across all economic sectors. *Energies*. 2024, Vol 17, Page 807 [Internet]. 2024 Feb 8 [cited 2025 Feb 2];17(4):807. Available from: https://www.mdpi.com/1996-1073/17/4/807/htm.

67 Alam SN, Khalid Z, Singh # Bhaskar, Guldhe A. Integration of Government Policies on the Global Level for Green Hydrogen Production. *ACS Symposium Series* [Internet]. 2024 Aug 21 [cited 2025 Feb 2];19:1–28. Available from: https://pubs.acs.org/doi/full/10.1021/bk-2024-1473.ch001.

68 Bade SO, Tomomewo OS, Meenakshisundaram A, Ferron P, Oni BA. Economic, social, and regulatory challenges of green hydrogen production and utilization in the US: A review. *Int J Hydrogen Energy*. 2024 Jan 2; 49:314–35.

69 Chaube A, Chapman A, Shigetomi Y, Huff K, Stubbins J. The role of hydrogen in achieving long term Japanese energy system goals. *Energies*. 2020, Vol 13, Page 4539 [Internet]. 2020 Sep 2 [cited 2025 Feb 2];13(17):4539. Available from: https://www.mdpi.com/1996-1073/13/17/4539/htm.

70 Harichandan S, Kar SK. Financing the hydrogen industry: Exploring demand and supply chain dynamics. *Environ Sci Pollut Res* [Internet]. 2023 Oct 9 [cited 2025 Feb 2];1:1–18. Available from: https://link.springer.com/article/10.1007/s11356-023-30262-9.

71 Lagioia G, Spinelli MP, Amicarelli V. Blue and green hydrogen energy to meet European Union decarbonisation objectives. An overview of perspectives and the current state of affairs. *Int J Hydrogen Energy*. 2023 Jan 12; 48(4):1304–22.

72 Lattin WC, Utgikar VP. Transition to hydrogen economy in the United States: A 2006 status report. *Int J Hydrogen Energy*. 2007 Oct 1; 32(15):3230–7.

73 Chai S, Zhang G, Li G, Zhang Y. Industrial hydrogen production technology and development status in China: A review. *Clean Technol Environ Policy* [Internet]. 2021 Sep 1 [cited 2025 Feb 2];23(7):1931–46. Available from: https://link.springer.com/article/10.1007/s10098-021-02089-w.

74 Kar SK, Sinha ASK, Harichandan S, Bansal R, Balathanigaimani MS. Hydrogen economy in India: A status review. *Wiley Interdiscip Rev Energy Environ* [Internet]. 2023 Jan 1 [cited 2025 Feb 2];12(1):e459. Available from: https://onlinelibrary.wiley.com/doi/full/10.1002/wene.459.

75 Wappler M, Unguder D, Lu X, Ohlmeyer H, Teschke H, Lueke W. Building the green hydrogen market – Current state and outlook on green hydrogen demand and electrolyzer manufacturing. *Int J Hydrogen Energy*. 2022 Sep 15; 47(79):33551–70.

76 Sapnken FE, Posso F, Kibong MT, Noumo PG, Fantah AC, Tamba JG. Green hydrogen demand in Cameroon's energy sectors by 2040. *Renew and Sustain Energy Rev*. 2024 Nov 1;205:114834.

77 Kharel S, Shabani B. Hydrogen as a long-term large-scale energy storage solution to support renewables. *Energies*. 2018, Vol 11, Page 2825 [Internet]. 2018 Oct 19 [cited 2025 Feb 2];11(10):2825. Available from: https://www.mdpi.com/1996-1073/11/10/2825/htm.

78 Arsad SR, Arsad AZ, Ker PJ, Hannan MA, Tang SGH, Goh SM, et al. Recent advancement in water electrolysis for hydrogen production: A comprehensive bibliometric analysis and technology updates. *Int J Hydrogen Energy*. 2024 Mar 22; 60:780–801.

79 Hassan Q, Sameen AZ, Salman HM, Jaszczur M, Al-Jiboory AK. RETRACTED: Hydrogen energy future: Advancements in storage technologies and implications for sustainability. *J Energy Storage*. 2023 Nov 20;72:108404.

80 Fan L, Tu Z, Chan SH. Recent development of hydrogen and fuel cell technologies: A review. *Energy Reports*. 2021 Nov 1; 7:8421–46.

81 Topolski K, Reznicek EP, Erdener BC, San Marchi CW, Ronevich JA, Fring L, et al. Hydrogen blending into natural gas pipeline infrastructure: Review of the state of technology. 2022 Oct 10 [cited 2025 Feb 2]; Available from: https://www.osti.gov/servlets/purl/1893355/.

82 Apostolou D, Xydis G. A literature review on hydrogen refuelling stations and infrastructure. Current status and future prospects. *Renew and Sustain Energy Rev.* 2019 Oct 1; 113:109292.

83 Moreno-Benito M, Agnolucci P, Papageorgiou LG. Towards a sustainable hydrogen economy: Optimisation-based framework for hydrogen infrastructure development. *Comput Chem Eng.* 2017 Jul 12; 102:110–27.

84 Shahriar MF, Khanal A, Khan MI, Pandey R. Current status of underground hydrogen storage: Perspective from storage loss, infrastructure, economic aspects, and hydrogen economy targets. *J Energy Storage.* 2024 Sep 1;97:112773.

85 *The hydrogen economy opportunities, costs, barriers, and R&D needs.* National Academy of Sciences; 2004. https://nap.nationalacademies.org/catalog/10922/the-hydrogen-economy-opportunities-costs-barriers-and-rd-needs.

86 Gordon JA, Balta-Ozkan N, Nabavi SA. Homes of the future: Unpacking public perceptions to power the domestic hydrogen transition. *Renew and Sustain Energy Rev.* 2022 Aug 1. Elsevier/Science Direct; 164: 112481.

87 Kim JW, Boo KJ, Cho JH, Moon I. Key challenges in the development of an infrastructure for hydrogen production, delivery, storage and use. *Advances in Hydrogen Production, Storage and Distribution.* 2014 Jan 1; 3–31.

88 Tseng P, Lee J, Friley P. A hydrogen economy: opportunities and challenges. *Energy.* 2005 Nov 1; 30(14):2703–20.

89 Nyangon J, Darekar A. Advancements in hydrogen energy systems: A review of levelized costs, financial incentives and technological innovations. *Innovation and Green Development.* 2024 Sep 1; 3(3):100149.

90 Shardeo V, Sarkar BD. Adoption of hydrogen-fueled freight transportation: A strategy toward sustainability. *Bus Strategy Environ* [Internet]. 2024 Feb 1 [cited 2025 Mar 17];33(2):223–40. Available from: https://onlinelibrary.wiley.com/doi/full/10.1002/bse.3482.

91 Zhang M, Lv H, Kang H, Zhou W, Zhang C. A literature review of failure prediction and analysis methods for composite high-pressure hydrogen storage tanks. *Int J Hydrogen Energy.* 2019 Oct 4; 44(47):25777–99.

92 Liu Z, Pan H, Liu Y, Li Y. Thermodynamic performance on the pressurized discharge process from a cryogenic fuel storage tank. *Int J Hydrogen Energy.* 2022 Mar 19; 47(24):12107–18.

93 Carpetis C. Estimation of storage costs for large hydrogen storage facilities. *Int J Hydrogen Energy.* 1982 Jan 1; 7(2):191–203.

94 Gandhi A, Zantye MS, Faruque Hasan MM. Integration of cryogenic energy storage with renewables and power plants: Optimal strategies and cost analysis. *Energy Convers Manag.* 2022 Oct 1;269:116165.

95 Schoots K, Rivera-Tinoco R, Verbong G, van der Zwaan B. Historical variation in the capital costs of natural gas, carbon dioxide and hydrogen pipelines and implications for future infrastructure. *International Journal of Greenhouse Gas Control.* 2011 Nov 1; 5(6):1614–23.

96 Maroufmashat A, Fowler M. Transition of future energy system infrastructure; through power-to-gas pathways. *Energies*. 2017, Vol 10, Page 1089 [Internet]. 2017 Jul 26 [cited 2025 Mar 17];10(8):1089. Available from: https://www.mdpi.com/1996-1073/10/8/1089/htm.

97 Reddi K, Elgowainy A, Rustagi N, Gupta E. Impact of hydrogen refueling configurations and market parameters on the refueling cost of hydrogen. *Int J Hydrogen Energy*. 2017 Aug 24; 42(34):21855–65.

98 Isaac N, Saha AK. A review of the optimization strategies and methods used to locate hydrogen fuel refueling stations. *Energies*. 2023, Vol 16, Page 2171 [Internet]. 2023 Feb 23 [cited 2025 Mar 17];16(5):2171. Available from: https://www.mdpi.com/1996-1073/16/5/2171/htm.

99 Simbeck DR, Chang E. Hydrogen supply: Cost estimate for hydrogen pathways-scoping analysis. 2002 [cited 2025 Mar 17]. Available from: http://www.osti.gov/bridge.

100 Khan KA, Hasan M, Islam MA, Alim MA, Asma U, Hassan L, et al. A study on conventional energy sources for power production. [cited 2025 Mar 17];4:2018. Available from: www.ijariie.com.

101 Karduri RKR, Ananth C. Hydrogen economy: Opportunities and challenges for a sustainable future. *International Journal of Advanced Research in Basic Engineering Sciences and Technology (IJARBEST)*. 2020;6(2): 69–76.

102 Kazi KSL. Advancing towards sustainable energy with hydrogen solutions: Adaptation and challenges. https://services.igi-global.com/resolvedoi/resolve.aspx?doi=104018/979-8-3693-8814-3.ch013 [Internet]. 1AD Jan 1 [cited 2025 Mar 18];357–94. Available from: https://www.igi-global.com/chapter/advancing-towards-sustainable-energy-with-hydrogen-solutions/370890.

103 Kamran M, Turzyński M. Exploring hydrogen energy systems: A comprehensive review of technologies, applications, prevailing trends, and associated challenges. *J Energy Storage*. 2024 Aug 15;96:112601.

104 Boretti A, Pollet BG. Hydrogen economy: Paving the path to a sustainable, low-carbon future. *Int J Hydrogen Energy*. 2024 Dec 3; 93:307–19.

105 Alsulaiman A. Review of hydrogen leakage along the supply chain: Environmental impact, mitigation, and recommendations for sustainable deployment. 2024 [cited 2025 Mar 18]; Available from: https://www.econstor.eu/handle/10419/306643.

106 Jumaat NAS, Khalid AAH. A comprehensive review of challenges, prospects, and future perspectives for hydrogen energy development in Malaysia. *Int J Hydrogen Energy*. 2024 Feb 15; 55:65–77.

107 Mittal H, Kushwaha OS. Policy implementation roadmap, diverse perspectives, challenges, solutions towards low-carbon hydrogen economy. *Green and Low-Carbon Economy* [Internet]. 2024 Mar 11 [cited 2025 Mar 18]; Available from: https://ojs.bonviewpress.com/index.php/GLCE/article/view/1846.

2

Innovative Hydrogen Production Technologies

2.1 Introduction to Hydrogen Production Technologies

Hydrogen has become a key energy carrier in the transition toward a low-carbon future. As industries and governments seek alternatives to fossil fuels, hydrogen offers a versatile and sustainable energy solution with applications in transportation, power generation, and industrial processes. However, the viability of hydrogen as a clean energy source depends significantly on how it is produced. Various hydrogen production technologies exist, ranging from fossil fuel-based methods to renewable-powered processes. The development of cost-effective and efficient hydrogen production techniques is essential for enabling large-scale adoption and reducing greenhouse gas (GHG) emissions [1].

Hydrogen is the most abundant element in the universe, yet it rarely exists in its pure molecular form on earth. Instead, it must be extracted from water or hydrocarbons through various chemical and electrochemical processes. Once produced, hydrogen can serve as an efficient energy carrier that can be stored, transported, and utilized in different sectors. The concept of the hydrogen economy envisions a future where hydrogen replaces fossil fuels in major industries, enabling deep decarbonization and sustainable energy management. Hydrogen production technologies can be broadly classified into three categories: fossil fuel-based, electrolysis-based, and renewable-driven methods. Fossil fuel-based production, particularly steam methane reforming (SMR) [2] and coal gasification, dominates the current hydrogen supply, but these methods generate significant carbon emissions. Electrolysis-based techniques, including alkaline electrolysis, PEM electrolysis [3], and solid oxide electrolysis (SOEC) [4], offer cleaner alternatives by using electricity to split water into hydrogen and oxygen. The sustainability of electrolysis depends on the energy source used; when powered by renewable electricity, it produces green hydrogen, a completely carbon-free fuel. Additionally, emerging renewable-driven methods, such as photocatalytic and photoelectrochemical (PEC) water splitting [5], thermochemical cycles, and biological hydrogen production, present innovative pathways for producing hydrogen with minimal environmental impact. Hydrogen's role in the future energy landscape extends beyond power generation and transportation.

Next-Generation Hydrogen Economy: Production, Storage, and Fuel Cell Technologies, First Edition.
Peeyush Phogat, Shreya Sharma and Satyam Rawat.

It enables the storage and transport of renewable energy, facilitates industrial decarbonization, and serves as a key feedstock for ammonia production, refining, and metallurgy. As illustrated in Figure 2.1, the hydrogen production landscape consists of multiple pathways, each contributing uniquely to the overall hydrogen economy. The shift toward green hydrogen, produced using renewable energy sources, is critical for reducing reliance on fossil fuels and achieving climate goals.

The development of efficient and sustainable hydrogen production technologies is essential for enabling a clean energy transition. With growing concerns about climate change and energy security, hydrogen presents a viable alternative to carbon-intensive fuels. The ability to produce hydrogen in an environmentally friendly manner is crucial for maximizing its benefits while minimizing negative ecological impacts. One of the most significant advantages of hydrogen is its potential to integrate with renewable energy systems. Solar and wind power are inherently intermittent, meaning energy generation fluctuates based on weather conditions. Hydrogen offers a solution by acting as an energy storage medium, enabling excess electricity to be converted into hydrogen through electrolysis. This stored hydrogen can later be used in fuel cells or combustion engines to generate power when renewable energy output is low. By serving as a buffer between energy supply and demand, hydrogen enhances the reliability and scalability of renewable energy systems. Hydrogen production technologies also play a crucial role in reducing GHG emissions. Traditional hydrogen production methods, such as SMR, produce significant amounts of carbon dioxide (CO_2), leading to the classification of gray hydrogen. However, by incorporating carbon capture and storage (CCS) technology,

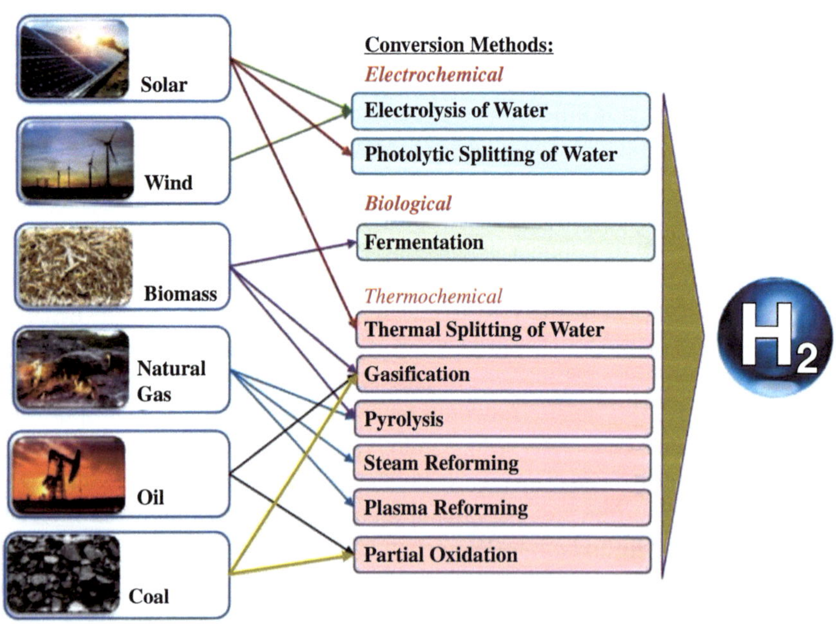

Figure 2.1 A diagram illustrating the various hydrogen production methods, including fossil-based, electrolysis-based, solar-driven, thermochemical, and biological processes, and their integration within the hydrogen economy. [9]/Taylor & Francis Group/CC BY 4.0.

blue hydrogen can be produced with reduced emissions. The cleanest form, green hydrogen, is generated through renewable-powered electrolysis, ensuring zero carbon emissions. Scaling up green hydrogen production is vital for achieving net-zero emissions and aligning with international climate commitments [6].

The economic and industrial applications of hydrogen production technologies are vast. In the industrial sector, hydrogen serves as a key feedstock for ammonia production, petroleum refining, and steel manufacturing. Transitioning to green hydrogen in these industries can drastically reduce their carbon footprints. In transportation, hydrogen fuel cells power buses, trucks, trains, ships, and even aircraft, providing a cleaner alternative to diesel and gasoline. Additionally, hydrogen-based synthetic fuels present a promising solution for decarbonizing aviation and maritime transport. In the power sector, hydrogen can be used in gas turbines or fuel cells to generate electricity, supporting grid stability and energy resilience. Residential and commercial applications, including hydrogen-based heating systems, further expand its role in the clean energy ecosystem. Despite its promising potential, hydrogen production faces several challenges. Cost competitiveness remains a major hurdle, as green hydrogen production is still more expensive than fossil-based alternatives due to the high cost of electrolyzers and electricity. Infrastructure development is another critical issue, as large-scale hydrogen production requires substantial investments in refueling stations, pipelines, and storage facilities. Moreover, technological advancements are needed to enhance the efficiency and durability of electrolyzers, photocatalysts, and thermochemical cycles [7].

In addition to technological and economic barriers, public perception and policy support play a crucial role in hydrogen adoption. Governments worldwide are introducing incentives, subsidies, and research grants to accelerate hydrogen production and deployment. Industry leaders and research institutions are also actively developing new materials and innovative techniques to reduce costs and improve efficiency. As investments and policies continue to drive progress, the future hydrogen landscape is set to evolve rapidly. Hydrogen production technologies will be at the heart of this transformation, shaping the future of clean energy and sustainable industrial practices. The continued advancement of electrolysis, photocatalysis, thermochemical processes, and biological hydrogen production will define the hydrogen economy in the coming decades. By addressing current challenges and leveraging cutting-edge innovations, hydrogen can fulfill its potential as a key pillar of the global energy transition [8].

2.2 Electrolysis: Advances in PEM, Alkaline, and Solid Oxide Electrolysis

2.2.1 Introduction to Electrolysis

Electrolysis is a fundamental process in hydrogen production that involves using electrical energy to split water into hydrogen and oxygen. As one of the cleanest methods for generating hydrogen, electrolysis is gaining significant attention due to its ability to produce green hydrogen when powered by renewable energy sources such as solar, wind, or hydroelectricity. This process plays a crucial role in the global shift toward a low-carbon hydrogen economy, reducing dependence on fossil fuel-based hydrogen production methods and significantly lowering GHG emissions [10].

At its core, electrolysis is an electrochemical reaction in which water molecules (H_2O) are decomposed into hydrogen gas (H_2) and oxygen gas (O_2) when an electric current is passed through an electrolyte solution. This reaction occurs within an electrolyzer, a device that consists of two electrodes (anode and cathode) submerged in water and separated by an ion-conducting membrane or electrolyte. When a voltage is applied across the electrodes, the water molecules undergo oxidation at the anode, releasing oxygen gas and positively charged hydrogen ions (protons). Simultaneously, at the cathode, the hydrogen ions gain electrons and are reduced to form hydrogen gas [11]. The overall reaction can be summarized as follows:

$$2H_2O(l) \rightarrow 2H_2(g) + O_2(g) \tag{2.1}$$

This simple yet powerful reaction underpins multiple electrolysis technologies, including alkaline electrolysis (AEL) [12], PEM electrolysis [13], and solid oxide electrolysis (SOEC) [14]. Each of these technologies has distinct advantages and is suited for different applications, depending on factors such as efficiency, cost, and scalability. One of the primary advantages of electrolysis is its ability to produce zero-emission hydrogen when coupled with renewable electricity sources. Unlike traditional SMR or coal gasification, which emit significant amounts of CO_2 and contribute to global warming, electrolysis-based hydrogen production generates only oxygen as a byproduct [15]. This makes it an attractive solution for industries looking to decarbonize and transition toward sustainable energy alternatives. Moreover, electrolysis supports the integration of renewable energy by serving as an energy storage medium. Since solar and wind power generation are inherently intermittent, excess electricity generated during peak production periods can be used to produce hydrogen through electrolysis. This hydrogen can then be stored and later converted back into electricity via fuel cells or utilized as a clean fuel for transportation, industry, and heating applications. By acting as a bridge between renewable energy generation and end-use applications, electrolysis helps enhance grid stability and energy security.

Despite its numerous benefits, electrolysis still faces challenges related to cost and efficiency. The high capital cost of electrolyzers, the energy-intensive nature of the process, and the need for large-scale infrastructure development remain barriers to widespread adoption. However, ongoing research and advancements in catalyst materials, system design, and operational efficiency are continuously improving the economic viability of electrolysis. With continued innovation and supportive government policies, electrolysis is poised to become a key enabler of a hydrogen-based energy future.

2.2.2 PEM Electrolysis

PEM electrolysis is a cutting-edge hydrogen production technology that utilizes a solid polymer electrolyte to facilitate the electrolysis of water. Unlike traditional alkaline electrolysis, PEM electrolysis offers higher efficiency, faster response times, and compact system design, making it particularly suitable for integration with renewable energy sources such as solar and wind power. As the global transition toward clean hydrogen accelerates, advancements in PEM technology are driving cost reductions, performance improvements, and scalability, making it a promising solution for large-scale hydrogen production [16].

PEM electrolyzers operate on the fundamental principle of electrochemical water splitting, wherein water (H_2O) is decomposed into hydrogen (H_2) and oxygen (O_2) by applying an external voltage across a PEM. The key components of a PEM electrolyzer include:

- **Anode (Oxygen Evolution Reaction – OER):** A catalyst-coated electrode where water molecules split into oxygen gas, protons (H^+), and electrons.
- **Cathode (Hydrogen Evolution Reaction – HER):** A catalyst-coated electrode where protons combine with electrons to form hydrogen gas.
- **PEM:** A specialized solid polymer electrolyte membrane that selectively conducts protons (H^+) from the anode to the cathode while preventing direct contact between hydrogen and oxygen.
- **Bipolar Plates and Gas Diffusion Layers:** Structural components that facilitate uniform distribution of reactants and removal of gas products.

The overall electrochemical reaction in a PEM electrolyzer is expressed as:

Anode Reaction (Oxidation):

$$2H_2O \rightarrow O_2 + 4H^+ + 4e^- \tag{2.2}$$

Cathode Reaction (Reduction):

$$4H^+ + 4e^- \rightarrow 2H_2 \tag{2.3}$$

Overall Reaction:

$$2H_2O \rightarrow 2H_2 + O_2 \tag{2.4}$$

As illustrated in Figure 2.2(a), the PEM electrolyzer consists of an electrolysis cell stack where water is fed to the anode side. Here, water molecules split into oxygen, protons, and electrons. The protons migrate through the PEM membrane to the cathode, where they combine with electrons to generate pure hydrogen gas. The oxygen gas produced is vented out as a byproduct. The ability of the PEM membrane to conduct only protons while blocking gases ensures high hydrogen purity, making PEM electrolysis ideal for fuel cell applications, industrial use, and energy storage.

PEM electrolysis requires specialized materials to achieve high efficiency, durability, and cost-effectiveness. The key material components include:

- **Catalysts:** The anode typically uses iridium oxide (IrO_2) or ruthenium oxide (RuO_2) due to their high activity for the OER. The cathode is coated with platinum (Pt) to enhance the HER. Ongoing research focuses on developing non-precious metal catalysts to reduce reliance on expensive platinum group metals.

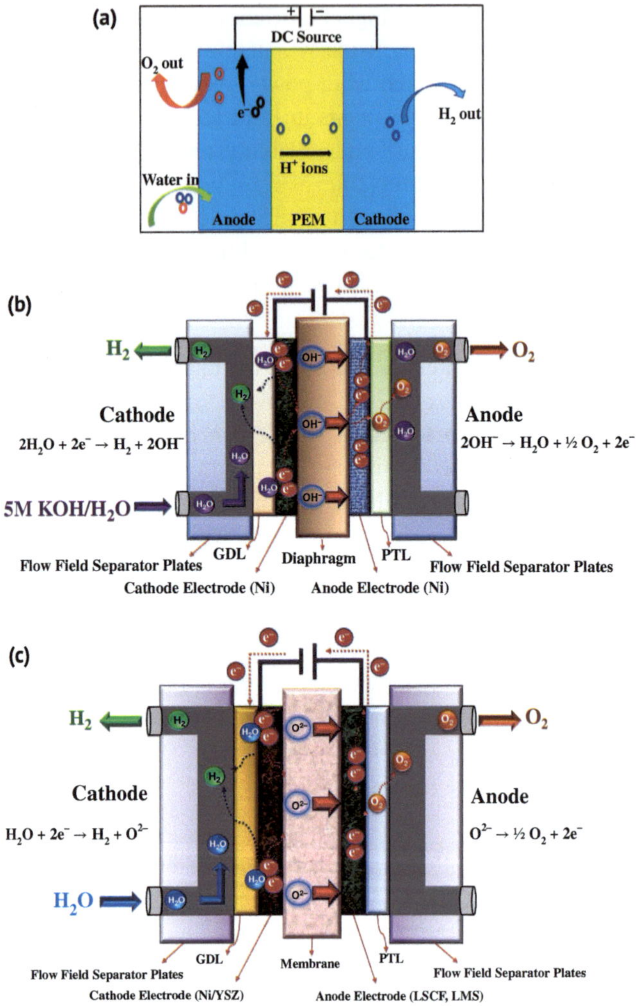

Figure 2.2 (a) Diagram of PEM processes – illustrating the working principle of PEM electrolysis, showing water splitting, proton migration, and hydrogen production. [22]/ Springer Nature/CC BY 4.0. (b) Diagram of alkaline electrolysis processes – showing the anode, cathode, electrolyte, ion movement, and hydrogen/oxygen production. [23]/Elsevier/ CC BY 4.0. (c) Diagram of SOEC electrolysis processes – illustrating the movement of ions, the role of electrodes, and high-temperature operation. [23]/Elsevier/CC BY 4.0.

- **PEM:** Nafion™ is the most widely used membrane material due to its high proton conductivity, chemical stability, and durability. Research is ongoing to develop cheaper and more durable PEM alternatives, such as polybenzimidazole (PBI) and sulfonated polyether ether ketone (sPEEK).
- **Bipolar Plates:** Made from stainless steel, titanium, or coated carbon composites to ensure structural integrity and corrosion resistance.

PEM electrolysis has witnessed significant advancements in recent years, primarily aimed at improving efficiency, reducing costs, and enhancing system durability. Some of the major breakthroughs are explained below.

Traditional PEM electrolyzers rely on iridium and platinum-based catalysts, which are expensive and scarce. Researchers are exploring non-precious metal catalysts (e.g., nickel, cobalt, and iron-based catalysts) to lower costs without compromising performance. Advances in thin-film membranes and nano-engineered electrodes have significantly improved proton conductivity and electrochemical stability, leading to higher energy efficiency. Enhanced bipolar plate coatings and corrosion-resistant materials have extended the operational life of PEM electrolyzers from 10,000 to over 60,000 hours. PEM electrolyzers have been successfully integrated with off-grid solar and wind farms, enabling dynamic operation to match fluctuating renewable power output. Companies like Siemens Energy, ITM Power, and Nel Hydrogen are developing megawatt-scale PEM electrolysis systems to support large-scale green hydrogen production for ammonia synthesis, refineries, and fuel cell vehicles. PEM electrolysis is poised to become a key enabler of the hydrogen economy, driven by continuous advancements in materials, system design, and cost efficiency. With increasing government support, investments, and industrial-scale projects, PEM electrolysis is expected to play a pivotal role in decarbonizing transportation, power generation, and industrial sectors. However, further cost reductions, efficiency improvements, and supply chain development will be critical for widespread adoption [17].

2.2.3 Alkaline Electrolysis

Alkaline electrolysis is one of the oldest and most widely used methods for hydrogen production, dating back to the early twentieth century. It is a mature and commercially available electrolysis technology that utilizes an alkaline electrolyte (typically potassium hydroxide, KOH, or sodium hydroxide, NaOH) to facilitate the electrochemical splitting of water into hydrogen and oxygen. Due to its low capital costs, high durability, and scalability, alkaline electrolysis remains a dominant hydrogen production method in industrial applications. However, ongoing research aims to address its efficiency limitations and improve electrode materials to enhance its competitiveness in the clean hydrogen economy [18].

Alkaline electrolyzers operate by passing an electric current through an electrolyte solution composed of 20–40% KOH or NaOH, which enables the transport of hydroxide ions (OH^-) between electrodes. The primary components of an alkaline electrolyzer include:

- **Anode (OER):** Water molecules react at the anode, producing O_2, electrons (e^-), and hydroxide ions (OH^-).
- **Cathode (HER):** Hydroxide ions migrate through the electrolyte to the cathode, where they react to form H_2 and regenerate water molecules.
- **Diaphragm/Separator:** A porous separator, typically made of asbestos, ceramic, or polymer-based materials, prevents the mixing of hydrogen and oxygen while allowing the transfer of hydroxide ions.

The overall chemical reactions in an alkaline electrolyzer are as follows:

Anode Reaction (Oxidation):

$$4OH^- \rightarrow O_2 + 2H_2O + 4e^- \tag{2.5}$$

Cathode Reaction (Reduction):

$$4H_2O + 4e^- \rightarrow 2H_2 + 4OH^- \tag{2.6}$$

Overall Reaction:

$$2H_2O \rightarrow 2H_2 + O_2 \tag{2.7}$$

As depicted in Figure 2.2(b), alkaline electrolyzers operate at relatively low current densities (100–400 mA/cm^2) and moderate temperatures (60–90 °C), making them reliable for industrial hydrogen production. They have been widely used in fertilizer production, oil refining, and steel manufacturing, where large-scale hydrogen demand exists.

Alkaline electrolysis is one of the most established and widely used hydrogen production technologies. Due to its long history, cost-effectiveness, and scalability, it remains a preferred choice for large-scale industrial hydrogen generation. However, despite these benefits, it faces certain efficiency and operational challenges that need to be addressed for its integration with renewable energy sources. Table 2.1 shows the benefits and limitations of alkaline electrolysis.

2.2.3.1 Recent Developments in Alkaline Electrolysis

- **Advanced Electrode Materials for Improved Performance:** Researchers are developing nickel-based, cobalt-based, and iron-based catalysts to enhance the HER and OER. New nickel-iron (Ni-Fe) alloys have demonstrated increased electrical conductivity and catalytic activity, improving system efficiency.
- **Innovative Separator Technologies:** Traditional asbestos diaphragms are being replaced with advanced polymer membranes that reduce hydrogen

Table 2.1 Benefits and limitations of alkaline electrolysis.

Aspect	Benefits	Limitations
Technology maturity	Established for over a century, widely used in industries	Requires advancements for better compatibility with renewables
Cost	No need for expensive PEM membranes or precious metal catalysts	Higher operational costs due to lower efficiency
Scalability	Can operate from kW to MW-scale, suited for large-scale hydrogen production	Less suitable for small-scale or decentralized applications
Energy efficiency	Achieves 60–80% efficiency (higher heating value, HHV)	Higher power consumption compared to PEM electrolysis
Durability	Operates for 60,000–90,000 hours with minimal degradation	Degradation of diaphragm separators affects long-term performance
Dynamic response	Suitable for continuous hydrogen production	Slower response to power fluctuations, not ideal for renewable energy
Gas separation	Uses diaphragm separator for gas separation	Higher risk of gas crossover, leading to hydrogen purity issues

crossover and enhance ionic conductivity. Nanostructured separator materials are being investigated to improve ion transport and system stability.

- **Integration with Renewable Energy Sources:** To address the slow dynamic response, researchers are exploring hybrid systems that integrate alkaline electrolyzers with battery storage to regulate power input fluctuations. Smart power management algorithms are being developed to enable alkaline electrolyzers to adapt to variable solar and wind power generation.
- **High-pressure Alkaline Electrolyzers:** Conventional alkaline electrolyzers operate at atmospheric pressure, requiring additional compressors for hydrogen storage and transport. New high-pressure alkaline electrolysis systems (30–50 bar) reduce the need for external compression, improving overall energy efficiency and reducing infrastructure costs.
- **Modular and Decentralized Electrolyzer Designs:** Compact and modular alkaline electrolyzers are being developed for localized hydrogen production in distributed energy systems. These small-scale units can be used in remote locations, hydrogen refueling stations, and off-grid renewable energy applications.

Alkaline electrolysis remains a reliable and cost-effective technology for hydrogen production, particularly in large-scale industrial applications. However, improving energy efficiency, reducing response time, and enhancing separator materials are key research priorities to ensure alkaline electrolysis remains competitive in the renewable hydrogen economy. With continued advancements, alkaline electrolysis is expected to play a significant role in the transition to green hydrogen, particularly in regions with abundant renewable energy resources and industrial hydrogen demand [19].

2.2.4 Solid Oxide Electrolysis (SOEC)

SOEC is an advanced high-temperature electrolysis technology that offers high efficiency and potential cost benefits for hydrogen production. Unlike PEM and alkaline electrolyzers, which operate at low to moderate temperatures, SOEC systems function at elevated temperatures (700–1,000 °C), leveraging solid oxide ceramic materials as electrolytes. This high-temperature operation allows SOEC to utilize waste heat from industrial processes or renewable sources to enhance efficiency and reduce electrical energy input requirements. By operating at such high temperatures, SOEC can convert steam (H_2O) into H_2 and O_2 more efficiently, reducing electricity consumption compared to conventional electrolysis methods. The integration of SOEC with renewable energy sources and industrial waste heat makes it a promising solution for large-scale green hydrogen production. However, significant challenges related to material degradation, cost, and system durability remain critical areas of ongoing research [20].

SOEC is based on the reverse operation of solid oxide fuel cells (SOFCs), meaning it takes in steam and electricity and produces H_2 and O_2. The system consists of three main components:

- **Cathode (HER):** Water molecules are reduced at the cathode to generate H_2 and oxygen ions (O^{2-}).

- **Solid Oxide Electrolyte:** O^{2-} migrate through a dense ceramic electrolyte, typically made of yttria-stabilized zirconia (YSZ) or ceria-based materials.
- **Anode (OER):** Oxygen ions reach the anode and release O_2 while returning electrons to the external circuit.

The electrochemical reactions occurring in SOEC can be summarized as follows:

Cathode Reaction (Reduction):

$$H_2O + 2e^- \rightarrow H_2 + O_2 \tag{2.8}$$

Anode Reaction (Oxidation):

$$O_2^- \rightarrow 12O_2 + 2e^- \tag{2.9}$$

Overall Reaction:

$$H_2O \rightarrow H_2 + 12O \tag{2.10}$$

As illustrated in Figure 2.2(c), SOEC operates in a manner similar to SOFCs, but in the reverse direction, using electricity and heat to split water instead of generating electricity from hydrogen.

Table 2.2 provides a comprehensive comparison of SOEC's key benefits and challenges, highlighting its potential in high-efficiency hydrogen production and its current research focus for commercial deployment.

Table 2.2 Advantages and challenges of SOEC.

Aspect	Advantages	Challenges
Efficiency	Achieves >80% efficiency (HHV) due to waste heat utilization	High-energy demand for maintaining elevated temperatures
Energy source utilization	Utilizes waste heat from industries, reducing energy costs	Requires specialized high-temperature materials, increasing manufacturing costs
CO_2 co-electrolysis	Enables co-electrolysis of CO_2 and H_2O to produce syngas for synthetic fuels	Still in the pilot phase, requiring scalability improvements for commercialization
Grid integration	Can operate in reverse mode as a fuel cell (SOFC) for energy storage	High thermal stress leads to material degradation, reducing long-term durability
Industrial integration	Can be directly integrated with solar thermal, geothermal, and nuclear systems	Sealing and containment issues at high temperatures limit system longevity
Durability	Research on ceria-based and perovskite oxide materials is improving stability	Nickel-YSZ electrodes undergo oxidation and structural degradation, reducing performance

2.2.4.1 Recent Developments and Future Outlook

- **Advanced Electrode Materials:** Researchers are developing nickel-free cathodes and ceria-based anodes to enhance stability and efficiency. The use of perovskite-based catalysts is being explored to minimize degradation and enhance OER efficiency.
- **Hybrid SOEC Technologies:** Efforts are being made to combine SOEC with renewable energy systems such as solar thermal and nuclear-assisted electrolysis for improved efficiency. Hybrid high-temperature fuel cell-electrolysis systems are also under investigation to provide flexibility between electricity generation and hydrogen production.
- **Large-scale Demonstration Projects:** The European Union and Japan are leading several pilot-scale SOEC initiatives to validate commercial hydrogen production at the megawatt scale. Companies such as Sunfire (Germany) and Bloom Energy (US) are actively developing scalable SOEC units for industrial hydrogen production.
- **Cost Reduction Strategies:** Developing lower-cost ceramic materials and optimizing manufacturing processes are key priorities. Increasing operating lifespan from the current 20,000–30,000 hours to >50,000 hours is a major research focus.

SOEC represents a highly promising hydrogen production technology due to its high efficiency, ability to utilize waste heat, and potential for co-electrolysis of CO_2 and H_2O. While challenges remain in material durability, cost, and large-scale implementation, ongoing research is focused on developing more robust materials, improving system efficiency, and reducing operational costs. With continued technological advancements, SOEC could become a key enabler of the hydrogen economy, particularly for industrial and renewable energy applications [21].

2.2.5 Comparison of Electrolysis Methods

Electrolysis is one of the most promising hydrogen production technologies, with three primary types: PEM electrolysis, alkaline electrolysis, and SOEC. Each method has unique characteristics, operational requirements, and advantages depending on the application. PEM electrolysis is known for its high efficiency and fast response time, making it suitable for intermittent renewable energy sources. Alkaline electrolysis, on the other hand, is a mature and cost-effective technology widely used in large-scale industrial applications. SOEC electrolysis, operating at high temperatures (700–1,000 °C), offers the highest efficiency by utilizing waste heat but faces challenges related to material durability and system scalability. To provide a clear comparison of these electrolysis methods, Table 2.3 presents a side-by-side evaluation of key parameters such as efficiency, cost, scalability, and operational conditions.

Table 2.3 Comparison of electrolysis technologies (PEM, alkaline, SOEC).

Parameter	PEM electrolysis	Alkaline electrolysis	SOEC
Operating temperature	Low (50–80 °C)	Moderate (60–90 °C)	High (700–1,000 °C)
Electrolyte	PEM	Liquid alkaline (KOH or NaOH)	Solid oxide ceramic (YSZ, ceria-based)
Efficiency (HHV)	60–75%	60–80%	>80% (utilizes waste heat)
Hydrogen purity	High (99.99%)	Moderate (requires gas separation)	High (99.99%)
Response time	Fast (suitable for renewable energy)	Slow (not ideal for fluctuating power)	Moderate
Material requirements	Requires precious metals (Pt, Ir, Ru)	Uses non-precious metals (Ni, Fe, Co)	High-temperature ceramics, perovskites
Durability/ lifespan	20,000–60,000 hours	60,000–90,000 hours	Still under research, durability concerns
Scalability	Moderate (ideal for small-to-medium scale)	High (well-established for large-scale use)	Limited (pilot projects, research phase)
Integration with renewables	Excellent	Limited	High potential with waste heat utilization
Capital cost	High (expensive catalysts and membranes)	Low (low-cost materials, well-developed)	Very high (high-temperature materials, R&D phase)
Industrial applications	Mobility, decentralized hydrogen production	Ammonia, methanol, and large-scale H_2 production	Industrial hydrogen, synthetic fuel production

2.3 Photocatalysis and Photoelectrochemical Water Splitting: Harnessing Solar Energy for Green Hydrogen

2.3.1 Introduction to Photocatalysis and PEC

Solar-driven hydrogen production has emerged as a promising solution for sustainable and clean energy generation, leveraging the abundant and renewable energy of the sun to split water into hydrogen and oxygen. Among the various solar-assisted hydrogen production methods, photocatalysis and PEC water splitting have gained significant attention due to their potential for direct conversion of solar energy into chemical fuel. These techniques provide a green pathway to hydrogen production, eliminating the need for fossil fuel-based energy sources and reducing GHG emissions. While both processes utilize light-absorbing materials (semiconductors) to drive water-splitting reactions, they operate under distinct mechanisms and configurations [24].

Photocatalysis involves the use of semiconductor materials that, when exposed to light, generate electron-hole pairs, which drive the redox reactions necessary for splitting water molecules into hydrogen and oxygen. This process occurs in a single-step reaction, where photocatalysts are suspended in water, absorbing sunlight to produce hydrogen directly. The efficiency of this method depends on the bandgap of the photocatalyst, the stability of the material, and the charge separation efficiency to prevent recombination of electrons and holes. Commonly used photocatalysts include titanium dioxide (TiO_2), metal oxides, perovskites, and graphitic carbon nitride (g-C_3N_4), often modified with co-catalysts such as platinum (Pt) or molybdenum sulfide (MoS_2) to improve activity. The major advantage of photocatalysis is its simple system design, which does not require external electrical inputs. However, challenges such as low solar-to-hydrogen (STH) efficiency, fast charge recombination, and limited visible light absorption hinder its large-scale implementation [25].

In contrast, PEC water splitting integrates semiconductor electrodes within an electrochemical cell to facilitate water-splitting reactions under light illumination. In a PEC system, a photoanode (often made of materials such as hematite (Fe_2O_3), bismuth vanadate ($BiVO_4$), or III-V semiconductors) absorbs sunlight, generating charge carriers that participate in oxidation and reduction reactions. Unlike photocatalysis, PEC systems have an external circuit that aids in charge separation, preventing recombination and improving hydrogen generation efficiency. A key advantage of PEC technology is its higher efficiency compared to pure photocatalysis, as the system allows for better control of reaction kinetics through applied bias voltage and material engineering. However, PEC systems require advanced electrode fabrication, protective coatings for stability, and optimized device architectures to overcome corrosion and energy loss [26]. The fundamental difference between photocatalysis and PEC lies in their system design and electron transport mechanisms. While photocatalysis operates as a single-step reaction in which suspended particles directly interact with water, PEC systems utilize a structured photoelectrode-electrolyte interface, allowing for separated oxidation and reduction reactions. Consequently, PEC technology exhibits greater control over reaction parameters and efficiency but comes with added complexity and cost. Both approaches continue to be actively researched, with material innovations and hybrid system developments aiming to improve the feasibility of STH conversion for large-scale applications.

2.3.2 Photocatalysis: Harnessing Sunlight for Water Splitting

Photocatalysis is a promising approach to solar-driven hydrogen production that exploits the energy of sunlight to split water molecules into hydrogen and oxygen. This method uses light-activated materials, known as photocatalysts, to initiate the required chemical reactions, and it holds considerable potential for sustainable hydrogen production. Photocatalysis operates through a light-induced process, where photons from sunlight are absorbed by semiconductor materials, generating electron-hole pairs. These charge carriers then participate in redox reactions, facilitating the dissociation of water molecules. The electrons reduce protons (H^+) to form H_2, while the holes oxidize water molecules to produce O_2. This process offers a direct, STH conversion pathway without the need for external electrical input, making it an appealing option for clean energy production [27].

The most commonly used photocatalysts are semiconductors, which absorb sunlight and enable the transfer of charge carriers. Among these, TiO_2 is one of the most studied materials due to its abundance, non-toxicity, and stability under light exposure. However, TiO_2 has limitations in its bandgap energy, which restricts its absorption of visible light, thereby limiting its efficiency. Other semiconductors, such as graphitic carbon nitride (g-C_3N_4), zinc oxide (ZnO), and perovskite materials, are also being explored to overcome these limitations. To improve the photocatalytic efficiency, these materials are often modified with various co-catalysts, such as Pt, MoS_2, or nickel-based catalysts, which help enhance charge separation and reduce the recombination of electrons and holes. These modifications improve the overall performance by increasing the reaction rate for hydrogen production [28].

In the photocatalytic process, the efficiency of STH conversion depends on several factors, including the bandgap of the photocatalyst, its surface area, and the charge carrier dynamics. When sunlight strikes the photocatalyst, electrons in the material's valence band absorb energy and jump to the conduction band, creating electron-hole pairs. The electrons then move to the surface of the material, where they reduce protons (H^+) to form hydrogen gas. Simultaneously, the holes move to the surface and oxidize water molecules to generate oxygen gas. This process is driven by the absorption of sunlight, particularly in the ultraviolet (UV) and visible light regions of the spectrum. However, in many cases, the recombination of electrons and holes before they can participate in the reactions results in low efficiency. To improve efficiency, researchers focus on enhancing the light absorption capabilities of the photocatalyst, typically through nanostructuring or doping to extend the absorption range into the visible light spectrum. Additionally, co-catalysts are added to the material to facilitate faster charge separation and improve reaction kinetics. For instance, Pt serves as an effective co-catalyst for hydrogen evolution, while MoS_2 helps in oxygen evolution. However, challenges in photocatalysis include charge recombination, poor conductivity, and the limited stability of photocatalysts under long-term solar exposure. These issues need to be addressed before photocatalysis can be scaled for commercial applications [29].

One of the major challenges faced by photocatalysis is its low efficiency, particularly in converting solar energy into hydrogen gas. The quantum efficiency of photocatalysts often falls short, as much of the sunlight absorbed by the catalyst does not contribute to water splitting. Additionally, charge recombination, where the electrons and holes reunite before they can participate in the reactions, further reduces the overall efficiency of the system. Moreover, achieving sufficient hydrogen production rates to meet the demands of large-scale energy production is another significant hurdle. This issue is compounded by the difficulty in scaling the process for industrial use. While photocatalysis can be applied in small, localized systems, scaling up to meet global hydrogen demands would require large surface areas of photocatalytic materials, which can be expensive and

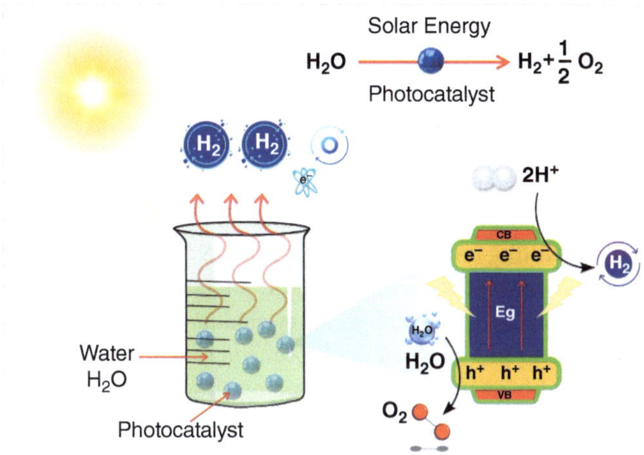

Figure 2.3 Mechanism of hydrogen production through sunlight using photocatalysis. This diagram shows the process of light absorption by photocatalysts, the formation of electron-hole pairs, and their role in driving the water-splitting reaction. [31]/Springer Nature/ CC BY 4.0.

inefficient to manufacture. Furthermore, material degradation is another obstacle that hinders the long-term viability of photocatalytic systems. Photocatalysts can degrade or become poisoned by prolonged exposure to environmental factors, such as moisture, oxygen, and solar radiation. This reduces their stability and efficiency over time, making it essential to develop robust materials that can maintain high performance over extended periods. Researchers are exploring the use of new, stable photocatalysts and advanced protective coatings to extend the lifespan of these materials in real-world conditions [30]. Figure 2.3 illustrates the mechanism of hydrogen production through sunlight using photocatalysis, showing how the absorption of photons results in the formation of electron-hole pairs and their subsequent participation in the redox reactions that split water molecules. The diagram provides a clear representation of how the electron-hole separation occurs and how they drive the hydrogen evolution reaction at the catalyst surface.

2.3.3 PEC Water Splitting

PEC water splitting is a highly promising technology that aims to harness solar energy to produce hydrogen from water. PEC is fundamentally a solar-driven electrochemical process where photovoltaic materials, typically semiconductors, are used to absorb sunlight and drive the electrolysis of water, resulting in hydrogen and oxygen. Unlike traditional electrolysis, which requires an external electric current, PEC splits water directly by utilizing the photogenerated electron-hole pairs produced when light strikes the semiconductor material. The process is both environmentally friendly and sustainable, as it uses only sunlight and water, with no emissions or the need for additional energy input, making it a key contender in the development of green hydrogen production technologies [32].

In a typical PEC system, the semiconductor material acts as both a photoanode and a photocathode, depending on whether it is an n-type or p-type semiconductor. When sunlight strikes the semiconductor, electron-hole pairs are generated in the material. The electrons flow toward the cathode to reduce protons (H^+) and produce hydrogen gas, while the holes flow toward the anode to oxidize water molecules and generate oxygen gas. The materials used for the electrodes must have high stability and efficient charge separation to ensure that the reaction proceeds efficiently. The efficiency of PEC water splitting is determined by several factors, including the bandgap of the semiconductor, its ability to absorb light, and the charge transfer kinetics at the semiconductor-electrolyte interface. A key challenge in PEC systems is to design semiconductors that can absorb a broad spectrum of light (particularly visible light) and have optimal bandgaps for the required reactions. Photocatalysts, often based on materials such as TiO_2, gallium arsenide (GaAs), copper indium gallium selenide, and perovskite-based materials, have shown promising potential in PEC applications [33].

Recent advancements in PEC technology have focused on improving the efficiency of light absorption and charge carrier separation. One approach involves the use of catalysts or co-catalysts that improve the kinetics of the OER and HER, respectively. Common co-catalysts include materials like Pt, ruthenium oxide (RuO_2), and nickel-based alloys. These co-catalysts reduce the overpotentials associated with both the OER and HER, which significantly increases the overall efficiency of the PEC system. In addition, recent innovations have led to the development of photoelectrodes with enhanced surface areas, which help to maximize light absorption and reduce electron recombination. To achieve high efficiency, nanostructuring and doping strategies have been employed to optimize the properties of the semiconductor materials. Nanostructured photoelectrodes have been found to improve the surface area and enhance charge separation by reducing the electron-hole recombination rates. Additionally, doping semiconductor materials with elements such as phosphorus (P) or nitrogen (N) can tune their electronic properties and enable better charge transport, thus improving performance [34].

Despite the potential of PEC water splitting, there are several challenges to overcome. The efficiency of PEC systems is still relatively low, with most systems achieving STH conversion efficiencies below 10%. To achieve large-scale adoption, it is essential to improve both efficiency and material stability, as photoelectrodes tend to degrade over time under solar irradiation and the harsh conditions of water splitting. Furthermore, current materials for photoanodes and photocathodes may face challenges related to cost, scalability, and durability. However, ongoing research in materials science and engineering is expected to provide innovative solutions to these challenges in the coming years. Figure 2.4 illustrates the redox mechanism involved in PEC water splitting, highlighting the different processes occurring at the n-type and p-type semiconductor interfaces. For the n-type semiconductor, photogenerated electrons move toward the cathode to reduce protons and generate hydrogen gas, while the holes flow toward the anode to oxidize water and produce oxygen gas. Similarly, in the case of a p-type semiconductor, the roles of the electrons and holes are reversed, with electrons moving toward the anode and holes toward the cathode, ensuring efficient water splitting at both electrodes [35].

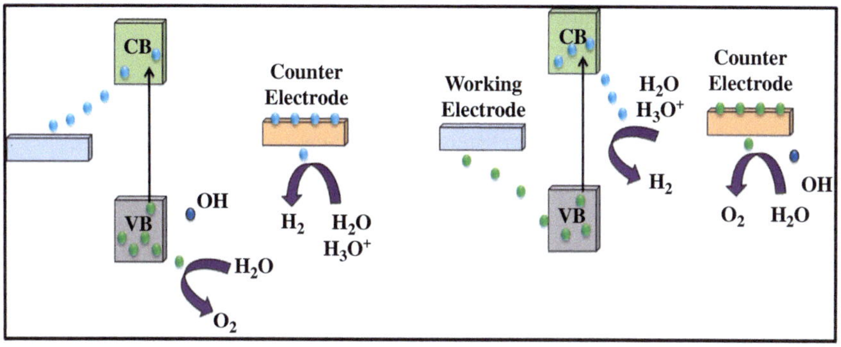

Figure 2.4 Redox mechanism for PEC water splitting. The diagram shows the redox processes at both n-type and p-type semiconductor photoelectrodes during photoelectrochemical water splitting, illustrating the flow of electrons and holes for hydrogen and oxygen production.

2.3.4 Challenges in and Potential Strategies for Solar Hydrogen Production

Table 2.4 summarizes some of the primary challenges in solar hydrogen production and potential strategies for overcoming them.

2.3.5 Future Directions in Photocatalysis and PEC

The future directions in photocatalysis and PEC water splitting are highly focused on addressing the existing challenges related to efficiency, scalability, and material stability. Current research is exploring innovative materials and systems to enhance the STH conversion efficiency, which remains a key barrier to large-scale implementation. Significant progress is expected in the development of next-generation photocatalysts, including nanoscale semiconductors and hybrid systems, designed to enhance light absorption and charge separation. Materials such as perovskites, graphene-based composites, and metal-organic frameworks are showing great promise for improving photocatalytic activity by offering more efficient pathways for electron transfer and hydrogen evolution. One of the most exciting areas of research is the use of tandem structures that combine multiple photocatalysts with complementary absorption spectra. These systems could harness a broader range of sunlight, enhancing overall efficiency. For example, pairing visible light absorbing semiconductors with UV-active materials could increase the efficiency of solar hydrogen production. Additionally, the development of multi-junction and photonic nanostructures holds the potential to significantly improve light harvesting and reduce energy losses during the conversion process [36, 37].

In the PEC space, much of the research is directed at improving the photoelectrode materials and electrocatalysts. High-performance photoanodes and cathodes made from earth-abundant, non-toxic materials like iron oxide, copper-based alloys, and tungsten trioxide are being investigated. These materials not only aim to enhance charge carrier mobility but also focus on improving long-term stability under the harsh operating conditions of PEC systems. Innovations in protective coatings and

Table 2.4 Challenges in solar hydrogen production.

Challenge	Description	Potential solutions
Efficiency limitations	Solar hydrogen production, particularly through photocatalysis and PEC, suffers from relatively low efficiency in converting solar energy to hydrogen. The STH conversion efficiency is often less than 10%, limiting the commercial viability	Development of high-efficiency photocatalysts Use of advanced materials such as perovskites and TiO_2 for better light absorption and charge separation Integration of nanostructured and doped semiconductors to reduce recombination rates
Material stability	Many photocatalysts and photoelectrodes degrade over time due to the harsh reaction conditions during water splitting (e.g., oxidation of electrodes and catalyst degradation)	Research into more stable materials such as perovskite-based semiconductors Introduction of co-catalysts for improving the stability of photoelectrodes Protective coatings and surface modifications to prevent degradation
Large-scale deployment	Scaling up solar hydrogen production requires addressing several issues related to the cost and efficiency of the systems, as well as material sourcing for large-scale production	Focus on modular, scalable designs for PEC systems Reduction of manufacturing costs through innovations in material selection Integration with renewable energy sources like solar and wind to ensure scalability and consistency in energy supply

surface modifications are also crucial to prevent degradation over time and improve the durability of PEC systems in real-world applications. Another critical research area is integrating PEC systems with renewable energy sources like solar and wind to provide consistent power for hydrogen production, ensuring that the systems are scalable for industrial applications. Recent breakthroughs in energy storage technologies like solid-state batteries and supercapacitors are also being explored for storing the hydrogen produced and integrating it into the grid. Furthermore, scalability is a major focus for future PEC and photocatalysis technologies. Researchers are exploring modular designs and automated fabrication techniques to lower the production costs of photocatalytic and PEC systems, making them viable for mass adoption. Advances in large-scale synthesis and low-cost manufacturing processes will be crucial in reducing the overall cost of these technologies, facilitating their deployment in industrial applications [38].

Overall, while significant hurdles remain in terms of efficiency, stability, and scalability, the rapid pace of research and development in photocatalysis and PEC water splitting provides hope for future breakthroughs that can transform these technologies into practical, large-scale solutions for sustainable hydrogen production. As these technologies continue to evolve, they will likely become an integral part of the global effort to transition to a hydrogen-based economy and achieve net-zero emissions.

2.4 Thermochemical and Hybrid Processes: Sulfur-iodine Cycle, Metal Oxide Cycles

2.4.1 Introduction to Thermochemical Hydrogen Production

Thermochemical hydrogen production refers to the generation of hydrogen gas through high-temperature chemical reactions, as opposed to traditional electrolysis methods that use electricity to split water [39]. These processes utilize heat, typically from concentrated solar energy or nuclear reactors, to drive endothermic reactions that produce hydrogen. One of the most studied and promising thermochemical cycles is the sulfur-iodine (SI) cycle, which involves a series of chemical reactions to produce hydrogen, sulfur dioxide, and oxygen at high temperatures (typically around 800–1,000 °C). The SI cycle consists of three main steps: the thermal decomposition of sulfuric acid, the reduction of iodine by sulfur dioxide, and the splitting of water to release hydrogen [40]. Another widely explored thermochemical method is the metal oxide cycle, which uses metal oxides, such as iron oxide (FeO/Fe_2O_3), to undergo cyclic reduction and oxidation reactions at high temperatures. These cycles typically operate in two stages: reduction (where oxygen is removed from the metal oxide using heat or carbon) and oxidation (where the reduced metal is reacted with steam to produce hydrogen). These reactions are highly efficient, especially when integrated with renewable or nuclear heat sources [41]. Thermochemical methods offer several advantages over conventional electrolysis. Unlike electrolysis, which relies on electricity and requires significant amounts of electrical energy to split water, thermochemical processes can leverage high-temperature heat, which is often more efficient and can be sourced from renewable energy, such as solar thermal or nuclear reactors. This thermal energy can drive endothermic reactions with less energy loss compared to traditional electrical input. Additionally, thermochemical cycles like the SI cycle can produce hydrogen at high temperatures without the need for expensive catalysts or electrodes, reducing material costs. Thermochemical hydrogen production also holds the potential for higher efficiency and scalability when compared to electrolysis, particularly for large-scale industrial applications. These methods can be coupled with concentrated solar power (CSP) systems, where solar heat directly powers the hydrogen production process, offering a sustainable pathway to green hydrogen production. Moreover, thermochemical cycles can be integrated with existing nuclear or industrial processes, providing a means for carbon-neutral hydrogen production at an industrial scale. Despite these advantages, thermochemical hydrogen production still faces challenges, particularly in terms of material durability at high temperatures, reactor design, and integration with heat sources. However, ongoing research is focused on overcoming these barriers, making thermochemical hydrogen production a promising avenue for clean hydrogen generation in the future.

2.4.2 SI Cycle

The SI cycle is a thermochemical process designed for hydrogen production through high-temperature reactions. It is one of the most researched methods for generating hydrogen, as it offers a promising pathway to producing hydrogen from water using heat, particularly from renewable or nuclear energy sources. The cycle is based on a series of chemical reactions that efficiently split water into hydrogen and oxygen at high temperatures, typically around 800–1,000 °C. The SI cycle involves three primary steps, each of which contributes to the overall production of hydrogen. The first step of the SI cycle is the thermal decomposition of sulfuric acid. This reaction occurs at temperatures around 850 °C, where sulfuric acid (H_2SO_4) is decomposed into sulfur dioxide (SO_2), water (H_2O), and oxygen (O_2) [42]. The reaction is as follows:

$$H_2SO_4(g) \rightarrow SO_2(g) + H_2O(g) + 21O_2(g) \qquad (2.11)$$

This reaction is endothermic, meaning it requires an external heat source to drive it. Heat sources such as CSP or nuclear reactors are typically employed in the SI cycle to provide the necessary high temperatures. The second step of the cycle involves the reaction between sulfur dioxide (SO_2) and iodine (I_2). This step is key to producing the iodine necessary for the third step, which involves hydrogen production. In this step, sulfur dioxide and iodine react to form sulfuric acid and hydrogen iodide (HI):

$$SO_2(g) + 2I_2(g) + 2H_2O(g) \rightarrow H_2SO_4(g) + 4HI(g) \qquad (2.12)$$

The final step in the SI cycle is the decomposition of HI to produce hydrogen (H_2) and I_2 through a high-temperature reaction. This reaction is crucial for generating hydrogen and closing the cycle:

$$2HI(g) \rightarrow H_2(g) + I_2(g) \qquad (2.13)$$

The hydrogen produced in this step is the desired output, and the I_2 can be recycled back into the cycle, making the process sustainable and reducing the need for external inputs. One of the primary advantages of the SI cycle is that it can operate efficiently at high temperatures, utilizing heat from nuclear reactors or solar thermal systems. Unlike electrolysis, which requires large amounts of electricity to split water, the SI cycle uses thermal energy, which can be more efficient and cost-effective. The cycle also requires fewer rare materials compared to other hydrogen production methods, making it potentially more economical for large-scale hydrogen production.

However, there are several challenges associated with the SI cycle. The main challenge is the high-temperature durability of the materials used in the process. The high temperatures required for the sulfuric acid decomposition and HI decomposition can cause significant wear and degradation of materials, including corrosion, erosion, and thermal stress. This has led to extensive research into improving the materials used in the cycle to enhance their stability and longevity. Recent developments in energy efficiency have focused on optimizing reaction conditions, improving the integration of the cycle with high-temperature heat sources, and developing advanced materials that can withstand the harsh conditions. For example, researchers have been investigating ceramic materials and metal alloys that exhibit better thermal stability and corrosion resistance under the high temperatures and aggressive chemical environments of the SI cycle. Moreover, catalysts are being optimized to increase reaction rates and improve overall efficiency. One of the key breakthroughs in improving the efficiency of the SI cycle is the development of advanced materials that can facilitate faster and more efficient reaction kinetics. Nanostructured materials, for example, offer enhanced surface areas for reactions and improved thermal conductivity, which help to enhance the efficiency of the overall process. Electrocatalysts that reduce the activation energy of reactions, such as the decomposition of HI, are also being investigated to improve the cycle's energy efficiency [43].

Additionally, advancements in process integration have led to more energy-efficient heat management strategies, enabling better recovery and reuse of heat within the system. This is particularly important for making the SI cycle more cost-effective and scalable for industrial hydrogen production. Integration with solar thermal plants and nuclear reactors is also being explored to make the SI cycle more sustainable and scalable [44, 45]. In Figure 2.5(a) and (b), FESEM (Field Emission Scanning

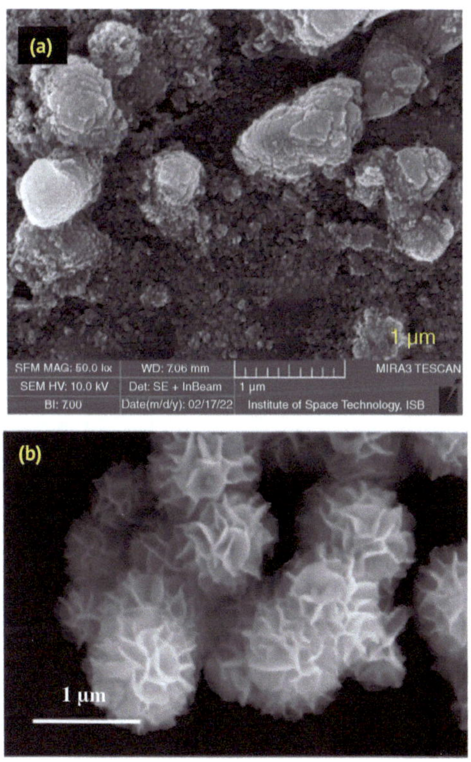

Figure 2.5 (a) FESEM images of materials used in the SI cycle, showing surface structure and properties. Reproduced with permission from [46]/Elsevier. (b) FESEM images depicting hydrogen reaction pathways and material interaction. [47]/MDPI/CC BY 4.0.

Electron Microscopy) images of materials used in the SI cycle show the structural properties and surface characteristics of materials that are being researched for improved performance in high-temperature reactions. These images highlight the advances in material design and surface engineering aimed at improving reaction efficiency and material durability under the extreme conditions of the SI cycle.

2.4.3 Metal Oxide Cycles

Metal oxide cycles are an emerging class of thermochemical processes that offer a promising pathway for hydrogen production. These cycles, like the SI cycle, utilize high-temperature reactions to split water and produce hydrogen. However, unlike the SI cycle, metal oxide cycles use metal oxides as oxygen carriers, which undergo reduction and oxidation reactions to facilitate hydrogen production. The most well-known metal oxide cycle for hydrogen production is the Cyclic Redox Reaction, which typically involves two steps: reduction and oxidation. In the first step, metal oxides are reduced at high temperatures, typically between 800 and 1,000 °C, by a source of heat such as CSP or nuclear heat [48]. The reduction process involves the metal oxide releasing oxygen to form a metal in its reduced state:

$$MO_x \rightarrow MO_{x-1} + (1/2)O_2 \tag{2.14}$$

In the second step, the reduced metal reacts with steam at lower temperatures, typically around 500–700 °C, to produce hydrogen. The reaction between the metal and water vapor generates hydrogen and regenerates the metal oxide:

$$MO_{x-1} + H_2O \rightarrow MO_x + H_2 \tag{2.15}$$

This two-step reaction process is repeated in cycles, making it a continuous method for hydrogen production.

One of the key advantages of metal oxide cycles is their high efficiency. Unlike conventional electrolysis, which requires large amounts of electricity to split water, metal oxide cycles leverage thermal energy to drive the reaction, making them more efficient when integrated with renewable heat sources like CSP or nuclear reactors. This high-temperature thermochemical approach allows for lower electricity consumption compared to electrolysis methods, which is a significant benefit in terms of energy efficiency. The key materials used in metal oxide cycles are typically metal oxides that can undergo reversible reduction and oxidation reactions. Commonly studied materials include cerium oxide (CeO_2), iron oxide (Fe_2O_3), titanium oxide (TiO_2), and chromium oxide (Cr_2O_3). These materials are selected for their ability to withstand high temperatures and their stability during multiple cycles. The metal oxides used must also exhibit high reactivity with water vapor, which is essential for efficient hydrogen production [49].

Cerium oxide is one of the most widely studied materials for metal oxide cycles due to its high oxygen storage capacity and good thermal stability. Iron oxide is another promising candidate, as it is abundant, non-toxic, and cost-effective. However, it requires high temperatures to achieve sufficient reduction, which can pose challenges in terms of heat management and material durability. Despite their potential, metal oxide cycles face several challenges. One of the major challenges is material durability. The metal oxides undergo significant thermal cycling during the reduction and oxidation processes, which can lead to material degradation over time. This includes thermal stress, volumetric expansion, and fracturing, which can reduce the efficiency of the cycle and limit the lifespan of the materials. Researchers are actively exploring advanced materials and coatings to improve the durability and stability of metal oxides under the extreme conditions of the cycle [50].

Another challenge is the efficiency of the reduction process. The temperature required for the reduction of metal oxides is relatively high, and maintaining such temperatures in a cost-effective and energy-efficient manner is difficult. To address this, researchers are exploring ways to lower the operational temperatures of the reduction step, such as by optimizing the use of solar energy or integrating waste heat from industrial processes. Scalability is another challenge for metal oxide cycles. While the technology shows promise in laboratory-scale experiments, scaling it up for large-scale hydrogen production remains a challenge. The design of reactors that can efficiently handle the cyclic processes and integrate with heat sources is an ongoing area of research. Figure 2.6 provides FESEM images of metal oxide materials used in the cycles. Figure 2.6(a) shows the surface morphology and properties of cerium oxide, while Figure 2.6(b) illustrates the interaction of metal oxides with hydrogen during the reaction process. These images highlight the material structures and the changes that occur during the thermochemical reactions, providing insights into how material properties affect hydrogen production efficiency [51].

2.4.4 Hybrid Thermochemical Processes

Hybrid thermochemical processes represent an advanced approach to hydrogen production that combines multiple energy sources – thermal, electrical, and chemical processes – to enhance overall efficiency and make the process more sustainable and cost-effective. These hybrid systems integrate various technologies, leveraging their respective strengths to overcome the limitations inherent in each individual method. By combining thermal energy with electricity or chemical reactions, hybrid systems can utilize lower-quality heat sources, like waste heat, alongside high-temperature processes to improve hydrogen production efficiency. The main advantage of hybrid thermochemical processes lies in their ability to integrate multiple energy inputs that operate synergistically. For instance, heat from industrial processes, solar concentrators, or nuclear reactors can be used to drive high-temperature chemical reactions, while electricity can power electrochemical reactions like electrolysis or contribute to controlling reaction conditions. This integration allows for a more efficient use of available energy and reduces the need for purely electrical or purely thermal energy inputs, which would otherwise require high operational costs [54].

An example of a hybrid thermochemical process is the combination of high-temperature electrolysis (HTE) with solar thermal power. In this setup, concentrated solar energy is used to heat water to extremely high temperatures (around 800–1,000 °C), providing the thermal energy necessary for electrolysis, while electricity is supplied to drive the electrochemical splitting of water into hydrogen and oxygen. The use of solar thermal energy decreases

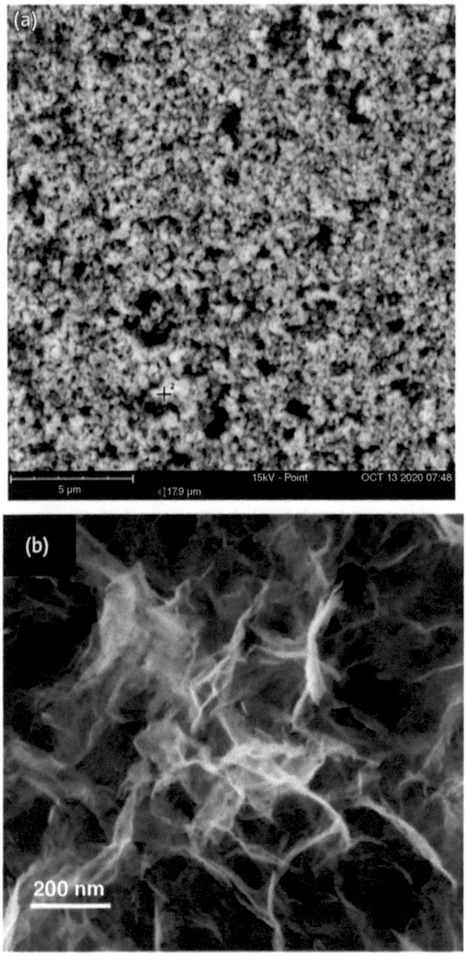

Figure 2.6 (a) FESEM images of cerium oxide material used in metal oxide cycles, showing surface structure and properties. [52]/MDPI/CC BY 4.0. (b) FESEM images illustrating metal oxide interaction during hydrogen production reactions. [53]/MDPI/CC BY 4.0.

the overall electricity consumption of the electrolysis process, making it more energy-efficient. Another example involves the combination of metal oxide cycles with electrical input to enhance the reduction reaction. By using electric current during the reduction phase, the system can operate at lower temperatures or provide additional energy for faster reaction kinetics, improving the overall hydrogen yield. Hybrid thermochemical systems are not limited to HTE and metal oxide cycles. Some approaches combine biological processes with chemical reactions, like biohybrid systems, which use microorganisms to catalyze hydrogen production from organic waste or biomass, supplemented by high-temperature or solar-driven chemical processes. These hybrid systems have the potential for large-scale hydrogen production, particularly in regions with abundant renewable heat sources or industrial waste heat, such as steel and cement production facilities, where excess heat can be repurposed for hydrogen generation [55].

The potential for large-scale hydrogen production with hybrid thermochemical processes is significant, as these systems can take advantage of the abundant and often underutilized waste heat produced in industrial settings. Moreover, integrating renewable energy sources, like solar and wind, with thermochemical processes can enable hydrogen production to be more closely aligned with local energy resources, reducing dependence on grid electricity and lowering the carbon footprint of the hydrogen production process. Despite the promise, challenges remain in the scalability and integration of hybrid systems. Efficiently coupling thermal, electrical, and chemical processes in a way that minimizes losses while optimizing reaction conditions is complex. Further advancements in system integration, material development, and process optimization will be essential for realizing the full potential of hybrid thermochemical processes for large-scale hydrogen production [56].

2.5 Biological Hydrogen Production: Algae, Bacteria, and Enzymatic Hydrogen Production

2.5.1 Introduction to Biological Hydrogen Production

Biological hydrogen production is a promising and environmentally friendly method for generating hydrogen, harnessing the natural capabilities of microorganisms to convert organic matter into biohydrogen. This process offers a potential solution to the increasing demand for clean and renewable hydrogen production methods, with minimal energy consumption and no direct emissions. Biological hydrogen production primarily relies on two types of microorganisms: algae and bacteria, which can use sunlight, organic substrates, or even waste materials as sources of energy to produce hydrogen. Unlike conventional methods, such as SMR or electrolysis, biological methods harness naturally occurring biochemical processes, making them potentially more sustainable and cost-effective [57].

Algae-based hydrogen production, often referred to as biophotolysis, involves the use of photosynthetic microorganisms, mainly green algae and cyanobacteria. These microorganisms can use light energy to convert water molecules into hydrogen and oxygen through a biochemical reaction. During photosynthesis, algae absorb sunlight and, using the energy, split water molecules (H_2O) into oxygen (O_2) and hydrogen (H_2). In some algae species, this hydrogen is released into the surrounding environment. A particularly promising group of algae are green algae and cyanobacteria, which can efficiently utilize sunlight to produce hydrogen in both aerobic and anaerobic conditions. The process in algae is highly like natural photosynthesis, with the added advantage that they can be cultivated in large-scale open ponds or photobioreactors, making them relatively low-cost compared to traditional electrolysis [58].

On the other hand, bacterial hydrogen production utilizes various bacteria species capable of producing hydrogen through fermentation or dark fermentation. In these processes, bacteria break down organic compounds (such as sugars or organic waste) in the absence of light to produce hydrogen. The biochemical reaction typically involves the conversion of carbohydrates into acids or alcohols, with hydrogen being a byproduct. Clostridium and Enterobacter are examples of bacteria that are commonly used for dark fermentation. Unlike algae, bacterial hydrogen production does not require sunlight, offering flexibility in terms of operating conditions and the potential for utilizing organic waste as feedstock. This ability to use waste materials, such as agricultural or industrial waste, further enhances the sustainability and economic viability of bacterial biohydrogen production [59].

Microorganisms play a central role in biohydrogen production, as they contain specific enzymes, such as hydrogenases and nitrogenases, which catalyze the conversion of substrates into hydrogen gas. These enzymes are the biological catalysts that enable the efficient splitting of water molecules or organic matter to release hydrogen. However, one of the main challenges in biological hydrogen production is maximizing the efficiency of these enzymes and optimizing the environmental conditions for microbial growth and hydrogen production. Factors such as temperature, pH, light intensity (for algae), and nutrient availability must be carefully controlled to achieve optimal hydrogen yields [60].

Biological hydrogen production offers several advantages, including the use of renewable resources, lower energy consumption compared to industrial hydrogen production methods, and the ability to recycle organic waste into valuable hydrogen gas. Despite these benefits, the technology faces several challenges, including low hydrogen production rates and the need for further advancements in microbial engineering and process optimization to make it commercially viable. However, with ongoing research and innovation, biological hydrogen production has the potential to become a key component of sustainable hydrogen production systems in the future [61].

2.5.2 Algae-based Hydrogen Production

Algae-based hydrogen production is a promising method for generating biohydrogen, particularly through the photosynthetic processes of green algae. This process, known as biophotolysis, involves the use of sunlight to split water molecules into hydrogen (H_2) and oxygen (O_2), similar to the natural photosynthesis that occurs in plants. Green algae, such as Chlamydomonas reinhardtii and Scenedesmus, are particularly well-suited for this process due to their efficient light absorption capabilities and their natural ability to produce hydrogen under specific environmental conditions. During the light phase of photosynthesis, these algae utilize sunlight to power the water-splitting reaction, which produces oxygen and protons. The protons are then used by hydrogenases and other enzymes within the algal cells to combine with electrons and produce hydrogen gas (H_2). The main advantage of algae-based hydrogen production is its reliance on renewable energy – sunlight – as the primary energy source. This makes it an environmentally friendly method with minimal carbon footprint, especially when compared to other industrial hydrogen production methods, such as SMR or electrolysis that require substantial electricity inputs. Algae's ability to convert solar energy directly into hydrogen gas through biophotolysis has been a subject of intensive research, as it holds significant promise for large-scale, sustainable hydrogen production [62].

However, algae-based hydrogen production faces several challenges, particularly in scaling up the process. While laboratory experiments have demonstrated the potential for high hydrogen yields from algae, achieving similar results on a commercial scale remains difficult. One of the primary challenges is the low hydrogen production rate, which is not yet sufficient to make algae-based systems economically competitive with other hydrogen production methods. The efficiency of hydrogen production in algae is also limited by several factors, including the rate of photosynthesis, the availability of suitable light and temperature conditions, and the need for specific growth environments. Another significant challenge is the activation and deactivation of hydrogenase enzymes. While these enzymes are critical for catalyzing the production of hydrogen, they are also highly sensitive to oxygen, which is released during photosynthesis. As a result, algae must enter a specialized anaerobic state to optimize hydrogen production, and maintaining such conditions in large-scale systems is a significant hurdle. Furthermore, the overall efficiency of algae-based hydrogen production depends heavily on the optimization of growth conditions, such as light intensity, temperature, and nutrient availability, all of which can vary significantly in open-air systems [63].

Despite these challenges, researchers have made significant progress in understanding the algae hydrogen production pathway and are exploring various ways to improve its efficiency and scalability. One promising approach is the genetic engineering of algae to enhance their ability to produce hydrogen under controlled conditions. Additionally, researchers are working on optimizing the growth medium, reactor designs, and improving enzyme stability to increase hydrogen production rates. Algae-based hydrogen production presents a viable and sustainable method for generating hydrogen fuel from renewable sources. By leveraging the natural ability of green algae to produce hydrogen through photosynthesis, this method holds promise for reducing dependence on fossil fuels and contributing to a cleaner energy future. However, to realize its full potential, significant improvements in efficiency, scalability, and system integration are necessary [64]. Figure 2.7(a) shows algae hydrogen production pathways through photosynthesis and biophotolysis.

2.5.3 Bacterial Hydrogen Production

Bacterial hydrogen production is a biological method of generating hydrogen gas through the metabolic processes of anaerobic bacteria. This process is distinct from algae-based hydrogen production in that it occurs in the absence of oxygen, utilizing the biochemical pathways of specific bacteria that thrive in anaerobic conditions. These microorganisms, such as Clostridium species, Enterobacter, Escherichia coli, and Rhodobacter, have evolved unique metabolic pathways that allow them to produce hydrogen gas as a byproduct of their cellular respiration. Bacterial hydrogen production is typically catalyzed by hydrogenase enzymes, which facilitate the conversion of protons (H^+) and electrons (e^-) into molecular hydrogen (H_2) under anaerobic conditions [65].

In nature, these bacteria utilize hydrogenase enzymes in their metabolism, which are crucial for converting organic substrates into hydrogen. The reactions are typically driven by the bacterial fermentation process, in which organic compounds like glucose or fatty acids are broken down to produce energy, releasing hydrogen gas as a byproduct. This process, known as dark fermentation, is most efficient under strictly anaerobic conditions. The potential of bacterial hydrogen production lies in its simplicity, as it does not require complex infrastructure or high-energy input. Moreover, the natural ability of bacteria to thrive on a variety of organic substrates makes this method adaptable and sustainable for producing hydrogen from waste materials, agricultural residues, or even industrial byproducts. One of the key advantages of bacterial hydrogen production is the diverse range of bacteria that can be utilized [66]. Among the most studied are the Clostridium species, which are known for their high hydrogen yields through fermentation processes. These bacteria can produce hydrogen efficiently under controlled conditions, making them suitable for bioreactor-based hydrogen production systems [67]. Other bacteria, such as Rhodobacter and Enterobacter, are studied for their ability to utilize light in addition to organic substrates, through a process known as photofermentation, further enhancing their hydrogen production capabilities [68].

However, despite the promising potential of bacterial hydrogen production, there are significant challenges in scaling up the process for industrial applications. One primary concern is the low hydrogen production rate. While some bacteria can produce hydrogen at relatively high rates in laboratory settings, the yields are still too low to make the process commercially viable without significant advancements in optimization. In addition, the production process is sensitive to environmental factors such as pH, temperature, and the concentration of organic substrates, which can fluctuate in large-scale systems. Furthermore, the efficiency of hydrogen production by bacteria is often limited by the hydrogenase enzymes' susceptibility to environmental inhibitors. Oxygen, for instance, can rapidly deactivate these enzymes, reducing the overall efficiency of the process. To overcome this, researchers are investigating ways to improve enzyme stability, either by genetic engineering or by developing more robust, oxygen-tolerant hydrogenase enzymes. Additionally, optimizing bioreactor design and maintaining ideal growth conditions are crucial factors that need to be addressed to improve the efficiency of bacterial hydrogen production [69].

Recent advancements have focused on improving the overall yield and efficiency of bacterial hydrogen production by genetically modifying bacteria to enhance their hydrogenase activity, as well as developing advanced bioreactor designs that optimize microbial growth and hydrogen production. Additionally, some researchers have explored using waste products, such as industrial wastewater or organic waste, to feed the bacteria, effectively creating a closed-loop system that could be both economically and environmentally beneficial [70]. Figure 2.7(b) illustrates bacterial hydrogen production pathways through anaerobic fermentation and photofermentation.

2.5.4 Enzymatic Hydrogen Production

Enzymatic hydrogen production involves the use of naturally occurring enzymes, specifically hydrogenases, to catalyze the production of hydrogen gas from protons and electrons. Hydrogenases are enzymes found in various microorganisms,

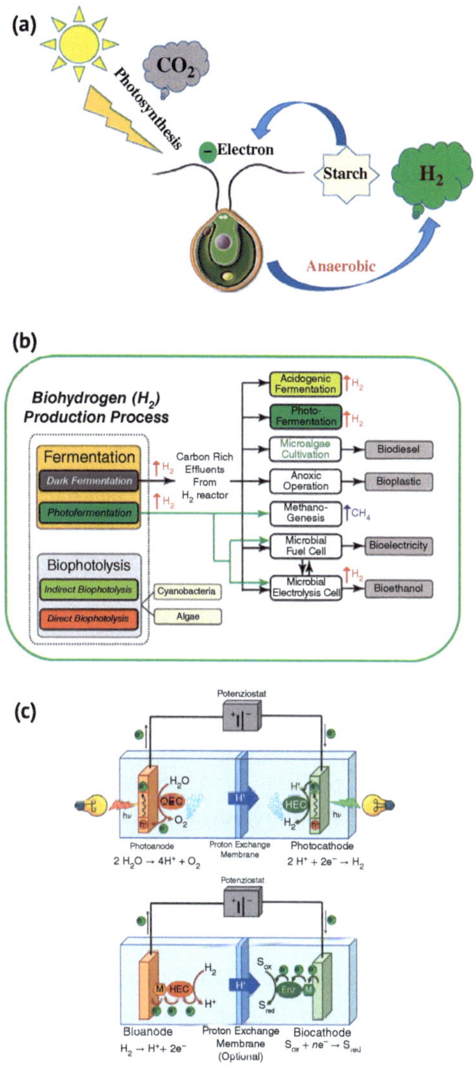

Figure 2.7 (a) Algae-based hydrogen production pathways, including the role of photosynthesis in hydrogen generation by green algae such as *Chlamydomonas reinhardtii*. [75]/Frontiers/CC BY 4.0. (b) Bacterial hydrogen production pathways, illustrating how anaerobic bacteria such as Clostridium and Rhodobacter generate hydrogen in bioreactors. [76]/MDPI/CC BY 4.0. (c) Enzymatic hydrogen production pathways, depicting the action of hydrogenase enzymes in the conversion of substrates to hydrogen at low temperatures and pressures. [77]/MDPI/CC BY 4.0.

including bacteria, archaea, and algae, that facilitate the reduction of protons (H^+) to hydrogen gas (H_2). These enzymes play a key role in the natural processes of hydrogen production in biological systems, such as during anaerobic respiration or fermentation. The ability of hydrogenases to efficiently catalyze this reaction at relatively low temperatures and pressures makes enzymatic hydrogen production an attractive option for sustainable hydrogen generation [71].

The primary function of hydrogenases is to catalyze the conversion of protons and electrons into molecular hydrogen in an oxygen-free environment. This process takes place within microbial cells, where hydrogenases are typically embedded in the cell membrane or found within specialized cellular compartments. The enzymatic reaction involves the transfer of electrons from an electron donor, often a biological substrate like organic compounds, to protons in the surrounding medium, resulting in the production of hydrogen gas. Notably, the reaction does not require high temperatures or extreme pressures, making it an energy-efficient method of hydrogen production. One of the most important advantages of enzymatic hydrogen production is the low operational requirements. Unlike traditional hydrogen production methods, such as SMR or electrolysis, which require high temperatures or electrical energy input, enzymatic hydrogen production can occur under ambient temperature and pressure conditions. This significantly reduces the energy demands of the process, offering a more sustainable and cost-effective pathway for hydrogen generation, especially in decentralized or small-scale applications [72].

However, despite its potential, enzymatic hydrogen production faces several challenges, primarily related to the stability and activity of the hydrogenase enzymes. Hydrogenases are highly sensitive to oxygen, which can deactivate the enzymes, significantly reducing their efficiency. Additionally, enzyme stability under long-term operational conditions remains a challenge for large-scale applications. Researchers are working to improve the stability and catalytic efficiency of hydrogenases by exploring enzyme engineering, as well as immobilization techniques to enhance their performance in industrial settings. One promising approach to overcoming these limitations is the genetic modification of microorganisms to produce more robust hydrogenases with enhanced activity and resistance to deactivation by oxygen. Additionally, innovations in enzyme encapsulation and immobilization techniques are being explored to protect the hydrogenases and improve their stability over extended periods of operation [73].

Enzymatic hydrogen production holds significant promise due to its low-energy requirements and the potential for hydrogen production at ambient conditions. Ongoing research to optimize enzyme activity, improve stability, and scale up production will be key to making this technology a viable option for sustainable hydrogen production. Figure 2.7(c) depicts enzymatic hydrogen production pathways catalyzed by hydrogenase enzymes, highlighting key steps in the reaction mechanism [74].

2.5.5 Comparison of Biological Hydrogen Production Methods

Table 2.5 presents a comparison of the primary biological hydrogen production methods, focusing on key parameters such as the type of organism used, hydrogen production efficiency, sustainability, scalability, and the current stage of research. It provides a comprehensive overview of algae-based, bacterial, and enzymatic hydrogen production methods, highlighting their respective advantages, challenges, and potential for large-scale implementation. By examining these parameters, the table offers insights into the future viability of biological hydrogen production as a sustainable alternative to conventional methods.

Table 2.5 Comparison of biological hydrogen production methods.

Method	Organism type	Efficiency	Sustainability	Scalability	Research stage
Algae-based hydrogen production	Green algae (e.g., *Chlamydomonas reinhardtii*)	Moderate (up to 0.2–0.5% STH efficiency)	High, renewable and CO_2-absorbing	Limited, high light and space requirements for scaling	Experimental/ Development
Bacterial hydrogen production	Anaerobic bacteria (e.g., *Clostridium, Rhodobacter*)	Moderate to high (up to 50% hydrogen yield)	High, utilizes organic waste and biofuels	Moderate, bioreactor systems for larger-scale use	Pilot scale/Advanced research
Enzymatic hydrogen production	Hydrogenase enzymes from bacteria/archaea	Low (limited by enzyme activity and stability)	High, renewable and low-energy consumption	Limited, enzyme instability at scale	Basic research/ Experimental

2.6 Comparative Analysis of Hydrogen Production Methods

2.6.1 Cost Analysis of Hydrogen Production Technologies

Hydrogen production technologies vary widely in terms of capital investment, operational costs, and economic feasibility, making cost considerations a crucial factor in determining the scalability and adoption of different methods. The production of green hydrogen (via electrolysis), blue hydrogen (via fossil fuels with carbon capture), and gray hydrogen (via fossil fuels without carbon capture) all present unique cost structures. While green hydrogen is the most sustainable, it remains significantly more expensive than blue and gray hydrogen due to high-energy input costs and infrastructure requirements. A comparative analysis of capital expenditures (CAPEX) and operational expenditures (OPEX) across different hydrogen production technologies helps in assessing their long-term viability [78].

Electrolysis, particularly PEM and alkaline electrolysis, is currently one of the most expensive hydrogen production methods due to high electricity costs, expensive catalyst materials (e.g., platinum and iridium for PEM electrolysis), and infrastructure investments. The CAPEX for electrolysis-based hydrogen production typically ranges between \$800 and \$1,500 per kW of installed electrolyzer capacity, with OPEX highly dependent on electricity prices. Since electricity accounts for 60–70% of total costs, the economic feasibility of electrolysis hinges on access to low-cost renewable energy sources. SOEC, a HTE method, shows potential for higher efficiency but faces challenges related to material degradation and high initial costs. The CAPEX for SOEC is currently higher than PEM and alkaline electrolysis, but ongoing advancements in ceramic electrolytes and system integration with waste heat sources could lower costs in the future [79]. Thermochemical hydrogen production, including SMR and coal gasification, remains the most cost-effective method for large-scale hydrogen production. Gray hydrogen from SMR costs around \$1.00–\$2.00 per kg, making it the dominant method for industrial hydrogen production. However, blue hydrogen, which incorporates CCS, increases production costs by \$0.50–\$1.00 per kg due to additional carbon capture infrastructure and energy requirements [80]. Despite these added costs, blue hydrogen remains significantly cheaper than green hydrogen, positioning it as a transitional solution in the shift toward a hydrogen economy. Biological hydrogen production methods, such as photofermentation, dark fermentation, and microbial electrolysis cells (MECs), are still in the early stages of commercialization [81]. While these methods offer low-energy input requirements, their low hydrogen yield, scalability limitations, and high cost of bioreactor materials result in CAPEX and OPEX that are currently not competitive with other hydrogen production technologies. Research in genetic engineering, microbial efficiency improvements, and bioreactor optimization is essential to reduce costs and improve feasibility.

The economic feasibility of green hydrogen depends on reducing electrolyzer costs, increasing efficiency, and lowering electricity prices. Currently, the cost of green hydrogen ranges from \$4.00 to \$6.00 per kg, making it significantly more expensive than gray and blue hydrogen. However, with improvements in electrolyzer technology, scale up of production, and declining renewable energy costs, green hydrogen is projected to fall below $2.00 per kg by 2030, making it competitive with fossil-fuel-based hydrogen. In contrast, blue hydrogen serves as an intermediate solution for decarbonization, with costs ranging from $2.00 to $3.00 per kg. While it reduces carbon emissions, it still relies on fossil fuels, and the efficiency of carbon capture technology remains a limiting factor. Some studies suggest that incomplete CO_2 capture rates (typically 85–90%) may undermine blue hydrogen's environmental benefits, reinforcing the push toward green hydrogen as a long-term solution. Gray hydrogen, while cheapest today, faces increasing regulatory and carbon pricing pressures. With many countries implementing carbon taxes and stricter emissions regulations, the cost of producing hydrogen from fossil fuels without carbon capture is expected to rise. Governments worldwide are promoting green hydrogen through subsidies, tax credits, and infrastructure investments, which will further accelerate its cost reductions and large-scale deployment [82].

Hydrogen production costs remain a critical factor in determining which technology will dominate the future energy landscape. Electrolysis-based green hydrogen, while currently expensive, is expected to become cost-competitive with fossil-based hydrogen by 2030 with advancements in renewable energy, electrolyzer efficiency, and manufacturing scale up. Blue hydrogen, with its reliance on carbon capture, serves as a temporary bridge to reduce emissions but is unlikely to be a long-term replacement for fossil fuels. Thermochemical and biological hydrogen production methods offer different advantages but face scalability and efficiency challenges. Government policies, technological breakthroughs, and energy pricing trends will ultimately determine how each production method contributes to the global hydrogen economy.

2.6.2 Efficiency and Scalability of Hydrogen Production

Hydrogen production methods vary significantly in terms of energy efficiency, resource consumption, and scalability. While electrolysis and thermochemical methods are the most widely researched and implemented at an industrial scale, biological hydrogen production remains in early development and presents unique challenges and limitations. A comprehensive comparison of these methods helps in identifying the most suitable technologies for large-scale hydrogen deployment [83].

Electrolysis-based hydrogen production, particularly PEM, alkaline, and SOEC electrolysis, relies on electricity to split water into hydrogen and oxygen [84]. The efficiency of these methods varies, with PEM electrolysis achieving 60–75% efficiency, alkaline electrolysis reaching 65–80% efficiency, and SOEC surpassing 80% when integrated with waste heat sources. However, the biggest drawback of electrolysis is its high electricity consumption, which makes it economically viable only when powered by low-cost renewable energy sources. Additionally, electrolysis requires large amounts of purified water, which can be a limiting factor in water-scarce regions. In contrast, thermochemical hydrogen production, including SMR [85] and coal gasification, is more energy-efficient in fossil-fuel-rich regions but generates significant carbon emissions unless coupled with CCS. SMR has an energy efficiency of 65–75%, while coal gasification is lower at around 50–60%. These methods require

Table 2.6 Electrolysis vs. thermochemical hydrogen production.

Parameter	Electrolysis (PEM, alkaline, SOEC)	Thermochemical (SMR, gasification, water-splitting cycles)
Energy efficiency	60–80% (depends on method) [86]	50–75% (SMR higher, gasification lower) [87]
Resource requirement	High electricity and purified water [88]	Fossil fuels or chemical catalysts [89]
Carbon emissions	Zero emissions if powered by renewables) [90]	high emissions (without CCS), moderate (with CCS) [91]
Feedstock	Water [92]	Natural gas, coal, or sulfur-based cycles [93]
Scalability	Limited by electricity availability [94]	High scalability but dependent on fossil resources [95]
Technology maturity	Commercially available, improving [96]	Established for fossil fuels, developing for water-splitting cycles [97]
Cost consideration	High electricity cost, declining with renewables [98]	Lower cost but subject to carbon pricing [95]
Future potential	Best suited for long-term green hydrogen economy [99]	Requires CCS for sustainability, promising for high-temperature processes [100]

hydrocarbon feedstocks (natural gas or coal), making them less sustainable than electrolysis. However, thermochemical water-splitting cycles, such as the SI cycle and metal oxide cycles, show promise for achieving high efficiency without direct carbon emissions. The following Table 2.6 provides a comparative analysis of these two dominant hydrogen production methods.

2.6.3 Scalability of Biological Hydrogen Production and Its Limitations

Biological hydrogen production methods, such as microbial fermentation, enzymatic hydrogen generation, and biophotolysis, offer a low-energy alternative to electrolysis and thermochemical processes. These methods leverage microorganisms (algae, bacteria, and archaea) to generate hydrogen under controlled conditions [101]. However, despite being environmentally friendly, biological hydrogen production has several scalability limitations that prevent it from becoming a dominant hydrogen source in the near future.

The key challenges include:

- Low Hydrogen Yields: Biological systems typically have low conversion efficiencies, making them impractical for high-volume production.
- Complex Bioreactor Design: Scaling up biological hydrogen production requires specialized bioreactors that must maintain optimal microbial conditions.

- Substrate Dependency: Many microbial hydrogen processes rely on organic feedstocks (biomass, sugars, wastewater), which limits feedstock availability and increases processing costs.
- Slow Reaction Rates: Compared to electrolysis and thermochemical methods, biological hydrogen production is significantly slower, leading to longer processing times and higher costs.

Hydrogen production technologies present a wide range of efficiency, resource requirements, and scalability considerations. Electrolysis offers a sustainable path but remains energy-intensive unless powered by low-cost renewables. Thermochemical processes are currently the most cost-effective but dependent on fossil fuels, requiring carbon capture solutions for sustainability. Biological hydrogen production, while environmentally promising, faces major scalability challenges due to low yields and slow reaction rates. Future developments in advanced catalysts, integrated hybrid systems, and AI-driven process optimizations will play a crucial role in making all hydrogen production methods more viable and cost-effective [102].

2.6.4 Environmental Impact and Sustainability

As hydrogen production scales globally, its environmental impact and sustainability must be carefully evaluated. While hydrogen itself is a clean energy carrier, its production processes require varying levels of water, energy, and natural resources, leading to different carbon footprints and environmental trade offs. The sustainability of hydrogen largely depends on feedstock availability, energy inputs, emissions, and land and water use, which influence its long-term viability as a low-carbon alternative to fossil fuels [103].

2.6.5 Water Usage and Energy Requirements for Different Hydrogen Production Technologies

Hydrogen production methods differ significantly in their water and energy consumption, which are critical factors in assessing their sustainability.

- Electrolysis consumes large volumes of purified water (H_2O) as a feedstock, making water availability a concern, particularly in arid regions. Alkaline and PEM electrolysis require approximately nine liters of water per kilogram of hydrogen produced, in addition to large energy inputs (50–55 kWh per kg H_2) [104].
- Thermochemical processes (SMR and coal gasification) use fossil fuels but also require water for steam generation. SMR uses four to six liters of water per kg H_2, while coal gasification can require more than 10 liters [105].
- Biological hydrogen production through algae and bacterial fermentation can operate using wastewater, reducing fresh water demand, but these processes have lower energy efficiencies, making them impractical for large-scale deployment [106].

2.6.6 Carbon Footprint and Sustainability of Large-scale Hydrogen Production

The carbon footprint of hydrogen production is directly linked to the energy source used and whether carbon capture technologies are implemented. Hydrogen is classified into three main types based on emissions:

1. Gray Hydrogen (from fossil fuels without carbon capture):
   - Produces 9–12 kg of CO_2 per kg of hydrogen.
   - Least sustainable but currently the most widely used method (~75% of global hydrogen) [107].
2. Blue Hydrogen (from fossil fuels with CCS):
   - Captures 60–90% of CO_2 emissions, reducing the carbon footprint to two to four kg CO_2 per kg H_2.
   - More sustainable than gray hydrogen but still relies on fossil fuels [108].
3. Green Hydrogen (from electrolysis using renewable energy):
   - Produces zero direct emissions, with an overall footprint of less than one kg CO_2 per kg H_2, depending on electricity source.
   - Most sustainable but currently expensive due to high electricity costs [109].
4. Biological Hydrogen Production:
   - Can be carbon-neutral when derived from organic waste but has low efficiency and scalability issues.

The sustainability of hydrogen production also depends on land use, biodiversity impact, and lifecycle emissions. Green hydrogen offers the best long-term solution, but scaling it requires significant advancements in renewable energy, electrolyzer efficiency, and cost reductions. Table 2.7 provides the comparative analysis of hydrogen production methods.

Table 2.7 Comparative analysis of hydrogen production methods (cost, efficiency, scalability, and environmental impact).

Parameter	Electrolysis (green H_2)	Thermochemical (SMR & gasification – gray & blue H_2)	Biological hydrogen
Energy efficiency	60–80%	50–75% (lower with CCS)	<40% (low yield)
Water usage (L/kg H_2)	9–10 L	4–10 L	Variable, can use wastewater
Carbon footprint (kg CO_2/kg H_2)	**<1 kg (green hydrogen)**	**Gray: 9–12 kg** / **Blue: 2–4 kg**	Low (varies based on feedstock)
Scalability	Medium (limited by electricity availability)	High (fossil fuel-dependent)	Low (due to slow microbial growth)
Technology maturity	Commercial, improving	Established for decades	Early-stage, experimental
Sustainability	High (renewable energy-based)	Moderate (depends on CCS efficiency)	Moderate (biodegradable feedstocks)
Cost ($/kg H_2)	**$4.00–$6.00** (declining with renewables)	**Gray: $1.00–$2.00** / **Blue: $2.00–$3.50**	**$5.00+ (due to low yields)**

2.7 Summary and Future Directions

In this chapter, we have explored various hydrogen production methods, each with its unique strengths and challenges. The primary techniques discussed include electrolysis, photocatalysis, thermochemical cycles, and biological processes. Electrolysis, particularly advancements in PEM, alkaline, and SOEC, has made significant strides in efficiency and cost reduction, positioning it as a clean and scalable method for hydrogen production. Photocatalysis and photoelectrochemical (PEC) water splitting have emerged as promising solar-driven technologies, harnessing sunlight to produce hydrogen directly from water, though scalability and efficiency remain key areas for development. Thermochemical cycles like the SI cycle and metal oxide cycles offer high-temperature methods for hydrogen generation, with the added potential for integrated systems that utilize industrial waste heat. Biological processes, including algae-based, bacterial, and enzymatic hydrogen production, hold great promise for sustainable biohydrogen, but challenges in scaling these methods remain significant.

Recent advancements in these technologies have focused on improving efficiency, reducing costs, and increasing the scalability of hydrogen production systems. In electrolysis, innovations such as the development of novel catalysts, materials, and improved electrolyzer designs have made hydrogen production from renewable sources more viable. For photocatalysis and PEC, ongoing research seeks to enhance the light absorption capacity of semiconductors and optimize reaction mechanisms. Similarly, thermochemical cycles are evolving to operate at lower temperatures and improve the stability of materials, while biological hydrogen production methods are being refined to improve yields and the efficiency of microbial processes.

Looking ahead, hybrid approaches that combine different hydrogen production technologies are poised to play a critical role in advancing hydrogen as a clean energy source. Hybrid systems could integrate electrolysis with solar energy, thermochemical cycles with waste heat recovery, and biological methods with industrial processes to maximize overall efficiency and sustainability. The integration of these diverse technologies could provide a more resilient and flexible hydrogen production infrastructure, capable of adapting to varying energy demands and resource availability. The future of hydrogen production is also closely tied to the evolution of global energy systems. As nations continue to shift toward renewable energy sources, hydrogen's role as an energy carrier and storage medium will become increasingly important. The ability to store excess energy from intermittent renewable sources like solar and wind in the form of hydrogen will be crucial for balancing energy supply and demand, particularly in regions with high renewable energy potential. The development of a hydrogen infrastructure, including production, storage, transport, and refueling stations, will be essential to unlock the full potential of hydrogen as a key player in global energy systems.

References

1 Holladay JD, Hu J, King DL, Wang Y. An overview of hydrogen production technologies. *Catal Today*. 2009 Jan 30;139(4):244–60.

2 Soltani R, Rosen MA, Dincer I. Assessment of CO_2 capture options from various points in steam methane reforming for hydrogen production. *Int J Hydrogen Energy*. 2014 Dec 3;39(35):20266–75.

3 Aouali FZ, Becherif M, Ramadan HS, Emziane M, Khellaf A, Mohammedi K. Analytical modelling and experimental validation of proton exchange membrane electrolyser for hydrogen production. *Int J Hydrogen Energy*. 2017 Jan 12;42(2):1366–74.

4 AlZahrani AA, Dincer I. Modeling and performance optimization of a solid oxide electrolysis system for hydrogen production. *Appl Energy*. 2018 Sep 1;225:471–85.

5 Pinaud BA, Benck JD, Seitz LC, Forman AJ, Chen Z, Deutsch TG, et al. Technical and economic feasibility of centralized facilities for solar hydrogen production via photocatalysis and photoelectrochemistry. *Energy Environ Sci* [Internet]. 2013 Jun 21 [cited 2025 Feb 2];6(7):1983–2002. Available from: https://pubs.rsc.org/en/content/articlehtml/2013/ee/c3ee40831k.

6 Ahmed SF, Mofijur M, Nuzhat S, Rafa N, Musharrat A, Lam SS, et al. Sustainable hydrogen production: Technological advancements and economic analysis. *Int J Hydrogen Energy*. 2022 Oct 30;47(88):37227–55.

7 Chai S, Zhang G, Li G, Zhang Y. Industrial hydrogen production technology and development status in China: A review. *Clean Technol Environ Policy* [Internet]. 2021 Sep 1 [cited 2025 Feb 2];23(7):1931–46. Available from: https://link.springer.com/article/10.1007/s10098-021-02089-w.

8 Akaev AA, Rudskoi AI, Korablev VV, Sarygulov AI. Technological and economic barriers to hydrogen energy growth. *Her Russ Acad Sci* [Internet]. 2022 Dec 1 [cited 2025 Feb 2];92(6):691–701. Available from: https://link.springer.com/article/10.1134/S1019331622050082.

9 Najafi B, Haghighatshoar F, Ardabili S, Band SS, Chau KW, Mosavi A. Effects of low-level hydroxy as a gaseous additive on performance and emission characteristics of a dual fuel diesel engine fueled by diesel/biodiesel blends. *Eng Appl of Comput Fluid Mech*. 2021 Jan 1;15(1):236–50.

10 Ursúa A, Gandía LM, Sanchis P. Hydrogen production from water electrolysis: Current status and future trends. *Proc IEEE*. 2012;100(2):410–26.

11 Li W, Shi Y, Luo Y, Cai N. Elementary reaction modeling of solid oxide electrolysis cells: Main zones for heterogeneous chemical/electrochemical reactions. *J Power Sources*. 2015 Jan 1;273:1–13.

12 Zeng K, Zhang D. Recent progress in alkaline water electrolysis for hydrogen production and applications. *Prog Energy Combust Sci*. 2010 Jun 1;36(3):307–26.

13 Goñi-Urtiaga A, Presvytes D, Scott K. Solid acids as electrolyte materials for proton exchange membrane (PEM) electrolysis: Review. *Int J Hydrogen Energy*. 2012 Feb 1;37(4):3358–72.

14 Hauch A, Küngas R, Blennow P, Hansen AB, Hansen JB, Mathiesen BV, et al. Recent advances in solid oxide cell technology for electrolysis. *Science* (1979) [Internet]. 2020 Oct 9 [cited 2025 Feb 2];370(6513):eaba6118. Available from: https://www.science.org/doi/10.1126/science.aba6118.

15 Guo K, Liu M, Wang B, Lou J, Hao Y, Pei G, et al. Hydrogen production and solar energy storage with thermo-electrochemically enhanced steam methane reforming. *Sci Bull (Beijing)*. 2024 Apr 30;69(8):1109–21.

16 Fahr S, Engel FK, Rehfeldt S, Peschel A, Klein H. Overview and evaluation of crossover phenomena and mitigation measures in proton exchange membrane (PEM) electrolysis. *Int J Hydrogen Energy*. 2024 May 28;68:705–21.

17 Bin S, Chen Z, Zhu Y, Zhang Y, Xia Y, Gong S, et al. High-pressure proton exchange membrane water electrolysis: Current status and challenges in hydrogen production. *Int J Hydrogen Energy*. 2024 May 20;67:390–405.

18 Sebbahi S, Assila A, Alaoui Belghiti A, Laasri S, Kaya S, Hlil EK, et al. A comprehensive review of recent advances in alkaline water electrolysis for hydrogen production. *Int J Hydrogen Energy*. 2024 Sep 11;82:583–99.

19 Emam AS, Hamdan MO, Abu-Nabah BA, Elnajjar E. A review on recent trends, challenges, and innovations in alkaline water electrolysis. *Int J Hydrogen Energy*. 2024 Apr 25;64:599–625.

20 Liu H, Yu M, Tong X, Wang Q, Chen M. High temperature solid oxide electrolysis for green hydrogen production. *Chem Rev* [Internet]. 2024 Sep 25;124(18):10509–10576. [cited 2025 Feb 2]; Available from: https://pubs.acs.org/doi/abs/10.1021/acs.chemrev.3c00795.

21 Xu Y, Cai S, Chi B, Tu Z. Technological limitations and recent developments in a solid oxide electrolyzer cell: A review. *Int J Hydrogen Energy*. 2024 Jan 2; 50:548–91.

22 Arunachalam M, Han DS. Efficient solar-powered PEM electrolysis for sustainable hydrogen production: An integrated approach. *Emergent Mater* [Internet]. 2024 Aug 1 [cited 2025 Feb 2];7(4):1401–15. Available from: https://link.springer.com/article/10.1007/s42247-024-00697-y.

23 Shiva Kumar S, Lim H. An overview of water electrolysis technologies for green hydrogen production. *Energy Rep*. 2022 Nov 1; 8:13793–813.

24 Li R. Latest progress in hydrogen production from solar water splitting via photocatalysis, photoelectrochemical, and photovoltaic-photoelectrochemical solutions. *Chinese J Catal*. 2017 Jan 1;38(1):5–12.

25 Faraji M, Yousefi M, Yousefzadeh S, Zirak M, Naseri N, Jeon TH, et al. Two-dimensional materials in semiconductor photoelectrocatalytic systems for water splitting. *Energy Environ Sci* [Internet]. 2019 Jan 16 [cited 2025 Feb 2];12(1):59–95. Available from: https://pubs.rsc.org/en/content/articlehtml/2019/ee/c8ee00886h.

26 Tayebi M, Lee BK. Recent advances in BiVO4 semiconductor materials for hydrogen production using photoelectrochemical water splitting. *Renew Sustain Energy Rev*. 2019 Sep 1;111:332–43.

27 Tao X, Zhao Y, Wang S, Li C, Li R. Recent advances and perspectives for solar-driven water splitting using particulate photocatalysts. *Chem Soc Rev* [Internet]. 2022 May 10 [cited 2025 Feb 2];51(9):3561–608. Available from: https://pubs.rsc.org/en/content/articlehtml/2022/cs/d1cs01182k.

28 Jiang C, Moniz SJA, Wang A, Zhang T, Tang J. Photoelectrochemical devices for solar water splitting—materials and challenges. *Chem Soc Rev* [Internet]. 2017 Jul 31 [cited 2025 Feb 2];46(15):4645–60. Available from: https://pubs.rsc.org/en/content/articlehtml/2017/cs/c6cs00306k.

29 Acar C, Dincer I, Naterer GF. Review of photocatalytic water-splitting methods for sustainable hydrogen production. *Int J Energy Res* [Internet]. 2016 Sep 1 [cited 2025 Feb 2];40(11):1449–73. Available from: https://onlinelibrary.wiley.com/doi/full/10.1002/er.3549.

30 Qureshi F, Tahir M. Photoelectrochemical water splitting with engineering aspects for hydrogen production: Recent advances, strategies and challenges. *Int J Hydrogen Energy*. 2024 Jun 5;69:760–76.

31 Kumar A, Sharma P, Sharma G, Dhiman P, Mola GT, Farghali M, et al. Simultaneous hydrogen production and photocatalytic pollutant removal: A review. *Environ Chem Lett* [Internet]. 2024 Jun 11 [cited 2025 Feb 2];22(5):2405–24. Available from: https://link.springer.com/article/10.1007/s10311-024-01756-w.

32 Yang W, Prabhakar RR, Tan J, Tilley SD, Moon J. Strategies for enhancing the photocurrent, photovoltage, and stability of photoelectrodes for photoelectrochemical water splitting. *Chem Soc Rev* [Internet]. 2019 Sep 30 [cited 2025 Feb 2];48(19):4979–5015. Available from: https://pubs.rsc.org/en/content/articlehtml/2019/cs/c8cs00997j.

33 Saraswat SK, Rodene DD, Gupta RB. Recent advancements in semiconductor materials for photoelectrochemical water splitting for hydrogen production using visible light. *Renew Sustain Energy Rev*. 2018 Jun 1;89:228–48.

34 Li D, Shi J, Li C. Transition-metal-based electrocatalysts as cocatalysts for photoelectrochemical water splitting: A mini review. *Small* [Internet]. 2018 Jun 1 [cited 2025 Feb 2];14(23):1704179. Available from: https://onlinelibrary.wiley.com/doi/full/10.1002/smll.201704179.

35 Vilanova A, Dias P, Lopes T, Mendes A. The route for commercial photoelectrochemical water splitting: A review of large-area devices and key upscaling challenges. *Chem Soc Rev* [Internet]. 2024 Mar 4 [cited 2025 Feb 2];53(5):2388–434. Available from: https://pubs.rsc.org/en/content/articlehtml/2024/cs/d1cs01069g.

36 Li Z, Luo W, Zhang M, Feng J, Zou Z. Photoelectrochemical cells for solar hydrogen production: Current state of promising photoelectrodes, methods to improve their properties, and outlook. *Energy Environ Sci* [Internet]. 2013 Jul 31 [cited 2025 Feb 2];6(2):347–70. Available from: https://pubs.rsc.org/en/content/articlehtml/2013/ee/c2ee22618a.

37 Zhou P, Navid IA, Ma Y, Xiao Y, Wang P, Ye Z, et al. Solar-to-hydrogen efficiency of more than 9% in photocatalytic water splitting. *Nature* [Internet]. 2023 Jan 4 [cited 2025 Feb 2];613(7942):66–70. Available from: https://www.nature.com/articles/s41586-022-05399-1.

38 Ye Y, Bruning H, Li X, Yntema D, Rijnaarts HHM. Significant enhancement of micropollutant photocatalytic degradation using a TiO2 nanotube array photoanode based photocatalytic fuel cell. *Chem Eng J*. 2018 Dec 15;354:553–62.

39 Lee JE, Shafiq I, Hussain M, Lam SS, Rhee GH, Park YK. A review on integrated thermochemical hydrogen production from water. *Int J Hydrogen Energy*. 2022 Jan 22;47(7):4346–56.

40 Yilmaz F, Selbaş R. Thermodynamic performance assessment of solar based sulfur-iodine thermochemical cycle for hydrogen generation. *Energy*. 2017 Dec 1;140:520–9.

41 Mao Y, Gao Y, Dong W, Wu H, Song Z, Zhao X, et al. Hydrogen production via a two-step water splitting thermochemical cycle based on metal oxide—A review. *Appl Energy*. 2020 Jun 1; 267:114860.

42 Zeng J, Zhang J, Ling B, He Y, Weng W, Wang Z. Hydrogen production by sulfur-iodine thermochemical cycle — Current status and recent advances. *Int J Hydrogen Energy*. 2024 Oct 11;86:677–702.

43 Dou B, Zhang H, Song Y, Zhao L, Jiang B, He M, et al. Hydrogen production from the thermochemical conversion of biomass: Issues and challenges. *Sustain Energy Fuels* [Internet]. 2019 Jan 29 [cited 2025 Feb 2];3(2):314–42. Available from: https://pubs.rsc.org/en/content/articlehtml/2019/se/c8se00535d.

44 Rosen MA. Advances in hydrogen production by thermochemical water decomposition: A review. *Energy*. 2010 Feb 1; 35(2):1068–76.

45 Vitart X, Carles P, Anzieu P. A general survey of the potential and the main issues associated with the sulfur–iodine thermochemical cycle for hydrogen production using nuclear heat. *Prog Nucl Energy*. 2008 Mar 1;50(2–6):402–10.

46 Hussain Z, Khan ZS, Khoja AH, Shabbir A, Al-Anazi A, Din IU. Corrosion behavior of SiC coated HX with MoSi2 interlayer to be utilized in iodine–sulfur cycle for hydrogen production. *Heliyon* [Internet]. 2023 Nov 1 [cited 2025 Feb 2];9(11):e21640. Available from: http://www.cell.com/article/S2405844023088485/fulltext.

47 Wu Y, Wang Z, Yan Y, Wei Y, Wang J, Shen Y, et al. Rational photodeposition of cobalt phosphate on flower-like ZnIn2S4 for Efficient photocatalytic hydrogen evolution. *Molecules* [Internet]. 2024 Jan 1 [cited 2025 Feb 2];29(2):465. Available from: https://www.mdpi.com/1420-3049/29/2/465/htm.

48 Wu S, Zhou C, Doroodchi E, Nellore R, Moghtaderi B. A review on high-temperature thermochemical energy storage based on metal oxides redox cycle. *Energy Convers Manag*. 2018 Jul 15;168:421–53.

49 Miller JE, McDaniel AH, Allendorf MD. Considerations in the design of materials for solar-driven fuel production using metal-oxide thermochemical cycles. *Adv Energy Mater* [Internet]. 2014 Jan 1 [cited 2025 Feb 2];4(2):1300469. Available from: https://onlinelibrary.wiley.com/doi/full/10.1002/aenm.201300469.

50 Charvin P, Abanades S, Beche E, Lemont F, Flamant G. Hydrogen production from mixed cerium oxides via three-step water-splitting cycles. *Solid State Ion*. 2009 Jun 25;180(14–16):1003–10.

51 Zheng K, Yu Z, Tan SC, Liu T, Kong H. Research progress on metal oxide oxygen carrier materials for two-step solar thermochemical cycles in the last five years. *Energy Convers Manag*. 2024 Mar 1;303:118116.

52 Tumkur PP, Gunasekaran NK, Lamani BR, Bayon NN, Prabhakaran K, Hall JC, et al. Cerium oxide nanoparticles: Synthesis and characterization for biosafe applications. *Nanomanufacturing* [Internet]. 2021 Dec 3 [cited 2025 Feb 2];1(3):176–89. Available from: https://www.mdpi.com/2673-687X/1/3/13/htm.

53 Rafique A, Zubair U, Serrapede M, Fontana M, Bianco S, Rivolo P, et al. Binder free and flexible asymmetric supercapacitor exploiting Mn3O4 and MoS2 nanoflakes on carbon fibers. *Nanomaterials* [Internet]. 2020 May 31 [cited 2025 Feb 2];10(6):1084. Available from: https://www.mdpi.com/2079-4991/10/6/1084/htm.

54 Shen Y, Jarboe L, Brown R, Wen Z. A thermochemical–biochemical hybrid processing of lignocellulosic biomass for producing fuels and chemicals. *Biotechnol Adv*. 2015 Dec 1;33(8):1799–813.

55 Chen C, Xia Q, Feng S, Liu Q. A novel solar hydrogen production system integrating high temperature electrolysis with ammonia based thermochemical energy storage. *Energy Convers Manag*. 2021 Jun 1;237:114143.

56 El-Emam RS, Özcan H. Comprehensive review on the techno-economics of sustainable large-scale clean hydrogen production. *J Clean Prod*. 2019 May 20; 220:593–609.

57 Akhlaghi N, Najafpour-Darzi G. A comprehensive review on biological hydrogen production. *Int J Hydrogen Energy*. 2020 Sep 3; 45(43):22492–512.

58 Chen Y. Global potential of algae-based photobiological hydrogen production. *Energy Environ Sci* [Internet]. 2022 Jul 13 [cited 2025 Feb 2];15(7):2843–57. Available from: https://pubs.rsc.org/en/content/articlehtml/2022/ee/d2ee00342b.

59 Maintinguer SI, Sakamoto IK, Adorno MAT, Varesche MBA. Bacterial diversity from environmental sample applied to bio-hydrogen production. *Int J Hydrogen Energy*. 2015 Mar 2;40(8):3180–90.

60 Sivaramakrishnan R, Shanmugam S, Sekar M, Mathimani T, Incharoensakdi A, Kim SH, et al. Insights on biological hydrogen production routes and potential microorganisms for high hydrogen yield. *Fuel*. 2021 May 1;291:120136.

61 Singh L, Wahid ZA. Methods for enhancing bio-hydrogen production from biological process: A review. *J Ind Eng Chem*. 2015 Jan 25;21:70–80.

62 Chen J, Li Y, Li M, Shi J, Wang L, Luo S, et al. Chemical flocculation-based green algae materials for photobiological hydrogen production. *ACS Appl Bio Mater* [Internet]. 2022 Feb 21 [cited 2025 Feb 2];5(2):897–903. Available from: https://pubs.acs.org/doi/abs/10.1021/acsabm.1c01281.

63 Goria K, Singh HM, Singh A, Kothari R, Tyagi VV. Insights into biohydrogen production from algal biomass: Challenges, recent advancements and future directions. *Int J Hydrogen Energy*. 2024 Jan 2; 52:127–51.

64 Vinayak V, Sirotiya V, Khandelwal P, Rai A, Jadhav DA, Pugazhendhi A, et al. Recent trends in engineering algae for biohydrogen production: State of art strategies. *Fuel*. 2023 Sep 15;348:128636.

65 Hallenbeck PC, Liu Y. Recent advances in hydrogen production by photosynthetic bacteria. *Int J Hydrogen Energy*. 2016 Feb 23; 41(7):4446–54.

66 Kim DH, Kim MS. Hydrogenases for biological hydrogen production. *Bioresour Technol*. 2011 Sep 1;102(18):8423–31.

67 Lin PY, Whang LM, Wu YR, Ren WJ, Hsiao CJ, Li SL, et al. Biological hydrogen production of the genus Clostridium: Metabolic study and mathematical model simulation. *Int J Hydrogen Energy*. 2007 Aug 1;32(12):1728–35.

68 Koku H, Erolu I, Gündüz U, Yücel M, Türker L. Aspects of the metabolism of hydrogen production by Rhodobacter sphaeroides. *Int J Hydrogen Energy*. 2002 Nov 1;27(11–12):1315–29.

69 Sagir E, Alipour S. Photofermentative hydrogen production by immobilized photosynthetic bacteria: Current perspectives and challenges. *Renew Sustain Energy Rev*. 2021 May 1;141:110796.

70 Nanda S, Li K, Abatzoglou N, Dalai AK, Kozinski JA. Advancements and confinements in hydrogen production technologies. In: *Bioenergy Systems for the Future: Prospects for Biofuels and Biohydrogen*. Elsevier. 2017 Jan 1;373–418.

71 Woodward J, Cordray KA, Edmonston RJ, Blanco-Rivera M, Mattingly SM, Evans BR. Enzymatic hydrogen production: Conversion of renewable resources for energy production. *Energy Fuels* [Internet]. 1999 Jan [cited 2025 Feb 2];14(1): 197–201. Available from: https://pubs.acs.org/doi/abs/10.1021/ef990126l.

72 Das D, Veziroğlu TN. Hydrogen production by biological processes: A survey of literature. *Int J Hydrogen Energy*. 2001 Jan 1; 26(1):13–28.

73 Percival Zhang YH. Production of biocommodities and bioelectricity by cell-free synthetic enzymatic pathway biotransformations: Challenges and opportunities. *Biotechnol Bioeng* [Internet]. 2010 Mar 1 [cited 2025 Feb 2];105(4):663–77. Available from: https://onlinelibrary.wiley.com/doi/full/10.1002/bit.22630.

74 Akram F, Fatima T, Ibrar R, ul Haq I. Biohydrogen: Production, promising progressions and challenges of a green carbon-free energy. *Sustain Energy Technol Assessments*. 2024 Sep 1;69:103893.

75 Xu L, Fan J, Wang Q. Omics application of bio-hydrogen production through green alga chlamydomonas reinhardtii. *Front Bioeng Biotechnol* [Internet]. 2019 Aug 21 [cited 2025 Feb 2];7(AUG):482439. Available from: www.frontiersin.org.

76 Chandrasekhar K, Lee YJ, Lee DW. Biohydrogen production: Strategies to improve process efficiency through microbial routes. *Int J Mol Sci* 2015 [Internet]. 2015 Apr 14 [cited 2025 Feb 2];16(4):8266–93. Available from: https://www.mdpi.com/1422-0067/16/4/8266/htm.

77 Leone L, Sgueglia G, La Gatta S, Chino M, Nastri F, Lombardi A. Enzymatic and bioinspired systems for hydrogen production. *Int J Mol Sci* 2023 [Internet]. 2023 May 11 [cited 2025 Feb 2];24(10):8605. Available from: https://www.mdpi.com/1422-0067/24/10/8605/htm.

78 Singla MK, Gupta J, Beryozkina S, Safaraliev M, Singh M. The colorful economics of hydrogen: Assessing the costs and viability of different hydrogen production methods - A review. *Int J Hydrogen Energy*. 2024 Apr 3;61:664–77.

79 Prosser JH, James BD, Murphy BM, Wendt DS, Casteel MJ, Westover TL, et al. Cost analysis of hydrogen production by high-temperature solid oxide electrolysis. *Int J Hydrogen Energy*. 2024 Jan 2;49:207–27.

80 Naqvi SR, kazmi B, Ammar Taqvi SA, Chen WH, Juchelková D. Techno economic analysis for advanced methods of green hydrogen production. *Curr Opin Green Sustain Chem*. 2024 Aug 1;48:100939.

81 Sharma A, Hussain Mehdi SE, Pandit S, Eun-Oh S, Natarajan V. Factors affecting hydrogen production in microbial electrolysis cell (MEC): A review. *Int J Hydrogen Energy*. 2024 Apr 3;61:1473–84.

82 Sayed K, Khamies M, Abokhalil AG, Aref M, Mossa MA, Metab Almalki M, et al. Feasibility study and economic analysis of PV/wind-powered hydrogen production plant. *IEEE Access*. 2024;12:76304–18.

83 Trangwachirachai K, Rouwenhorst K, Lefferts L, Faria Albanese JA. Recent progress on ammonia cracking technologies for scalable hydrogen production. *Curr Opin Green Sustain Chem*. 2024 Oct 1;49:100945.

84 Dash S, K AS, S J, D VHW, D E, Surapraraju SK, et al. Advances in green hydrogen production through alkaline water electrolysis: A comprehensive review. *Int J Hydrogen Energy*. 2024 Sep 19;83:614–29.

85 Alotaibi FN, Berrouk AS, Salim IM. Scaling up dry methane reforming: Integrating computational fluid dynamics and machine learning for enhanced hydrogen production in industrial-scale fluidized bed reactors. *Fuel*. 2024 Nov 15; 376:132673.

86 Sun H, Xu X, Fei L, Zhou W, Shao Z. Electrochemical oxidation of small molecules for energy-saving hydrogen production. *Adv Energy Mater* [Internet]. 2024 Aug 1 [cited 2025 Mar 18];14(30):2401242. Available from: https://onlinelibrary.wiley.com/doi/full/10.1002/aenm.202401242.

87 Darabadi Zare AA, Yari M. Techno economic analysis of efficient and environmentally friendly methods for hydrogen, power, and heat production using chemical looping combustion integrating plastic waste gasification and thermochemical copper chlorine cycles. *Energy*. 2024 Feb 15;289:129941.

88 Muhammad HA, Naseem M, Kim J, Kim S, Choi Y, Lee YD. Solar hydrogen production: Technoeconomic analysis of a concentrated solar-powered high-temperature electrolysis system. *Energy*. 2024 Jul 1;298:131284.

89 Azizimehr B, Armaghani T, Ghasemiasl R, Kaabi Nejadian A, Javadi MA. A comprehensive review of recent developments in hydrogen production methods using a new parameter. *Int J Hydrogen Energy*. 2024 Jun 27;72:716–29.

90 Noh W, Cho S, Lee I. Carbon utilization in natural gas-based hydrogen production via carbon dioxide electrolysis: Towards cost-competitive clean hydrogen. *Energy Convers Manag*. 2024 Aug 15;314:118719.

91 Batgi SU, Dincer I. A study on comparative environmental impact assessment of thermochemical cycles and steam methane reforming processes for hydrogen production processes. *Comput Chem Eng*. 2024 Jan 1;180:108514.

92 Afanasev P, Askarova A, Alekhina T, Popov E, Markovic S, Mukhametdinova A, et al. An overview of hydrogen production methods: Focus on hydrocarbon feedstock. *Int J Hydrogen Energy*. 2024 Aug 12; 78:805–28.

93 Cavaliere P. Hydrogen: Energy vector and fuel. *Hydrogen Embrittlement in Metals and Alloys* [Internet]. 2025 [cited 2025 Mar 18];1–155. Available from: https://link.springer.com/chapter/10.1007/978-3-031-83681-7_1.

94 Ju HK, Yoon D, Bong S, Lee J. Challenge and opportunity in scaling-up hydrogen production via electrochemical ammonia electrolysis process. *Curr Opin Electrochem*. 2025 Feb 1;49:101609.

95 Ramteke SM, Meshram SA, Chelladurai H, Choudhary AK. Thermochemical hydrogen production methods: A review. *Hydrogen Energy: Production, Storage, and Utilization* [Internet]. 2024 Jan 1 [cited 2025 Mar 18];130–47. Available from: https://www.taylorfrancis.com/chapters/edit/10.1201/9781003537816-8/thermochemical-hydrogen-production-methods-sangharatna-ramteke-sachin-akoji-meshram-chelladurai-akhilesh-kumar-choudhary.

96 Zheng C, Zhang L, Li Q, Yang Y, Zhang C, Dou Z. Review of electric hydrogen production technology combined with technology maturity analysis. *Int J Hydrogen Energy*. 2025 Jan 13;98:429–40.

97 Elsapagh MR, Sultan NS, Mohamed FA, Fahmy HM. The role of nanocatalysts in green hydrogen production and water splitting. *Int J Hydrogen Energy*. 2024 May 20;67:62–82.

98 Hill SJP, Bamisile O, Hatton L, Staffell I, Jansen M. The cost of clean hydrogen from offshore wind and electrolysis. *J Clean Prod*. 2024 Mar 15;445:141162.

99 Chang SH, Rajuli MF. An overview of pure hydrogen production via electrolysis and hydrolysis. *Int J Hydrogen Energy*. 2024 Sep 26; 84:521–38.

100 Jankovic J, Wilkinson DP, Wilkinson DP. Towards a more sustainable future: Transitioning from thermochemical to electrochemical processes in clean energy technologies relevant to hydrogen-containing fuels. *Can J Chem Eng* [Internet]. 2025 Apr 1 [cited 2025 Mar 18];103(4):1602–22. Available from: https://onlinelibrary.wiley.com/doi/full/10.1002/cjce.25549.

101 Karim S, Tanwar N, Das S, Ranjit R, Banerjee A, Gulafshan N, et al. Shaping the future of green hydrogen production: Overcoming conventional challenges with molecular catalysts, immobilization, and scalable electrolyzers. *ACS Catal* [Internet]. 2025 Jan 17 [cited 2025 Mar 18];15(2):1073–96. Available from: https://pubs.acs.org/doi/abs/10.1021/acscatal.4c05986.

102 Snousy MG, Abouelmagd AR, Moustafa YM, Gamvroula DE, Alexakis DE, Ismail E, et al. The potential role of Africa in green hydrogen production: A short-term roadmap to protect the world's future from climate crisis. *Water* [Internet]. 2025 Feb 2 [cited 2025 Mar 18];17(3):416. Available from: https://www.mdpi.com/2073-4441/17/3/416/htm.

103 Martínez de León C, Molina P, Ríos C, Brey JJ. Green hydrogen production's impact on sustainable development goals. *Int J Hydrogen Energy*. 2025 Jan 4, 142:642–653.

104 Anand C, Chandraja B, Nithiya P, Akshaya M, Tamizhdurai P, Shoba G, et al. Green hydrogen for a sustainable future: A review of production methods, innovations, and applications. *Int J Hydrogen Energy*. 2025 Mar 20;111:319–41.

105 Kafle S, Sapkota S, Higgins BT, Adhikari S. Research trends in life cycle assessment of hydrogen production: Methodological review on thermochemical conversion processes. *Int J Hydrogen Energy*. 2025 Mar 6;106:432–43.

106 Pandey AK, Park KD, Morya R, Joo HH, Kim SH. Exploring sustainable biohydrogen production from dried food waste: Optimization strategies and environmental implications. *J Environ Manage*. 2025 Feb 1;375:124195.

107 Beigi Kheradmand A, Heidari Soureshjani M, Jahangiri M, Hamawandi B. Comparative techno-economic analysis of gray hydrogen production costs: A case study. *Sustainability* [Internet]. 2025 Jan 12 [cited 2025 Mar 18];17(2):547. Available from: https://www.mdpi.com/2071-1050/17/2/547/htm.

108 Yoon HJ, Choi J, Cherif A, Lee CJ. Integrated hydrogen liquefaction with autothermal reforming utilizing liquefied oxygen as a pre-coolant. *J Clean Prod*. 2025 Jan 1;486:144563.

109 Hassan NS, Jalil AA, Rajendran S, Khusnun NF, Bahari MB, Johari A, et al. Recent review and evaluation of green hydrogen production via water electrolysis for a sustainable and clean energy society. *Int J Hydrogen Energy*. 2024 Jan 2;52:420–41.

3

Advanced Materials for Hydrogen Production

3.1 Introduction to Advanced Materials for Hydrogen Production

3.1.1 Importance of Material Science in Hydrogen Production

Material science plays a crucial role in hydrogen production by directly influencing the efficiency, stability, and economic feasibility of various production methods. The choice of materials affects reaction kinetics, energy consumption, durability, and scalability across electrolysis, photocatalysis, and thermochemical cycles. As hydrogen emerges as a key energy carrier for a sustainable future, advancements in material science are essential to overcoming the limitations of traditional methods and enabling large-scale, cost-effective hydrogen generation [1].

Efficiency in hydrogen production is largely dictated by the properties of the materials used in catalytic and reaction processes. For electrolysis, the efficiency of the hydrogen evolution reaction (HER) and oxygen evolution reaction (OER) is determined by the electrocatalyst's ability to lower activation energy and enhance reaction kinetics [2]. Traditional catalysts like platinum (Pt) and iridium oxide (IrO_2) offer high efficiency but are limited by their high cost and scarcity [3]. In contrast, emerging materials such as transition metal dichalcogenides (TMDs), metal-organic frameworks (MOFs), and perovskite oxides are being explored for their ability to achieve similar efficiency at a lower cost [4]. Similarly, in photocatalytic hydrogen production, semiconductor materials play a pivotal role in absorbing solar energy and driving water-splitting reactions. Conventional materials like titanium dioxide (TiO_2) have been widely used, but they suffer from limited visible-light absorption. Advanced photocatalysts, including modified perovskites, carbon-based materials like graphene, and doped metal oxides, are being developed to improve light absorption, charge separation, and overall efficiency [5]. In thermochemical hydrogen production, materials must withstand extreme temperatures while facilitating redox reactions efficiently. Metal oxides, particularly cerium oxide (CeO_2) and ferrites, are promising candidates for thermochemical cycles due to their high oxygen-exchange capacity and thermal stability. The advancement of these materials is critical for improving efficiency and ensuring

Next-Generation Hydrogen Economy: Production, Storage, and Fuel Cell Technologies, First Edition.
Peeyush Phogat, Shreya Sharma and Satyam Rawat.

sustained hydrogen production at high temperatures. As illustrated in Figure 3.1, material applications in different hydrogen production technologies involve a variety of advanced catalysts, semiconductors, and metal oxides, each optimized for specific reaction conditions and efficiency goals [6].

The longevity of materials used in hydrogen production directly affects operational costs and large-scale deployment feasibility. Traditional electrocatalysts such as Pt and IrO_2 degrade over time due to corrosion, dissolution, and phase

Figure 3.1 Schematic representation of material applications in different hydrogen production technologies.

changes under harsh electrochemical conditions. Research is now focused on designing stable alternatives like cobalt-nickel alloys, molybdenum sulfides (MoS_2), and tungsten-based catalysts that exhibit prolonged durability under acidic and alkaline conditions. In photocatalysis, the long-term stability of semiconductor materials is a major concern. Many photocatalysts suffer from photocorrosion, where prolonged exposure to light and water degrades the material's structure. Strategies such as surface modification, protective coatings, and heterojunction formation have been explored to enhance material stability while maintaining high photocatalytic activity. For thermochemical cycles, material degradation due to repeated high-temperature cycling remains a key challenge. Metal oxides used in these processes experience sintering, grain growth, and phase segregation, which reduce their reactivity over time. Novel synthesis techniques, such as nanostructuring and doping, are being investigated to enhance the thermal stability and redox performance of these materials.

One of the biggest barriers to widespread hydrogen adoption is the high cost of production, which is largely influenced by the materials used. Precious metal catalysts like Pt, Ru, and Ir provide excellent catalytic activity but are too expensive for large-scale applications. Therefore, researchers are developing cost-effective alternatives based on earth-abundant elements like nickel, iron, cobalt, and manganese. Nanomaterial-based solutions have also gained attention for reducing material costs while maintaining high catalytic performance. By using nanostructured catalysts, the surface area and active sites are maximized, allowing for lower material consumption while achieving high reaction rates. Furthermore, advancements in machine learning and computational material discovery are accelerating the identification of new low-cost materials with optimal catalytic properties [7].

A direct comparison between traditional and advanced materials across different hydrogen production technologies reveals significant progress in efficiency, stability, and cost-effectiveness:

a. **Electrolysis:**
   - Traditional: Pt and IrO_2 catalysts offer high efficiency but are expensive and scarce.
   - Advanced: Nickel-based catalysts, transition metal carbides/nitrides, and perovskite oxides are promising alternatives that reduce costs while maintaining performance.

b. **Photocatalysis:**
   - Traditional: TiO_2 is widely used but has a limited absorption range in the ultraviolet spectrum.
   - Advanced: Doped semiconductors, carbon-based photocatalysts (graphene-TiO_2 hybrids), and perovskite-based materials improve light absorption and charge transfer efficiency.

c. **Thermochemical Cycles:**
   - Traditional: Ferrite-based oxides and cerium oxides offer high-temperature stability but suffer from degradation over multiple cycles.
   - Advanced: Nanostructured metal oxides with doping strategies and improved oxygen transport capabilities enhance reaction kinetics and cycle stability.

Material science is a cornerstone of hydrogen production, impacting efficiency, stability, and cost-effectiveness across various production methods. While traditional materials have laid the foundation for hydrogen generation, ongoing research into advanced catalysts, nanomaterials, and cost-effective alternatives is driving significant improvements. By leveraging material innovations, the hydrogen economy can transition toward a more sustainable and economically viable future [8–10].

3.1.2 Key Challenges in Material Development

The advancement of hydrogen production technologies relies significantly on the development of high-performance materials that enhance efficiency, durability, and cost-effectiveness. However, materials used in electrolysis, photocatalysis, and thermochemical processes face critical challenges that hinder large-scale commercialization. These challenges primarily include the degradation of catalysts and electrodes, the high cost and scarcity of noble metals, and the difficulty in scaling up novel materials for industrial applications. Addressing these issues is essential for making hydrogen production a viable solution for sustainable energy [11].

One of the most pressing concerns in material development is durability and degradation under extreme operating conditions. Electrolyzers, for instance, require stable catalysts that can withstand prolonged exposure to water, high current densities, and fluctuating voltages [12]. In proton exchange membrane (PEM) electrolysis, Pt and IrO_2 are the most effective catalysts, but they suffer from dissolution and sintering, leading to a decline in performance over time [13]. Similarly, alkaline electrolyzers use nickel-based electrodes, which experience oxidation and mechanical wear, reducing their efficiency. The membranes used in electrolysis also degrade due to gas crossover and loss of ionic conductivity, further limiting system longevity [14]. In photocatalysis, semiconductors like TiO_2 and tungsten oxide (WO_3) degrade due to photochemical instability, causing efficiency losses over extended periods [15]. Thermochemical cycles, such as those based on CeO_2, face structural breakdown at high temperatures (800–1,500 °C), reducing their long-term viability [16]. To address these issues, researchers are developing protective coatings, doped materials, and nanostructured catalysts to improve thermal, chemical, and electrochemical stability.

Another major challenge is the cost and scarcity of noble metals like platinum, iridium, and ruthenium, which are widely used in hydrogen production technologies. Platinum is the preferred catalyst for hydrogen evolution and oxygen reduction reactions in electrolysis and fuel cells, but its high cost (approximately $30,000 per kg) makes large-scale adoption difficult. Iridium and ruthenium oxides are critical for oxygen evolution in PEM electrolyzers, but their global availability is limited, with iridium production estimated at only seven tons per year. This scarcity poses a serious obstacle to widespread deployment. To mitigate this issue, significant efforts have been made to develop earth-abundant alternatives such as nickel, cobalt, iron, molybdenum, and manganese-based catalysts. Transition metal carbides, nitrides, and phosphides have also been explored as potential substitutes due to their comparable catalytic activity. Additionally, single-atom catalysts and nanostructured materials are being designed to maximize catalyst efficiency while reducing material costs. Despite these advancements, matching the long-term stability and efficiency of noble metals remains a challenge, requiring further research and optimization [17].

The scalability of novel materials for large-scale hydrogen production is another critical hurdle. Many high-performance materials that show promising results in laboratory studies face difficulties in transitioning to industrial applications. Advanced catalysts, such as MOFs and perovskite oxides, require complex synthesis methods, making large-scale production costly and inefficient. In addition, scaling from milligram-scale synthesis in research labs to kilogram-scale industrial production often results in material inconsistencies, such as variations in particle size, morphology, and defect density, which negatively impact performance. Electrolysis systems, in particular, require new materials that can integrate seamlessly into existing commercial setups. While alternatives to platinum and iridium are being developed, they must function efficiently in acidic or alkaline conditions without compromising performance. In photocatalysis, large-scale deployment requires cost-effective deposition techniques for semiconductor coatings, while thermochemical materials need to withstand repeated heating and cooling cycles without losing activity. Addressing these scalability issues requires innovations in automated synthesis, artificial intelligence (AI)-driven material discovery, and advanced manufacturing techniques. Material challenges play a pivotal role in determining the success of hydrogen production technologies. Degradation of catalysts and electrodes limits long-term performance, while the high cost and scarcity of noble metals create economic barriers to widespread adoption. Additionally, the transition from lab-scale materials to industrial applications remains a major challenge due to difficulties in scaling up production. Overcoming these obstacles will require a multi-disciplinary approach, integrating advances in material engineering, nanotechnology, and computational modeling to develop cost-effective, durable, and scalable materials. Continued research, industrial collaboration, and government support will be crucial in making hydrogen a mainstream energy solution for a sustainable future [18].

3.2 High-performance Catalysts for Electrolysis

3.2.1 Role of Catalysts in HER and OER

Hydrogen production through electrolysis relies on two fundamental half-reactions: the HER at the cathode and the OER at the anode. These reactions are crucial for splitting water molecules into hydrogen and oxygen, but they require highly efficient catalysts to minimize energy loss and maximize reaction rates. The efficiency of electrolysis largely depends on the catalytic activity, stability, and cost-effectiveness of the materials used in HER and OER. Traditional catalysts, such as Pt for HER and IrO_2 and ruthenium oxide (RuO_2) for OER, have long been considered the gold standard due to their exceptional performance. However, their high cost and scarcity limit their widespread adoption, necessitating the development of alternative catalysts that are both cost-effective and durable [19].

3.2.1.1 Mechanisms of HER and OER in Electrolysis

The HER occurs at the cathode, where protons (H^+) or water molecules are reduced to form molecular hydrogen (H_2). The reaction mechanism varies depending on the electrolyte used:

In acidic media, HER follows two key steps:

- **Volmer step (Electrochemical adsorption):**

$$H^+ + e^- \rightarrow H_{ads} \tag{3.1}$$

- **Tafel step (Recombination) or Heyrovsky step (Electrochemical desorption):**

$$\text{Tafel: } 2H_{ads} \rightarrow H_2 \tag{3.2}$$

$$\text{Heyrovsky: } H_{ads} + H^+ + e^- \rightarrow H_2 \tag{3.3}$$

In alkaline media, HER is slightly more complex due to the involvement of hydroxyl ions (OH^-):

- **Volmer step:**

$$H_2O + e^- \rightarrow H_{ads} + OH^- \tag{3.4}$$

- **Tafel or Heyrovsky step:** (Similar to acidic media)

Since HER requires minimal overpotential on high-performance catalysts, the choice of materials significantly influences reaction efficiency. Figure 3.2(a) illustrates a comparison of catalyst efficiency in HER and OER, highlighting the overpotentials and kinetic barriers associated with different catalyst materials.

On the other hand, the OER occurs at the anode and involves the oxidation of water molecules to generate oxygen gas (O_2). This reaction is inherently more complex and sluggish compared to HER due to the involvement of a four-electron transfer process:

- **Electrochemical oxidation of water molecules:**

$$H_2O \rightarrow OH_{ads} + H^+ + e^- \tag{3.5}$$

- **Formation of superoxide species:**

$$OH_{ads} \rightarrow O_{ads} + H^+ + e^- \tag{3.6}$$

- **Oxygen-oxygen bond formation:**

$$O_{ads} + OH_{ads} \rightarrow OOH_{ads} \tag{3.7}$$

- **Release of molecular oxygen:**

$$OOH_{ads} \rightarrow O_2 + H^+ + e^- \tag{3.8}$$

OER is significantly more energy-intensive than HER, requiring higher overpotentials even with the best-known catalysts. This is one of the main reasons why electrolysis processes, particularly in PEM electrolyzers, require efficient and durable catalysts to facilitate OER [20].

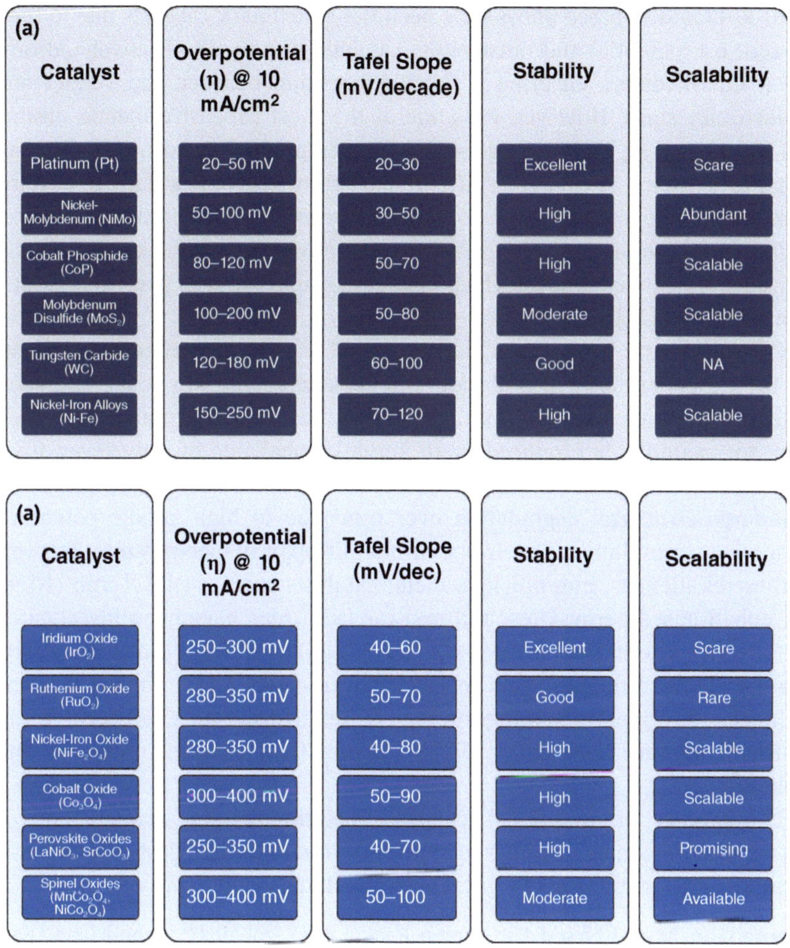

(a)

Catalyst	Overpotential (η) @ 10 mA/cm2	Tafel Slope (mV/decade)	Stability	Scalability
Platinum (Pt)	20–50 mV	20–30	Excellent	Scare
Nickel-Molybdenum (NiMo)	50–100 mV	30–50	High	Abundant
Cobalt Phosphide (CoP)	80–120 mV	50–70	High	Scalable
Molybdenum Disulfide (MoS_2)	100–200 mV	50–80	Moderate	Scalable
Tungsten Carbide (WC)	120–180 mV	60–100	Good	NA
Nickel-Iron Alloys (Ni-Fe)	150–250 mV	70–120	High	Scalable

(a)

Catalyst	Overpotential (η) @ 10 mA/cm2	Tafel Slope (mV/dec)	Stability	Scalability
Iridium Oxide (IrO_2)	250–300 mV	40–60	Excellent	Scare
Ruthenium Oxide (RuO_2)	280–350 mV	50–70	Good	Rare
Nickel-Iron Oxide ($NiFe_2O_4$)	280–350 mV	40–80	High	Scalable
Cobalt Oxide (Co_3O_4)	300–400 mV	50–90	High	Scalable
Perovskite Oxides ($LaNiO_3$, $SrCoO_3$)	250–350 mV	40–70	High	Promising
Spinel Oxides ($MnCo_2O_4$, $NiCo_2O_4$)	300–400 mV	50–100	Moderate	Available

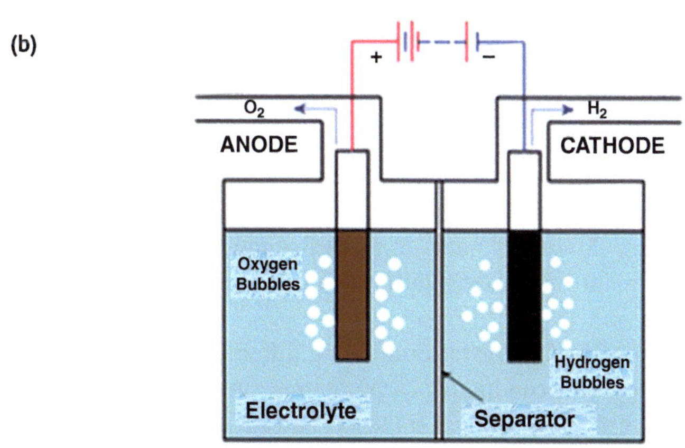

Figure 3.2 (a) Comparison of catalyst efficiency in HER and OER. (b) Schematic diagram of a high-performance catalyst in an electrolysis setup. Reproduced with permission of [29]/ The Royal Society of Chemistry.

For HER, Pt and Pt-based alloys have been the benchmark catalysts due to their nearly zero overpotential and outstanding catalytic activity. Pt effectively adsorbs hydrogen intermediates, ensuring a smooth transition between the Volmer and Tafel/Heyrovsky steps. However, Pt is one of the most expensive metals, costing around $30,000 per kg, making large-scale hydrogen production economically challenging. Furthermore, Pt catalysts suffer from stability issues, especially in alkaline environments where the metal tends to dissolve over prolonged operation. To address these limitations, researchers are exploring alternatives such as nickel (Ni), MoS_2, and tungsten-based compounds, which offer reasonable catalytic performance at a fraction of the cost [21].

For OER, IrO_2 and RuO_2 are widely regarded as the most effective catalysts due to their excellent oxygen evolution kinetics and stability in acidic and alkaline media. However, like platinum, these materials are scarce and expensive. Iridium, for example, is produced at an annual global rate of only seven tons, making it one of the least abundant elements on earth. Moreover, these catalysts undergo structural degradation over time due to high anodic potentials and oxidative stress, limiting their operational lifespan in electrolyzers. To overcome these challenges, non-precious metal catalysts such as nickel-iron (NiFe) oxides, cobalt-based perovskites, and manganese oxides have been investigated as cost-effective alternatives for OER. These materials exhibit promising activity in alkaline media, although their performance in acidic environments remains a challenge. Additionally, doping strategies, heterostructure engineering, and nanostructuring have been employed to enhance their catalytic efficiency and durability [22].

The limitations of traditional catalysts necessitate the development of next-generation materials that offer high activity, durability, and cost-effectiveness. Some promising directions in catalyst research include:

a. **Transition Metal-based Catalysts:** Nickel (Ni), cobalt (Co), and iron (Fe)-based catalysts, such as NiFe layered double hydroxides (LDHs), have shown excellent OER activity in alkaline electrolytes. Similarly, Mo-based materials, including MoS_2 and $MoSe_2$, are emerging as efficient HER catalysts.
b. **Single-atom Catalysts (SACs):** SACs, where isolated metal atoms are dispersed on a conductive support, provide high atomic efficiency and enhanced catalytic performance. These catalysts exhibit remarkable stability and activity due to their unique electronic structures.
c. **Perovskite Oxides:** Materials such as $LaNiO_3$ and $SrCoO_3$ are gaining attention for their ability to catalyze OER with improved stability and lower overpotentials.
d. **Hybrid and Composite Catalysts:** Combining different catalytic materials, such as metal oxides with conductive carbon-based supports (e.g., graphene, carbon nanotubes), enhances electron transport and stability.

Catalysts play a vital role in determining the efficiency and feasibility of hydrogen production via electrolysis. While platinum, iridium oxide, and ruthenium oxide remain the benchmark catalysts for HER and OER, their high cost and limited availability necessitate the search for alternatives. HER benefits from TMDs (e.g., MoS_2), while OER research focuses on NiFe-based oxides and perovskites. The development of cost-effective, durable, and scalable catalysts remains a crucial challenge in making green hydrogen a commercially viable energy source. Future advancements in catalyst design, leveraging nanotechnology, computational modeling, and AI-driven material discovery, will be key to optimizing hydrogen production technologies and ensuring their large-scale adoption.

3.2.2 Emerging Catalysts for Electrolysis

The development of advanced catalysts is crucial for improving the efficiency, cost-effectiveness, and long-term stability of electrolysis-based hydrogen production. Traditional catalysts, such as Pt for the HER and IrO_2 for the OER, offer exceptional catalytic performance but suffer from high costs and scarcity. To overcome these challenges, researchers are exploring non-precious metal catalysts, TMDs, and hybrid catalyst systems that combine multiple materials to enhance efficiency and durability. These emerging catalysts have the potential to significantly reduce costs while maintaining or even surpassing the performance of conventional noble-metal-based materials.

Non-precious metal catalysts, including Ni, Fe, and Co-based materials, have emerged as promising alternatives for HER and OER due to their abundance, lower cost, and reasonable catalytic activity [23].

- **Nickel-based catalysts** (e.g., NiFe hydroxides and NiMo alloys) exhibit excellent OER activity in alkaline electrolytes, making them ideal for use in alkaline water electrolysis. NiFe LDHs are particularly effective due to their synergistic interactions between Ni and Fe, which enhance charge transfer and active site exposure.
- **Cobalt-based catalysts**, such as cobalt phosphides (CoP) and cobalt oxides, have shown promising results in both HER and OER, with CoP demonstrating efficient hydrogen adsorption-desorption kinetics comparable to Pt in alkaline media.
- **Iron-based catalysts**, including Fe-based perovskites and Fe-nitride composites, have also been explored for OER, benefiting from their strong electronic interactions with oxygen species and high intrinsic activity.

Despite their advantages, non-precious metal catalysts often suffer from stability issues, particularly under acidic conditions. Strategies such as doping, alloying, and nanostructuring are being employed to improve their durability and enhance their catalytic efficiency [24]. Figure 3.2(b) illustrates a schematic diagram of a high-performance non-precious metal catalyst integrated into an electrolysis setup, highlighting its active sites and reaction pathways.

TMDs, such as MoS_2, tungsten disulfide (WS_2), and selenium-based compounds, have gained significant attention for HER applications due to their unique layered structures and tunable electronic properties [25].

- **MoS_2**, one of the most extensively studied TMDs, possesses active edge sites that promote hydrogen adsorption and desorption. Engineering MoS_2 with

defects, heteroatom doping, and conductive supports like graphene can further improve its catalytic performance.
- **WS_2** and **$MoSe_2$** also demonstrate promising HER activity, with researchers focusing on optimizing their basal plane activation to enhance reaction kinetics.

For OER, perovskite oxides (ABO_3-type structures) such as $LaNiO_3$, $SrCoO_3$, and $BaFeO_3$ have shown high oxygen evolution activity, particularly in alkaline environments. These materials offer strong metal-oxygen interactions, enabling efficient electron transfer and oxygen formation. Additionally, perovskites are highly tunable, allowing for modifications that enhance their conductivity, stability, and catalytic performance [26].

Hybrid catalysts, which combine multiple materials to optimize performance, represent a cutting-edge approach in electrolysis research. These catalysts integrate non-precious metals, TMDs, perovskites, and carbon-based materials to achieve superior activity and stability.

- **MOFs** serve as an effective platform for hybrid catalyst design, allowing for precise structural control and high surface area exposure. MOF-derived materials, such as Co/Ni-MOFs, have shown exceptional performance in both HER and OER.
- **Core-shell structures**, where a catalytically active core (e.g., Ni, Co, or Fe) is coated with a protective or conductive shell (e.g., carbon, MoS_2, or phosphides), enhance stability while maintaining high catalytic efficiency.
- **Heterostructured catalysts**, such as NiFe-LDH coupled with MoS_2 or graphene-supported CoP, leverage synergistic effects between different materials to enhance charge transfer and reaction kinetics.

Hybrid catalysts not only improve efficiency but also provide enhanced durability under harsh electrolysis conditions. Their potential for large-scale implementation makes them a strong candidate for future hydrogen production technologies [27]. Table 3.1 presents a comparative analysis of traditional noble metal catalysts and emerging catalyst materials in terms of efficiency, cost, and stability, highlighting the

Table 3.1 Comparison of traditional and emerging catalysts for electrolysis.

Catalyst type	Examples	Reaction (HER/OER)	Efficiency	Cost	Stability
Traditional noble metal catalysts	Pt, IrO_2, RuO_2	HER (Pt), OER (IrO_2, RuO_2)	High	Very high	Excellent
Non-precious metal catalysts	Ni, Fe, Co, NiFe-LDH	HER/OER	Moderate to high	Low	Moderate to high
TMDs	MoS_2, WS_2, $MoSe_2$	HER	Moderate to high	Low to moderate	Moderate
Perovskite catalysts	$LaNiO_3$, $SrCoO_3$, $BaFeO_3$	OER	High	Moderate	High

Catalyst type	Examples	Reaction (HER/OER)	Efficiency	Cost	Stability
Hybrid catalysts	NiFe-LDH/ MoS_2, CoP/ Graphene, MOF-derived catalysts	HER/OER	High	Moderate	High
Core-shell catalysts	Ni@MoS_2, Co@C, Fe@P	HER/OER	High	Moderate	High
MOFs	Ni-MOF, Co-MOF, Fe-MOF	HER/OER	High	Moderate	High

advantages of newer approaches. Emerging catalysts for electrolysis focus on cost reduction, enhanced performance, and improved long-term stability. Non-precious metal catalysts (e.g., Ni, Fe, Co) offer viable alternatives to noble metals, while TMDs and perovskites provide promising pathways for high-efficiency hydrogen production. Hybrid catalysts, combining different materials, are revolutionizing catalyst design by optimizing both HER and OER processes. The ongoing advancements in catalyst development are paving the way for large-scale, economically viable, and sustainable hydrogen production, accelerating the transition toward a clean energy future [28].

3.3 Nanomaterials and Composites for Enhanced Hydrogen Production

3.3.1 Role of Nanomaterials in Hydrogen Evolution

Nanomaterials have revolutionized the field of hydrogen evolution by significantly enhancing the efficiency and stability of catalytic processes [30]. Due to their high surface area, unique electronic properties, and tunable reactivity, nanomaterials serve as excellent candidates for improving HER kinetics. Traditional catalysts like Pt exhibit high catalytic activity but suffer from high costs and scarcity. In contrast, nanostructured materials offer cost-effective and scalable alternatives while maintaining excellent catalytic performance. By engineering nanomaterials at the atomic and molecular levels, researchers have been able to optimize the efficiency of HER, making hydrogen production via electrolysis more feasible for large-scale applications. One of the key advantages of nanomaterials is their dramatically increased surface area, which provides more active sites for hydrogen evolution. Unlike bulk materials, nanostructures expose a higher fraction of catalytic sites, leading to improved interaction between the reactant species and the catalyst. Additionally, the unique quantum effects observed in nanomaterials enable better charge transfer, reducing the energy barrier for hydrogen formation. Moreover, nanostructured catalysts exhibit superior mechanical and chemical stability, making them more durable under harsh electrolysis conditions.

Several classes of nanomaterials have been widely explored for HER applications. Among them, nanostructured metal oxides, graphene-based materials, and carbon nanotubes (CNTs) have demonstrated exceptional catalytic properties. Nanostructured metal oxides such as WO_3, TiO_2, and molybdenum trioxide (MoO_3) have been investigated extensively for their ability to act as efficient electrocatalysts. These materials exhibit excellent conductivity, stability, and synergistic interactions when combined with other transition metals. For instance, doping metal oxides with non-precious metals such as Ni, Fe, or Co can further enhance their electrocatalytic activity by modifying the electronic structure and improving charge transfer kinetics [31].

Graphene-based materials have also gained significant attention in the field of HER due to their excellent electrical conductivity, high surface area, and tunable chemical properties. Graphene, a two-dimensional (2D) material composed of a single layer of carbon atoms, can act as a support structure to stabilize active catalytic sites. When combined with transition metals, graphene derivatives such as reduced graphene oxide (rGO) and nitrogen-doped graphene exhibit remarkable hydrogen evolution performance. The unique electronic interactions between the catalyst and the graphene substrate help in lowering the overpotential and improving hydrogen production efficiency. Another promising class of nanomaterials for HER is CNTs. These cylindrical structures, composed of rolled-up graphene sheets, possess extraordinary mechanical strength, high electrical conductivity, and excellent chemical stability. CNTs provide an ideal platform for dispersing metal nanoparticles, ensuring uniform distribution of active sites and minimizing catalyst aggregation. Functionalized CNTs, modified with metal oxides or heteroatoms (e.g., nitrogen, sulfur), show enhanced electrocatalytic activity by altering the electronic structure and improving proton adsorption. Furthermore, CNTs enable efficient electron transport, reducing charge recombination and increasing overall HER kinetics [32].

Recent advancements in nanotechnology have led to the development of hybrid nanomaterials, which integrate multiple components to achieve synergistic effects in HER catalysis. For example, hybrid structures combining metal oxides with graphene or CNTs offer a unique combination of conductivity, stability, and catalytic efficiency. These hybrid materials exhibit improved charge transfer dynamics and enhanced electrochemical stability under prolonged operational conditions. Additionally, alloyed nanostructures, such as Ni-Mo or Fe-Co nanoparticles, provide cost-effective alternatives to precious metal catalysts while maintaining comparable performance. Despite their numerous advantages, nanomaterials face several challenges that need to be addressed for practical implementation in large-scale hydrogen production. One of the main concerns is the scalability of nanomaterial synthesis. Many nanomaterials require complex and expensive fabrication techniques, making large-scale production economically challenging. Developing scalable and cost-effective synthesis methods, such as green chemistry approaches or scalable chemical vapor deposition techniques, will be crucial for the widespread adoption of nanomaterials in HER applications. Another critical issue is the stability and degradation of nanostructured catalysts. While nanomaterials exhibit high initial catalytic performance, prolonged exposure to harsh electrolysis conditions can lead to degradation, agglomeration, or leaching of active sites. To mitigate these effects, researchers are exploring strategies such as surface passivation, core-shell nanostructures, and protective coatings. Additionally, the incorporation of self-healing properties in nanocatalysts could further enhance their long-term stability and durability [33].

The future of nanomaterials in HER research looks promising, with ongoing efforts focused on designing next-generation catalysts with ultrahigh efficiency and long-term durability. Emerging techniques such as AI-assisted material discovery and computational modeling are being employed to accelerate the identification of optimal nanomaterial compositions. AI-driven approaches allow researchers to predict the catalytic performance of new materials before experimental validation, significantly reducing the time and cost associated with traditional trial-and-error methods [34]. Nanomaterials have transformed the landscape of hydrogen evolution by offering enhanced catalytic activity, improved charge transfer kinetics, and greater stability. The integration of nanostructured metal oxides, graphene-based materials, and CNTs has opened new avenues for developing high-performance electrocatalysts. However, challenges related to scalability, stability, and cost must be addressed for the successful commercialization of nanomaterial-based HER technologies [35]. Future research should focus on optimizing material properties, developing scalable synthesis techniques, and exploring hybrid nanostructures to achieve sustainable and cost-effective hydrogen production. Figure 3.3(a) and (b) depicts the FESEM and TEM image of CNTs.

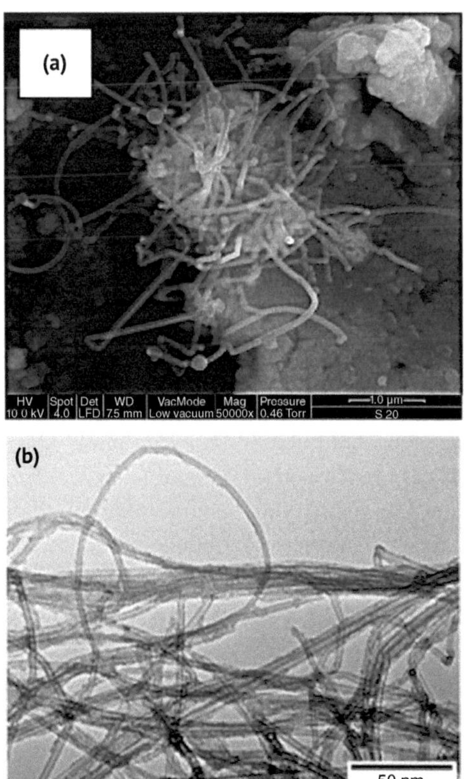

Figure 3.3 (a) FESEM image of CNT. [36]/IOP Publishing/CC BY 3.0. (b) TEM image of CNT. Reproduced with permission from [37]/Elsevier.

3.3.2 Advanced Composites for Electrolysis and Photocatalysis

The development of advanced composite materials has emerged as a promising strategy to enhance the efficiency, durability, and scalability of hydrogen production technologies such as electrolysis and photocatalysis. By combining different material classes, researchers have been able to design composites with superior catalytic properties, improved charge transfer capabilities, and enhanced stability. Among these, MOFs and composites integrating metal oxides with carbon-based supports have garnered significant attention due to their tunable structures, high surface areas, and synergistic effects [38].

MOFs are a class of hybrid materials composed of metal ions or clusters coordinated with organic ligands, forming porous, crystalline structures. Their high surface area, tunable porosity, and diverse chemical functionalities make them ideal candidates for catalytic applications in hydrogen production. Unlike conventional catalysts, MOFs offer precise control over active sites, enabling efficient charge transport and enhanced catalytic performance. In electrolysis, MOFs serve as promising electrocatalysts for the HER and OER. Transition metal-based MOFs, such as Co-, Ni-, and Fe-based MOFs, have demonstrated remarkable electrocatalytic activity due to their intrinsic electronic properties. These MOFs can facilitate proton adsorption and charge transfer, reducing the overpotential required for water splitting. Additionally, MOFs with bimetallic or multimetallic centers exhibit enhanced stability and reactivity by leveraging the synergistic interactions between different metal species. For photocatalysis, MOFs have shown great potential in improving light absorption and charge separation efficiency. By incorporating photoactive metal centers such as Ti or Mo, MOFs can effectively utilize solar energy to drive hydrogen production through photocatalytic water splitting. The ordered porous structures of MOFs also act as charge transport channels, minimizing charge recombination and improving overall photocatalytic efficiency. One of the main challenges associated with MOFs is their limited conductivity, which can hinder electron transfer in electrochemical applications. To address this, researchers have explored the integration of conductive nanomaterials such as graphene or CNTs with MOFs to enhance charge transport properties [39].

Another significant advancement in composite materials for electrolysis and photocatalysis is the combination of metal oxides with carbon-based supports. Metal oxides such as TiO_2, MoO_3, Fe_2O_3, and Co_3O_4 are known for their excellent catalytic properties, but they often suffer from low conductivity and charge recombination issues. By integrating these metal oxides with conductive carbon-based materials, researchers have successfully enhanced their performance [40]. Graphene-based composites have been widely explored due to the exceptional electrical conductivity, high surface area, and flexibility of graphene. When combined with metal oxides, graphene serves as an electron-conducting bridge, facilitating charge transport and improving catalytic efficiency. For example, TiO_2-graphene composites exhibit superior photocatalytic activity due to graphene's ability to enhance light absorption and charge separation [41]. Similarly, CNT-based composites provide a robust platform for high-performance electrocatalysts. CNTs, owing to their unique tubular nanostructure and excellent conductivity, enhance the dispersion of metal oxide nanoparticles and improve electron mobility. Studies have shown that NiO-CNT and Co_3O_4-CNT composites demonstrate high HER and OER efficiency with enhanced stability in alkaline and acidic electrolytes [42]. Another promising approach is the development of porous carbon-metal oxide composites, where porous carbon structures (e.g., activated carbon, carbon aerogels) are combined with metal oxides to create materials with high surface areas and enhanced catalytic activity. These composites offer significant advantages in terms of stability, conductivity, and adsorption capabilities, making them suitable for large-scale hydrogen production. Figure 3.4 illustrates the FESEM and TEM images of these advanced composites, highlighting their structural morphology and catalytic performance. These images provide insight into the nanostructure of composite materials, showcasing how the integration of metal oxides with carbon-based supports can enhance hydrogen production.

While advanced composite materials have shown significant promise in enhancing electrolysis and photocatalysis, several challenges remain:

a. **Synthesis Complexity:** The fabrication of MOFs and metal oxide-carbon composites often involves multistep synthesis processes, which can be time-consuming and expensive.
b. **Long-term Stability:** Many MOF-based catalysts suffer from structural degradation in harsh electrochemical environments, requiring further research on stabilization strategies.
c. **Scalability:** The integration of advanced composites into large-scale water-splitting systems remains a challenge due to limitations in production techniques and cost.

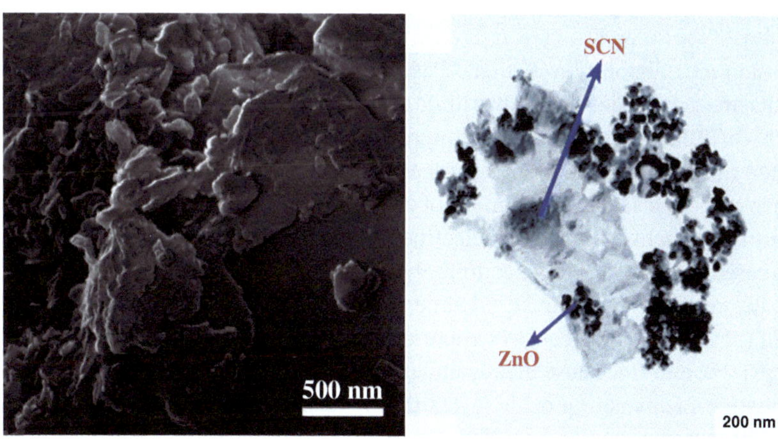

Figure 3.4 FESEM and TEM images of advanced composites for electrolysis and photocatalysis, highlighting their structural morphology and catalytic performance. [44]/ The Royal Society of Chemistry/CC BY 3.0.

Overall, the integration of MOFs and metal oxide-carbon composites represents a promising direction for advancing hydrogen production technologies, offering a pathway toward more efficient and sustainable energy solutions [43].

3.4 Durability and Degradation Challenges in Catalyst Materials

3.4.1 Factors Affecting Catalyst Durability

Catalyst durability is a critical concern in hydrogen production, as catalysts are exposed to extreme electrochemical, thermal, and mechanical stresses during long-term operation. These stresses lead to gradual degradation, reducing the efficiency and lifespan of catalysts used in electrolysis, photocatalysis, and thermochemical processes. Several factors influence catalyst stability, including electrochemical cycling, corrosion, and structural changes that occur over time. Understanding these degradation mechanisms is crucial for improving catalyst materials and developing more sustainable hydrogen production technologies [45].

One of the primary factors affecting catalyst durability is electrochemical cycling and prolonged operation in HER and OER systems. In processes such as PEM electrolysis and alkaline electrolysis, catalysts undergo continuous oxidation-reduction cycles, which place significant stress on the active sites. This repeated cycling leads to structural fatigue and loss of catalytic activity, primarily due to metal dissolution, nanoparticle detachment, and catalyst sintering [46]. In PEM electrolysis, for instance, Pt and IrO_2 catalysts degrade over time due to electrochemical stress, leading to a reduction in active surface area and increased overpotential. This results in higher energy consumption and reduced hydrogen production efficiency. Similarly, in alkaline electrolysis, nickel-based catalysts suffer from hydroxide formation, which limits their ability to sustain long-term operation without efficiency losses [47].

Corrosion and dissolution of metal-based catalysts in acidic and alkaline environments are major challenges that limit the durability of electrodes and catalytic surfaces. In PEM electrolysis, the strongly acidic operating conditions (pH ~0–1) accelerate catalyst dissolution and ion leaching, particularly for noble metal catalysts such as Pt and Ir. The dissolution of these metals results in loss of active sites and eventual detachment from the electrode surface. Additionally, the corrosion of carbon-based catalyst supports, commonly used in PEM electrolysis, further exacerbates degradation. In contrast, alkaline electrolysis operates in a high-pH environment (pH >10), which causes structural instability in transition metal catalysts such as Ni, Fe, and Co. These metals undergo passivation and oxidation, leading to the formation of insulating oxide layers that reduce charge transfer efficiency. To address these issues, researchers are developing corrosion-resistant catalyst coatings and stable metal alloys to extend catalyst lifetimes in both acidic and alkaline conditions [48].

Structural and morphological changes in catalysts also play a significant role in degradation and performance loss over time. During long-term operation, catalyst nanoparticles tend to agglomerate and sinter, leading to a reduction in active surface area. In high-temperature applications such as solid oxide electrolysis (SOECs), metal catalysts experience particle coalescence, which reduces their catalytic activity and efficiency [49]. Similarly, in low-temperature electrolysis, catalyst restructuring due to Ostwald ripening causes smaller particles to dissolve and redeposit onto larger ones, further decreasing overall reactivity [50]. Another common issue is supporting material degradation, where materials such as carbon black used in PEM electrodes experience oxidation and collapse, reducing mechanical stability and leading to catalyst detachment. Apart from thermal and electrochemical stresses, mechanical stress also contributes to catalyst degradation. Frequent start-stop cycles and fluctuating current densities in hydrogen production systems induce expansion and contraction in catalyst layers, leading to the formation of cracks, delamination, and eventual failure. This is especially problematic in high-pressure and high-temperature environments, where mechanical fatigue accelerates catalyst degradation [51]. Moreover, in photocatalytic hydrogen production, photocorrosion of semiconductor materials such as TiO_2 and CdS limits their long-term effectiveness. Exposure to UV light and oxidizing conditions degrades the surface structure of photocatalysts, reducing their efficiency in water-splitting applications. The combined effects of electrochemical cycling, corrosion, structural degradation, and mechanical fatigue ultimately determine the lifespan and reliability of catalysts in hydrogen production. Addressing these challenges requires the development of more robust catalyst materials, such as high-entropy alloys, core-shell nanostructures, and protective coatings that can resist degradation while maintaining high activity. Future research must focus on enhancing catalyst durability through nanostructuring, surface modification, and self-repairing materials, ensuring long-term efficiency in scalable hydrogen energy applications [52].

3.4.2 Strategies to Improve Catalyst Stability

The long-term efficiency and durability of catalysts used in hydrogen production depend on their ability to withstand electrochemical cycling, corrosion, and structural degradation. Several strategies have been developed to enhance catalyst stability, ensuring longer operational lifetimes while maintaining high efficiency. These strategies include protective coatings, support materials, alloying strategies, nanostructuring, and self-healing catalysts, all of which help mitigate degradation mechanisms such as metal dissolution, particle sintering, and surface restructuring. Implementing these approaches is crucial for the economic viability of large-scale hydrogen production technologies [53].

3.4.2.1 Protective Coatings and Support Materials to Prevent Degradation

Protective coatings and stable support materials play a vital role in enhancing catalyst stability by preventing corrosion, nanoparticle detachment, and material leaching. One of the most effective solutions is the use of carbon-based materials, such as graphene, CNTs, and mesoporous carbon supports, which provide high conductivity while reducing metal catalyst oxidation. As shown in Figure 3.5, catalysts coated

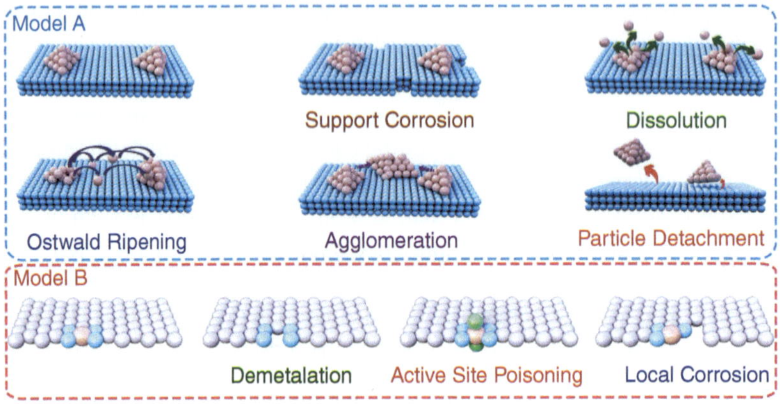

Figure 3.5 Schematic representation of degradation mechanisms in hydrogen evolution catalysts [46]/John Wiley & Sons/CC BY 4.0.

with carbon layers exhibit lower degradation rates due to the hydrophobic nature of carbon, which reduces exposure to corrosive electrolytes [54]. In addition to carbon supports, metal oxide coatings such as TiO_2, Al_2O_3, and ZrO_2 serve as protective barriers against acidic and alkaline degradation. These oxides create a thin, stable layer that prevents metal dissolution and catalyst leaching in PEM and alkaline electrolysis environments. Another promising approach is the use of perovskite oxide supports, which exhibit self-repairing properties and can withstand high-temperature electrolysis conditions in SOECs. The selection of appropriate support materials is crucial in minimizing performance loss and ensuring long-term stability in hydrogen production systems [55].

3.4.2.2 Alloying Strategies and Nanostructuring to Enhance Durability

Alloying catalysts with transition metals has emerged as an effective method to enhance catalyst durability while reducing noble metal content. By incorporating elements such as Ni, Co, Fe, and Mo, alloyed catalysts exhibit greater resistance to dissolution and oxidation, making them more suitable for long term operation. For instance, Pt-Ni and Ir-Ru alloys have shown improved electrochemical stability in HER and OER applications due to synergistic effects between metal atoms, which enhance catalytic performance while preventing excessive leaching [56]. Nanostructuring is another widely explored strategy that helps increase surface area, optimize charge transfer, and prevent catalyst aggregation. Core-shell structures, where a noble metal shell (e.g., Pt, Ir) is deposited over a non-precious metal core (e.g., Ni, Fe, Co), significantly improve stability and catalytic efficiency while reducing material costs. As indicated in Table 3.2, core-shell catalysts exhibit lower degradation rates compared to conventional metal nanoparticles, making them an economically viable alternative for scalable hydrogen production. Additionally, high-entropy alloys (HEAs), which consist of multiple metallic elements in a single-phase structure, demonstrate exceptional durability due to their ability to resist corrosion and maintain catalytic activity under extreme conditions [57].

Table 3.2 Comparison of catalyst degradation rates and stability enhancement strategies.

Degradation mechanism	Causes	Effects on catalyst performance	Stability enhancement strategies
Electrochemical cycling degradation	Repeated oxidation-reduction reactions in HER and OER	Loss of active sites, reduced catalytic efficiency, and increased overpotential	Use of HEAs, core-shell nanostructures, and doped metal catalysts to minimize oxidation stress [58]
Corrosion in acidic environments (PEM electrolysis)	Leaching of metal ions (Pt, Ir) under low-pH conditions	Catalyst dissolution, structural degradation, and efficiency loss	Protective metal oxide coatings (TiO_2, ZrO_2) and carbon-based encapsulation for corrosion resistance [59]
Corrosion in alkaline environments (alkaline electrolysis)	Formation of hydroxide layers on Ni, Fe, and Co catalysts	Passivation of active sites, lower charge transfer efficiency	Alloying with Co, Mo, and Ru, using perovskite oxides for higher stability [60]
Metal dissolution and leaching	Loss of metal atoms due to ion migration in electrolyte	Catalyst thinning, performance decline, and shorter lifespan	Development of self-healing catalysts with in situ metal redeposition [61]
Particle agglomeration and sintering	Nanoparticle coalescence due to high temperatures	Reduction in surface area, loss of active catalytic sites	Using mesoporous catalyst supports, plasma-treated nanomaterials, and graphene-based reinforcements [62]
Support material degradation	Oxidation of carbon-based supports (e.g., carbon black) in PEM fuel cells	Weakening of electrode structure, catalyst detachment	Using metal oxide supports (Nb_2O_5, TiO_2) or graphene-modified conductive supports [63]
Photocorrosion in photocatalytic hydrogen production	UV-induced degradation of semiconductor catalysts (TiO_2, CdS)	Surface restructuring, loss of charge carrier mobility	Doping with noble metals (Au, Pt), passivation layers, and core-shell structures [64]
Mechanical stress and cracking	Expansion/contraction cycles in variable operating conditions	Crack formation, delamination, loss of conductivity	Flexible binder-free electrodes, nanostructured catalyst layers, and self-repairing coatings [65]

3.4.2.3 Self-healing and Regenerable Catalysts for Prolonged Efficiency

Recent advancements in self-healing catalysts have opened new possibilities for prolonging catalyst lifespan and maintaining efficiency over extended periods. These catalysts possess the ability to restructure themselves or regenerate lost active sites during operation, ensuring consistent performance. One approach involves the use of redox-active metal oxides, such as MnO_2 and CeO_2, which can capture and redeposit catalyst nanoparticles to prevent active site depletion. Another promising avenue is the development of plasma-treated catalysts, which undergo in situ surface modification to repair structural damage and enhance electrocatalytic activity. These plasma-assisted techniques introduce defects and vacancies, which self-adjust during reaction cycles, thereby improving mechanical stability and electronic conductivity. Additionally, hydrogel-based catalyst matrices have been explored for continuous reactivation of immobilized nanoparticles, further improving the longevity of catalyst systems in hydrogen production [66]. As shown in Figure 3.5, these self-healing mechanisms contribute to significantly slower degradation rates, extending the operational lifetime of catalysts in hydrogen evolution and oxygen evolution reactions. As summarized in Table 3.2, self-healing catalysts exhibit up to 50% longer durability compared to conventional materials, making them an attractive option for next-generation hydrogen technologies.

The durability of catalysts in hydrogen production remains a key challenge for achieving long-term stability and cost-effectiveness. Strategies such as protective coatings, alloying, nanostructuring, and self-healing mechanisms have shown significant improvements in catalyst longevity while maintaining high catalytic activity. The integration of novel support materials and regenerable catalysts will be instrumental in reducing degradation rates, ultimately leading to more sustainable and commercially viable hydrogen production systems. Future research should focus on scalable fabrication techniques for these advanced catalysts to ensure their real-world application in electrolysis and photocatalytic hydrogen generation.

3.5 Cost-effective and Non-precious Metal Alternatives

3.5.1 Challenges with Precious Metal Catalysts

Precious metal catalysts, such as Pt and Ir, have long been considered the gold standard for hydrogen production, particularly for the HER and OER in electrolysis. These metals are highly efficient, providing fast kinetics and excellent stability under reaction conditions, making them ideal for use in alkaline electrolysis, PEM electrolysis, and photocatalysis. However, despite their outstanding performance, there are several critical challenges that hinder their widespread adoption for large-scale hydrogen production [67].

One of the primary issues is the high cost and scarcity of platinum and iridium. Both metals are rare and expensive, limiting the economic viability of hydrogen production technologies that rely on them. For example, Pt is not only rare but also faces supply chain challenges due to geopolitical factors, further driving up costs. As hydrogen production moves toward commercialization, the high demand for these precious metals could lead to price volatility and further strain economic feasibility. Furthermore, stability and performance limitations pose additional hurdles in the use of Pt and Ir-based catalysts under real-world conditions. While these metals perform well under ideal laboratory conditions, electrolyzers operating at high currents, especially during intermittent renewable energy integration, can cause rapid degradation of catalyst materials due to surface oxidation, corrosion, and loss of active sites. In PEM electrolysis systems, for example, platinum catalysts can suffer from particle sintering and leaching, both of which diminish their catalytic activity and overall lifespan. In addition, high-temperature and high-pressure conditions that are common in industrial-scale hydrogen production can exacerbate these issues, leading to reduced performance and higher maintenance costs over time [68].

While platinum and iridium are irreplaceable in some high-performance applications, these limitations have prompted researchers to explore alternative catalytic materials that offer similar efficiency at a fraction of the cost. Table 3.3 provides a performance comparison of precious and non-precious metal catalysts, highlighting the efficiency and durability gaps between the two categories.

3.5.2 Development of Non-precious Metal Catalysts

Given the challenges associated with precious metal catalysts, there has been increasing interest in developing non-precious metal catalysts for alkaline and PEM electrolysis. These catalysts are not only more abundant but also offer significant cost advantages, making them a more sustainable option for large-scale hydrogen production. Ni, Co, and Fe-based catalysts have gained attention as promising alternatives to platinum and iridium. These metals are widely available and less expensive, yet they can still exhibit decent catalytic activity, particularly in alkaline electrolysis. Ni, for instance, is a well-established catalyst for the HER in alkaline environments due to its high electrical conductivity and good stability. Co and Fe also demonstrate solid catalytic properties, though they typically require support materials to enhance their efficiency and durability. By using oxide-based catalysts or bimetallic systems, researchers have been able to tune the electronic structure of these metals to optimize their performance for water splitting [69].

In addition to these transition metals, there is growing interest in sulfides, phosphides, and nitrides as cost-effective alternatives to precious metals. These materials exhibit promising electrocatalytic activity for both HER and OER while being far less expensive than platinum or iridium. For example, nickel sulfide (NiS) and cobalt phosphide (CoP) have shown high efficiency and good stability in alkaline electrolyzers, making them attractive candidates for large-scale applications. Additionally, transition metal nitrides, such as Mo_2N and Fe-N, have demonstrated good performance in PEM electrolyzers, offering higher efficiency and longer durability compared to traditional noble metal catalysts. One of the main advantages of non-precious metal catalysts is their low cost and earth abundance, making them suitable for scalable hydrogen production. These materials can be incorporated

Table 3.3 Performance comparison of precious and non-precious metal catalysts.

Catalyst type	Material	Catalytic activity	Efficiency (HER/OER)	Stability	Cost	Limitations
Precious metal catalysts	**Pt**	Excellent for both HER and OER due to its high conductivity and low overpotentials	High	High (under controlled conditions)	Very high	Expensive, limited availability, suffers from corrosion and sintering under extreme conditions [71]
	IrO_2	Excellent for OER, particularly in acidic environments	High	Moderate to high (but prone to oxidation in acidic media)	High	Expensive, scarce, limited stability in high-temperature conditions [72]
	RuO_2	Efficient for OER in acidic solutions, especially in PEM electrolysis	Moderate to high	Moderate (leaching and corrosion under harsh conditions)	High	Expensive, poor stability in alkaline conditions [73]
Non-precious metal catalysts	**Ni**	Good activity for HER, particularly in alkaline electrolysis	Moderate to high	Moderate (long-term stability improves with alloys or composites)	Low to moderate	Relatively low activity for OER, corrosion at high currents [74]
	Co	Effective for both HER and OER, particularly in alkaline electrolyzers	Moderate to high	Moderate (susceptible to oxidation)	Low to moderate	Lower efficiency for OER compared to precious metals, needs improvements for stability [75]
	Fe	Good HER activity in alkaline electrolyzers, useful in combination with carbon supports	Moderate	Moderate (better stability with composites)	Low	Limited efficiency for OER, stability concerns under harsh conditions [76]

	NiS	Good electrocatalytic activity, particularly for HER in alkaline media	High	High (improved stability in alkaline solutions)	Low	Lower performance for OER compared to precious metals [77]
	CoP	High efficiency for both HER and OER in alkaline electrolysis	High	High (excellent stability under alkaline conditions)	Low	Requires improved efficiency for commercialization [78]
	Mo_2N	High activity for HER and OER, particularly in PEM electrolysis	High	High (good stability in both acidic and alkaline environments)	Low	Needs further optimization for industrial applications
	Fe-N	Effective for HER in acidic and alkaline environments	Moderate to high	High (stable in alkaline solutions)	Low	Needs improved performance for OER, scalability issues
Hybrid catalysts	**Nickel/Graphene composites**	Synergistic effect, enhanced charge transfer and catalytic activity	High	High (improved mechanical stability)	Low	Needs further optimization for cost-effectiveness and scalability [79]
	Cobalt/Carbon nanotubes (CNTs)	Enhanced electrocatalytic activity and stability	High	High (good for both HER and OER)	Moderate	High-performance under specific conditions, scalability challenges [80]
	NiO/Fe_2O_3 composites	Enhanced electrocatalytic properties for both HER and OER	High	High (excellent stability in alkaline electrolyzers)	Low	Requires optimization for industrial-scale applications [81]

into advanced composite catalysts or used in combination with carbon-based supports (e.g., graphene or carbon nanotubes) to enhance their electrocatalytic activity and mechanical stability. Hybrid catalysts, which combine metal oxides with graphene or carbon-based materials, have been shown to exhibit synergistic effects, improving charge transfer and catalytic efficiency. Despite these advantages, non-precious metal catalysts still face challenges in terms of stability and efficiency under real-world conditions. To address these issues, ongoing research is focused on optimizing material synthesis to improve the catalytic properties and long-term durability of these materials. Moreover, much of the current research is aimed at developing bifunctional catalysts capable of efficiently catalyzing both HER and OER, which is essential for overall system efficiency in electrolysis applications. Table 3.3 compares the performance of precious and non-precious metal catalysts, illustrating how non-precious metals, while still in the early stages of development, offer promising potential for efficient and cost-effective hydrogen production [70].

3.6 Role of AI and Machine Learning in Material Discovery

3.6.1 How AI Accelerates Material Discovery

AI is revolutionizing the field of material discovery, particularly in the development of catalysts for hydrogen production. Traditional methods of catalyst development often involve time-consuming trial-and-error experiments that can be resource-intensive. AI accelerates this process by enabling high-throughput screening of new materials, which dramatically reduces both time and cost in the discovery phase. One of the key ways AI facilitates material discovery is through high-throughput screening. This technique uses machine learning algorithms to analyze vast databases of materials and predict their properties. By leveraging AI, researchers can quickly identify promising materials that may exhibit the desired catalytic activity for hydrogen production. This approach is especially valuable for screening thousands of potential catalysts at once, filtering through materials based on various performance indicators such as catalytic efficiency, cost, stability, and environmental impact. AI-powered screening tools can also suggest modifications to existing materials to enhance their properties, leading to novel catalysts that may not have been discovered through conventional methods [82].

In addition to high-throughput screening, AI also plays a crucial role in predictive modeling. Predictive modeling uses machine learning to simulate how a material will behave under different conditions, such as during hydrogen evolution or oxygen evolution reactions. This enables researchers to predict the stability and performance of materials without needing to conduct extensive physical experiments. For example, AI can simulate the degradation of a catalyst over time or model its performance under varying temperature and pressure conditions, which is vital for assessing its viability in real-world applications. By using predictive modeling, researchers can select materials that are likely to maintain their catalytic properties over prolonged periods, reducing the need for trial-and-error testing [83].

Furthermore, AI can integrate data from multiple sources, such as experimental results, literature, and computational simulations, to provide a more comprehensive understanding of material properties. Machine learning algorithms can analyze these large datasets to uncover hidden patterns and relationships between material structure, composition, and catalytic performance. This kind of deep learning approach can yield new insights into the design of catalysts, guiding researchers toward the most promising materials with higher precision. AI-driven techniques like high-throughput screening and predictive modeling are accelerating the discovery and optimization of catalysts for hydrogen production. By reducing the time and cost associated with material discovery, AI is helping researchers identify high-performance, stable, and cost-effective catalysts more efficiently than traditional methods. These advancements have the potential to revolutionize hydrogen production technologies, making them more scalable and economically viable for large-scale applications [84].

3.6.2 Examples of AI-optimized Hydrogen Production Materials

The integration of AI with materials science has led to the accelerated discovery and optimization of catalysts for hydrogen production. By combining AI with computational materials science, researchers are now able to rapidly design, test, and refine new electrocatalysts and photocatalysts with unprecedented efficiency. Several case studies demonstrate the transformative impact of AI in hydrogen production research, showcasing how AI-optimized materials are paving the way for more efficient and cost-effective hydrogen production methods [85].

One significant case study involves the development of AI-designed electrocatalysts for water-splitting reactions, particularly the HER. Traditionally, Pt has been the catalyst of choice due to its high catalytic activity and stability. However, its high cost and limited availability have motivated researchers to explore non-precious metal alternatives. In one example, AI-driven high-throughput screening was used to explore a vast library of transition metal alloys and identify candidates that could mimic the performance of platinum while being significantly more affordable. Through AI, researchers discovered that alloys of nickel and iron exhibited remarkable catalytic properties for HER, with efficiency comparable to Pt-based catalysts. AI algorithms analyzed data on atomic configurations, electronic structures, and catalytic activity, rapidly narrowing down potential candidates for experimental validation. This approach has accelerated the transition from platinum to more sustainable, cost-effective materials for electrocatalysis in hydrogen production [86].

Another prominent case study is the development of AI-optimized photocatalysts for solar-driven hydrogen production through water splitting. Photocatalysis harnesses the energy of sunlight to drive the chemical reaction that splits water into hydrogen and oxygen. While TiO_2 has been extensively studied as a photocatalyst, its relatively low efficiency limits its practical applications. AI has been employed to identify modifications to TiO_2 and other photocatalysts, optimizing their band gap, light absorption capabilities, and charge carrier dynamics. For example, AI algorithms have successfully designed composite photocatalysts by combining TiO_2

with other materials such as carbon-based nanomaterials or MOFs, enhancing the overall performance of the photocatalytic system. These AI-guided designs have resulted in significant improvements in photocatalyst efficiency, opening up new possibilities for sustainable hydrogen production through solar energy [87].

The use of computational materials science has further propelled AI-driven innovation in hydrogen production materials. By simulating the interactions between materials and molecules at the atomic scale, AI models can predict how materials will behave in real-world conditions, such as during electrolysis or photocatalysis. One example is the development of perovskite-based materials for photocatalysis. Using AI-powered simulations, researchers identified new compositions of perovskites that enhanced the efficiency of photocatalytic water splitting. These simulations provided valuable insights into the electronic structure of perovskites, leading to the design of materials with optimized properties for solar-driven hydrogen production. This AI-enabled approach allows for the rapid screening of novel materials without the need for time-consuming experimental testing, significantly reducing the development time for new photocatalysts [85].

AI has also been applied to hybrid catalysts, which combine multiple materials to take advantage of their complementary properties. For instance, AI has been used to design hybrid materials that combine metal oxides with carbon-based supports, such as graphene or carbon nanotubes, for use in electrolysis and photocatalysis. These hybrid materials benefit from enhanced charge transfer, increased surface area, and improved stability, all of which are critical for efficient hydrogen production. AI algorithms have optimized the composition and architecture of these hybrid materials, enabling the development of catalysts with higher performance and longer lifetimes than traditional single-material catalysts. The integration of AI with computational materials science is not limited to the design of individual materials. AI has also been used to optimize the reaction conditions and system configurations for hydrogen production. By analyzing large datasets from experimental studies, AI can predict the best operating conditions for a particular catalyst or reaction system, such as temperature, pressure, or electrolyte composition. This capability has been particularly useful in electrolysis, where AI has helped fine-tune the parameters for maximum efficiency and stability.

3.7 Summary and Future Directions

In the realm of hydrogen production, material innovations have played a pivotal role in enhancing efficiency, sustainability, and cost-effectiveness. The development of advanced catalysts, nanomaterials, composites, and AI-driven discoveries has accelerated the progression toward scalable hydrogen production technologies. Catalysts, both traditional and emerging, are crucial in processes like electrolysis, photocatalysis, and thermochemical hydrogen production. Traditional precious metals such as Pt and Ir have long been considered the gold standard due to their excellent catalytic properties. However, the high cost and scarcity of these materials have spurred the search for non-precious metal catalysts, such as Ni, Co, and Fe. These non-precious metal catalysts, along with novel alternatives like sulfides, phosphides, and nitrides, have demonstrated promising performance while offering a significant reduction in cost.

The role of nanomaterials in improving hydrogen production efficiency cannot be overstated. Nanostructured materials, with their high surface area and enhanced reactivity, have been shown to greatly improve catalytic activity. Materials like graphene, CNTs, and nanostructured metal oxides have emerged as important components in next-generation catalysts for electrolysis and photocatalysis. These nanomaterials enhance the performance of hydrogen production processes by providing active sites that facilitate more efficient electron and ion transfer during the reaction. Additionally, composite materials – such as MOFs and carbon-based supports combined with metal oxides – have shown substantial promise in increasing the efficiency and stability of hydrogen production. The synergy between different materials in composites helps to enhance the overall performance of electrolysis and photocatalysis, making these systems more durable and efficient.

Moreover, the integration of AI into material discovery is rapidly transforming the landscape of catalyst development. AI-driven approaches, such as high-throughput screening and predictive modeling, enable the identification and optimization of new materials at an unprecedented rate. AI is helping to overcome the limitations of traditional trial-and-error methods, accelerating the discovery of low-cost, high-performance materials and providing invaluable insights into the stability and longevity of catalysts. These material innovations have significant implications for the future of the hydrogen economy. As the demand for green hydrogen grows, the availability of cost-effective, efficient, and sustainable materials will play a crucial role in making hydrogen production technologies more scalable and commercially viable. The advancements in materials science will not only lower production costs but also enhance the integration of hydrogen into global energy systems, contributing to a cleaner and more sustainable future.

Looking ahead, several promising research directions in materials science hold the potential for further breakthroughs in hydrogen production. One of the most critical areas is the development of low-cost, high-performance catalysts. While significant progress has been made in finding non-precious metal catalysts that perform well in electrolysis and photocatalysis, further research is needed to improve their stability and durability under industrial conditions. Catalysts capable of withstanding harsh operational environments, such as high temperatures and pressures, while maintaining high catalytic efficiency, will be essential for large-scale hydrogen production. Hybrid catalysts combining multiple material types (such as metal oxides with carbon-based supports or MOFs) are also an exciting avenue for future research. By leveraging the unique properties of different materials, these hybrid catalysts can offer enhanced performance, making them a potential game-changer in hydrogen production technologies.

Another key area for research is the scalability of novel materials. While many of the emerging catalysts and nanomaterials show promise in laboratory-scale experiments, their performance and cost-effectiveness need to be evaluated at larger scales. Scaling up the synthesis and integration of these materials for industrial applications will be a significant challenge, particularly for materials that require complex production processes or rare components. Efforts to develop more cost-effective synthesis methods and improve the overall scalability of these materials are critical for the widespread adoption of hydrogen production technologies. The role of AI and machine learning in material discovery will continue to expand, enabling researchers to identify new materials with greater efficiency. AI models trained on vast datasets of material properties can predict the performance and stability of potential catalysts, allowing researchers to focus their efforts on the most promising candidates. Furthermore, AI can assist in optimizing the design of hydrogen production systems, such as improving the architecture of electrolysis cells or designing better composite materials for photocatalysis.

Finally, the industrial application of these materials will be crucial to realizing the full potential of hydrogen as a clean energy source. As advancements in material science continue, industries such as renewable energy, transportation, and chemicals will increasingly integrate hydrogen production technologies into their operations. The development of cost-effective and scalable materials will facilitate the integration of hydrogen into broader energy systems, enabling its use as a clean, sustainable energy carrier across various sectors.

In summary, the future of hydrogen production is closely tied to ongoing advancements in material science. The next generation of catalysts, nanomaterials, and composites will drive the efficiency, scalability, and affordability of hydrogen production technologies. By continuing to explore new materials and leveraging AI to accelerate discovery, researchers are paving the way for a sustainable hydrogen economy that can play a crucial role in addressing global energy challenges.

References

1 Alamiery A. Advancements in materials for hydrogen production: A review of cutting-edge technologies. *ChemPhysMater*. 2024 Jan 1;3(1):64–73.

2 Xu Y, Zhang X, Liu Y, Wang R, Yang Y, Chen J. A critical review of research progress for metal alloy materials in hydrogen evolution and oxygen evolution reaction. *Environ Sci Pollut Res* [Internet]. 2023 Jan 1 [cited 2025 Feb 3]; 30(5):11302–20. Available from: https://link.springer.com/article/10.1007/s11356-022-24728-5.

3 Zhang J, Li X, Chen H, Qi M, Zhang G, Hu H, et al. Hydrogen production by catalytic methane decomposition: Carbon materials as catalysts or catalyst supports. *Int J Hydrogen Energy*. 2017 Aug 3;42(31):19755–75.

4 Tiu ZC, Ooi SI, Guo J, Zhou G, Ji J, et al. Review—Recent developments in the applications of 2D transition metal dichalcogenides as electrocatalysts in the generation of hydrogen for renewable energy conversion. *J Electrochem Soc* [Internet]. 2022 Jun 6 [cited 2025 Feb 3];169(6):064504. Available from: https://iopscience.iop.org/article/10.1149/1945-7111/ac7172.

5 Kumar R, Kumar A, Pal A. An overview of conventional and non-conventional hydrogen production methods. *Mater Today Proc*. 2021 Jan 1;46:5353–9.

6 Bhosale RR, AlMomani F. Hydrogen production via solar driven thermochemical cerium oxide—Cerium sulfate water splitting cycle. *Int J Hydrogen Energy*. 2020 Mar 27;45(17):10381–90.

7 Rönnebro ECE, Oelrich RL, Gates RO. Recent Advances and prospects in design of hydrogen permeation barrier materials for energy applications—A review. *Molecules* [Internet]. 2022 Oct 2 [cited 2025 Feb 3];27(19):6528. Available from: https://www.mdpi.com/1420-3049/27/19/6528/htm.

8 Godula-Jopek A. *Hydrogen production: By electrolysis*. Wiley; 2015.

9 Preethi V, Kanmani S. Photocatalytic hydrogen production. *Mater Sci Semicond Process* 2013 Jun 1;16(3):561–75.

10 Ji G, Yao JG, Clough PT, Da Costa JCD, Anthony EJ, Fennell PS, et al. Enhanced hydrogen production from thermochemical processes. *Energy Environ Sci* [Internet]. 2018 Oct 10 [cited 2025 Feb 3];11(10):2647–72. Available from: https://pubs.rsc.org/en/content/articlehtml/2018/ee/c8ee01393d.

11 Jeje SO, Marazani T, Obiko JO, Shongwe MB. Advancing the hydrogen production economy: A comprehensive review of technologies, sustainability, and future prospects. *Int J Hydrogen Energy* 2024 Aug 12;78:642–61.

12 Jacobsson TJ, Fjällström V, Edoff M, Edvinsson T. Sustainable solar hydrogen production: From photoelectrochemical cells to PV-electrolyzers and back again. *Energy Environ Sci* [Internet]. 2014 Jun 19 [cited 2025 Feb 3];7(7):2056–70. Available from: https://pubs.rsc.org/en/content/articlehtml/2014/ee/c4ee00754a.

13 Bessarabov D, Wang H, Li H, Zhao N. *PEM Electrolysis for Hydrogen Production: Principles and Applications*. CRC press England & Wales. 2016 Feb 3;1–382.

14 Dincer IE, Agelin-Chaab M. Experimental investigation and evaluation of newly designed electrodes for hydrogen production in alkaline water electrolysis. *J Power Sources* [Internet]. 2025 Mar 15 [cited 2025 Feb 3];632:236326. Available from: https://linkinghub.elsevier.com/retrieve/pii/S0378775325001624.

15 Colón G. Towards the hydrogen production by photocatalysis. *Appl Catal A Gen*. 2016 May 25;518:48–59.

16 Karaca AE, Qureshy AMMI, Dincer I. An overview and critical assessment of thermochemical hydrogen production methods. *J Clean Prod*. 2023 Jan 20; 385:135706.

17 Younas M, Shafique S, Hafeez A, Javed F, Rehman F. An overview of hydrogen production: Current status, potential, and challenges. *Fuel*. 2022 May 15; 316:123317.

18 Mani P, Shenoy S, Sagayaraj PJJ, Agamendran N, Son S, Bernaurdshaw N, et al. Scaling up of photocatalytic systems for large-scale hydrogen generation. *Appl Phys Rev* [Internet]. 2025 Mar 1 [cited 2025 Feb 3];12(1):011303. Available from: /aip/apr/article/12/1/011303/3330351/Scaling-up-of-photocatalytic-systems-for-large.

19 Biradha K, Goswami A, Moi R. Coordination polymers as heterogeneous catalysts in hydrogen evolution and oxygen evolution reactions. *Chem Commun* [Internet]. 2020 Sep 17 [cited 2025 Feb 3];56(74):10824–42. Available from: https://pubs.rsc.org/en/content/articlehtml/2020/cc/d0cc04236f.

20 Alobaid A, Wang C, Adomaitis RA. Mechanism and kinetics of HER and OER on NiFe LDH films in an alkaline electrolyte. *J Electrochem Soc* [Internet]. 2018 Dec 5 [cited 2025 Feb 3];165(15):J3395–404. Available from: https://iopscience.iop.org/article/10.1149/2.0481815jes.

21 Huang L, Zaman S, Tian X, Wang Z, Fang W, Xia BY. Advanced platinum-based oxygen reduction electrocatalysts for fuel cells. *Acc Chem Res* [Internet]. 2021 Jan 19 [cited 2025 Feb 3];54(2):311–22. Available from: https://pubs.acs.org/doi/abs/10.1021/acs.accounts.0c00488.

22 Fabbri E, Schmidt TJ. Oxygen evolution reaction - The enigma in water electrolysis. *ACS Catal* [Internet]. 2018 Oct 5 [cited 2025 Feb 3];8(10):9765–74. Available from: https://pubs.acs.org/doi/full/10.1021/acscatal.8b02712.

23 Zhou D, Li P, Xu W, Jawaid S, Mohammed-Ibrahim J, Liu W, et al. Recent advances in non-precious metal-based electrodes for alkaline water electrolysis. *ChemNanoMat* [Internet]. 2020 Mar 1 [cited 2025 Feb 3];6(3):336–55. Available from: https://onlinelibrary.wiley.com/doi/full/10.1002/cnma.202000010.

24 Kim HJ, Kim HY, Joo J, Joo SH, Lim JS, Lee J, et al. Recent advances in non-precious group metal-based catalysts for water electrolysis and beyond. *J Mater Chem A Mater* [Internet]. 2021 Dec 21 [cited 2025 Feb 3];10(1):50–88. Available from: https://pubs.rsc.org/en/content/articlehtml/2022/ta/d1ta06548c.

25 Upadhyay SN, Satrughna JAK, Pakhira S. Recent advancements of two-dimensional transition metal dichalcogenides and their applications in electrocatalysis and energy storage. *Emerg Mater* [Internet]. 2021 Jun 25 [cited 2025 Feb 3];4(4):951–70. Available from: https://link.springer.com/article/10.1007/s42247-021-00241-2.

26 Wang Y, Wang L, Zhang K, Xu J, Wu Q, Xie Z, et al. Electrocatalytic water splitting over perovskite oxide catalysts. *Chinese J Catal*. 2023 Jul 1;50:109–25.

27 Wang HY, Sun ML, Ren JT, Yuan ZY. Circumventing challenges: Design of anodic electrocatalysts for hybrid water electrolysis systems. *Adv Energy Mater* [Internet]. 2023 Jan 1 [cited 2025 Feb 3];13(4):2203568. Available from: https://onlinelibrary.wiley.com/doi/full/10.1002/aenm.202203568.

28 Zhang X, Cao C, Ling T, Ye C, Lu J, Shan J. Developing practical catalysts for high-current-density water electrolysis. *Adv Energy Mater* [Internet]. 2024 Dec 1 [cited 2025 Feb 3];14(45):2402633. Available from: https://onlinelibrary.wiley.com/doi/full/10.1002/aenm.202402633.

29 Hoang AL, Balakrishnan S, Hodges A, Tsekouras G, Al-Musawi A, Wagner K, et al. High-performing catalysts for energy-efficient commercial alkaline water electrolysis. *Sustain Energy Fuels* [Internet]. 2022 Dec 20 [cited 2025 Feb 2];7(1):31–60. Available from: https://pubs.rsc.org/en/content/articlehtml/2023/se/d2se01197b.

30 Chen Y, Yang K, Jiang B, Li J, Zeng M, Fu L. Emerging two-dimensional nanomaterials for electrochemical hydrogen evolution. *J Mater Chem A Mater* [Internet]. 2017 May 10 [cited 2025 Feb 3];5(18):8187–208. Available from: https://pubs.rsc.org/en/content/articlehtml/2017/ta/c7ta00816c.

31 He Z, Que W. Molybdenum disulfide nanomaterials: Structures, properties, synthesis and recent progress on hydrogen evolution reaction. *Appl Mater Today*. 2016 Jun 1;3:23–56.

32 Huang H, Yan M, Yang C, He H, Jiang Q, Yang L, et al. Graphene nanoarchitectonics: Recent advances in graphene-based electrocatalysts for hydrogen evolution reaction. *Adv Mater* [Internet]. 2019 Nov 1 [cited

2025 Feb 3];31(48):1903415. Available from: https://onlinelibrary.wiley.com/doi/full/10.1002/adma.201903415.

33 Zhu J, Hu L, Zhao P, Lee LYS, Wong KY. Recent advances in electrocatalytic hydrogen evolution using nanoparticles. *Chem Rev* [Internet]. 2020 Jan 22 [cited 2025 Feb 3];120(2):851–918. Available from: https://pubs.acs.org/doi/abs/10.1021/acs.chemrev.9b00248.

34 Baghban A, Habibzadeh S, Zokaee Ashtiani F. On the evaluation of hydrogen evolution reaction performance of metal-nitrogen-doped carbon electrocatalysts using machine learning technique. *Sci Rep* [Internet]. 2021 Nov 9 [cited 2025 Feb 3];11(1):1–12. Available from: https://www.nature.com/articles/s41598-021-00031-0.

35 Li J, Wu N, Zhang J, Wu HH, Pan K, Wang Y, et al. Machine learning-assisted low-dimensional electrocatalysts design for hydrogen evolution reaction. *Nano-Micro Lett* [Internet]. 2023 Oct 13 [cited 2025 Feb 3];15(1):1–27. Available from: https://link.springer.com/article/10.1007/s40820-023-01192-5.

36 Suriani AB, Asli NA, Salina M, Mamat MH, Aziz AA, Falina AN, et al. Effect of iron and cobalt catalysts on the growth of carbon nanotubes from palm oil precursor. *IOP Conf Ser Mater Sci Eng* [Internet]. 2013 Jun 6 [cited 2025 Feb 2];46(1):012014. Available from: https://iopscience.iop.org/article/10.1088/1757-899X/46/1/012014.

37 Landois P, Peigney A, Laurent C, Frin L, Datas L, Flahaut E. CCVD synthesis of carbon nanotubes with W/Co–MgO catalysts. *Carbon N Y*. 2009 Mar 1;47(3):789–94.

38 Deshmukh MA, Park SJ, Thorat HN, Bodkhe GA, Ramanavicius A, Ramanavicius S, et al. Advanced energy materials: Current trends and challenges in electro- and photo-catalysts for H_2O splitting. *J Ind Eng Chem*. 2023 Mar 25;119:90–111.

39 Gupta A, Puri N. Recent advances in hydrogen production using metal organic frameworks and their composites. *Int J Hydrogen Energy*. 2024 Aug 12;78:303–24.

40 Kannan K, Chanda D, Gautam J, Behera A, Meshesha MM, Gwon Jang S, et al. Hydrothermally synthesized mixed metal oxide nanocomposites for electrochemical water splitting and photocatalytic hydrogen production. *Int J Hydrogen Energy*. 2023 Dec 1;48(93):36412–26.

41 Kuang P, Sayed M, Fan J, Cheng B, Yu J. 3D graphene-based H_2-production photocatalyst and electrocatalyst. *Adv Energy Mater* [Internet]. 2020 Apr 1 [cited 2025 Feb 3];10(14):1903802. Available from: https://onlinelibrary.wiley.com/doi/full/10.1002/aenm.201903802.

42 Ahmed SF, Kumar PS, Ahmed B, Mehnaz T, Shafiullah GM, Nguyen VN, et al. Carbon-based nanomaterials: Characteristics, dimensions, advances and challenges in enhancing photocatalytic hydrogen production. *Int J Hydrogen Energy*. 2024 Jan 2;52:424–42.

43 Abdelazeez AAA, Rabia M, Hasan F, Mahanta V, Adly ER. Polymer nanocomposites: Catalysts for Sustainable hydrogen production from challenging water sources. *Adv Energy Sustain Res* [Internet]. 2024 Oct 1 [cited 2025 Feb 3];5(10):2400077. Available from: https://onlinelibrary.wiley.com/doi/full/10.1002/aesr.202400077.

44 Kalisamy P, Lallimathi M, Suryamathi M, Palanivel B, Venkatachalam M. ZnO-embedded S-doped g-C3N4 heterojunction: Mediator-free Z-scheme mechanism for enhanced charge separation and photocatalytic degradation. *RSC Adv* [Internet]. 2020 Jul 27 [cited 2025 Feb 2];10(47):28365–75. Available from: https://pubs.rsc.org/en/content/articlehtml/2020/ra/d0ra04642f.

45 Abdalla AM, Hossain S, Nisfindy OB, Azad AT, Dawood M, Azad AK. Hydrogen production, storage, transportation and key challenges with applications: A review. *Energy Convers Manag*. 2018 Jun 1;165:602–27.

46 Zhai W, Ma Y, Chen D, Ho JC, Dai Z, Qu Y. Recent progress on the long-term stability of hydrogen evolution reaction electrocatalysts. *InfoMat* [Internet]. 2022 Sep 1 [cited 2025 Mar 18];4(9):e12357. Available from: https://onlinelibrary.wiley.com/doi/full/10.1002/inf2.12357.

47 Salehmin MNI, Husaini T, Goh J, Sulong AB. High-pressure PEM water electrolyser: A review on challenges and mitigation strategies towards green and low-cost hydrogen production. *Energy Convers Manag*. 2022 Sep 15;268:115985.

48 Liu M, Yao Z, Gu J, Li C, Huang X, Zhang L, et al. Issues and opportunities facing hydrolytic hydrogen production materials. *Chem Eng J*. 2023 Apr 1;461:141918.

49 Subotić V, Hochenauer C. Analysis of solid oxide fuel and electrolysis cells operated in a real-system environment: State-of-the-health diagnostic, failure modes, degradation mitigation and performance regeneration. *Prog Energy Combust Sci*. 2022 Nov 1;93:101011.

50 Gao M, Zhu L, Ong WL, Wang J, Ho GW. Structural design of TiO2-based photocatalyst for H_2 production and degradation applications. *Catal Sci Technol* [Internet]. 2015 Sep 21 [cited 2025 Mar 18];5(10):4703–26. Available from: https://pubs.rsc.org/en/content/articlehtml/2015/cy/c5cy00879d.

51 Dubau L, Castanheira L, Maillard F, Chatenet M, Lottin O, Maranzana G, et al. A review of PEM fuel cell durability: Materials degradation, local heterogeneities of aging and possible mitigation strategies. *Wiley Interdiscip Rev Energy Environ* [Internet]. 2014 Nov 1 [cited 2025 Mar 18];3(6):540–60. Available from: https://onlinelibrary.wiley.com/doi/full/10.1002/wene.113.

52 Xu F, Weng B. Photocatalytic hydrogen production: An overview of new advances in structural tuning strategies. *J Mater Chem A Mater* [Internet]. 2023 Feb 27 [cited 2025 Mar 18];11(9):4473–86. Available from: https://pubs.rsc.org/en/content/articlehtml/2023/ta/d2ta09614e.

53 Kumar R, Singh R, Dutta S. Review and outlook of hydrogen production through catalytic processes. *Energy Fuels* [Internet]. 2024 Feb 15 [cited 2025 Mar 18];38(4):2601–29. Available from: https://pubs.acs.org/doi/abs/10.1021/acs.energyfuels.3c04026.

54 Faisal NH, Prathuru A, Ahmed R, Rajendran V, Hossain M, Venkatachalapathy V, et al. Application of thermal spray coatings in electrolysers for hydrogen production: Advances, challenges, and opportunities. *ChemNanoMat* [Internet]. 2022 Dec 1 [cited 2025 Mar 18];8(12):e202200384. Available from: https://onlinelibrary.wiley.com/doi/full/10.1002/cnma.202200384.

55 Liu W, Sun L, Li Z, Fujii M, Geng Y, Dong L, et al. Trends and future challenges in hydrogen production and storage research. *Environ Sci Pollut Res* [Internet].

2020 Sep 1 [cited 2025 Mar 18];27(25):31092–104. Available from: https://link.springer.com/article/10.1007/s11356-020-09470-0.

56 Kumar N, Aepuru R, Lee SY, Park SJ. Advances in catalysts for hydrogen production: A comprehensive review of materials and mechanisms. *Nanomaterials* [Internet]. 2025 Feb 8 [cited 2025 Mar 18];15(4):256. Available from: https://www.mdpi.com/2079-4991/15/4/256/htm.

57 Zhu J, Zäch M. Nanostructured materials for photocatalytic hydrogen production. *Curr Opin Colloid Interface Sci*. 2009 Aug 1; 14(4):260–9.

58 Luo H, Barrio J, Sunny N, Li A, Steier L, Shah N, et al. Progress and perspectives in photo- and electrochemical-oxidation of biomass for sustainable chemicals and hydrogen production. *Adv Energy Mater* [Internet]. 2021 Nov 1 [cited 2025 Mar 18];11(43):2101180. Available from: https://onlinelibrary.wiley.com/doi/full/10.1002/aenm.202101180.

59 Overview of PEM Electrolysis for Hydrogen Production. PEM Electrolysis for Hydrogen Production [Internet]. 2016 Feb 3 [cited 2025 Mar 18];13–22. Available from: https://www.taylorfrancis.com/chapters/edit/10.1201/b19096-5/overview-pem-electrolysis-hydrogen-production-nicola-briguglio-vincenzo-antonucci.

60 Amores E, Sánchez-Molina M, Sánchez M. Effects of the marine atmosphere on the components of an alkaline water electrolysis cell for hydrogen production. *Results Eng*. 2021 Jun 1;10:100235.

61 Anantharaj S, Sugime H, Noda S. Chemical leaching of inactive Cr and subsequent electrochemical resurfacing of catalytically active sites in stainless steel for high-rate alkaline hydrogen evolution reaction. *ACS Appl Energy Mater* [Internet]. 2020 Dec 28 [cited 2025 Mar 18];3(12):12596–606. Available from: https://pubs.acs.org/doi/abs/10.1021/acsaem.0c02505.

62 Wang J, Kang D, Shen B, Sun H, Wu C. Enhanced hydrogen production from catalytic biomass gasification with in-situ CO_2 capture. *Environ Pollut*. 2020 Dec 1;267:115487.

63 Song B, Chen M, Zeng G, Gong J, Shen M, Xiong W, et al. Using graphdiyne (GDY) as a catalyst support for enhanced performance in organic pollutant degradation and hydrogen production: A review. *J Hazard Mater*. 2020 Nov 5; 398:122957.

64 Gonuguntla S, Jaksani B, Jamma A, Vennapoosa CS, Chatterjee D, Pal U. Design principle of anti-corrosive photocatalyst for large-scale hydrogen production. *Wiley Interdiscip Rev Energy Environ* [Internet]. 2024 Jul 1 [cited 2025 Mar 18]; 13(4):e530. Available from: https://onlinelibrary.wiley.com/doi/full/10.1002/wene.530.

65 Li W, Cao R, Xu L, Qiao L. The role of hydrogen in the corrosion and cracking of steels - A review. *Corros Commun*. 2021 Dec 1;4:23–32.

66 Kinik FP, Nguyen TN, Oveisi E, Valizadeh B, Ebrahim FM, Gładysiak A, et al. Discovery of a self-healing catalyst for the hydrolytic dehydrogenation of ammonia borane. *J Mater Chem A Mater* [Internet]. 2019 Oct 22 [cited 2025 Mar 18];7(41):23830–7. Available from: https://pubs.rsc.org/en/content/articlehtml/2019/ta/c9ta02123j.

67 Tong Y, Ma H, Xiao F, Bohm S, Fu H, Luo Y. Precision engineering of precious metal catalysts for enhanced hydrogen production efficiency. *Process Saf Environ Prot*. 2023 Oct 1;178:559–79.

68 Zhao W, Chen Z, Yang X, Qian X, Liu C, Zhou D, et al. Recent advances in photocatalytic hydrogen evolution with high-performance catalysts without precious metals. *Renew Sustain Energy Rev*. 2020 Oct 1;132:110040.

69 Sukackienė Z, Valeckytė G, Kepenienė V, Stalnionienė I, Jasulaitienė V, Vaičiūnienė J, et al. Non-precious metals catalysts for hydrogen generation. *Coatings* [Internet]. 2023 Oct 7 [cited 2025 Feb 3];13(10):1740. Available from: https://www.mdpi.com/2079-6412/13/10/1740/htm.

70 Chang K, Hai X, Ye J. Transition metal disulfides as noble-metal-alternative co-catalysts for solar hydrogen production. *Adv Energy Mater* [Internet]. 2016 May 1 [cited 2025 Feb 3];6(10):1502555. Available from: https://onlinelibrary.wiley.com/doi/full/10.1002/aenm.201502555.

71 Irmak S, Meryemoglu B, Ozsel BK, Hasanoglu A, Erbatur O. Improving activity of Pt supported metal catalysts by changing reduction method of Pt precursor for hydrogen production from biomass. *Int J Hydrogen Energy*. 2015 Nov 16; 40(43):14826–32.

72 Du Y, Liu J, Chen J, Wang S, Tang Y, Wang AL, et al. Design principle and regulation strategy of noble metal-based materials for practical proton exchange membrane water electrolyzer. *Adv Energy Mater* [Internet]. 2024 [cited 2025 Feb 3];15(10): 2404113. Available from: https://onlinelibrary.wiley.com/doi/full/10.1002/aenm.202404113.

73 Zhiani M, Balali Dehkordi H, Xiao J, Yan CF, Tan HY. Optimized $Irruo_x$ mixed oxide nanocatalyst for enhanced oxygen evolution electrocatalysis in acidic water splitting: A comparative study with Iro_2 and Ruo_2 nanoparticles. 2024 [cited 2025 Feb 3]; Available from: https://papers.ssrn.com/abstract=4791663.

74 Sher F, Al-Shara NK, Iqbal SZ, Jahan Z, Chen GZ. Enhancing hydrogen production from steam electrolysis in molten hydroxides via selection of non-precious metal electrodes. *Int J Hydrogen Energy*. 2020 Oct 30;45(53):28260–71.

75 Liu J, Wang Z, Yan X, Jian P. Metallic cobalt nanoparticles imbedded into ordered mesoporous carbon: A non-precious metal catalyst with excellent hydrogenation performance. *J Colloid Interface Sci*. 2017 Nov 1;505:789–95.

76 Liu J, Wu H, Li F, Feng X, Zhang P, Gao L. Recent progress in non-precious metal single atomic catalysts for solar and non-solar driven hydrogen evolution reaction. *Adv Sustain Syst* [Internet]. 2020 Nov 1 [cited 2025 Feb 3];4(11):2000151. Available from: https://onlinelibrary.wiley.com/doi/full/10.1002/adsu.202000151.

77 Sudeep M, Vijay Rajyaguru Y, Kumar Rastogi C, MP SA, Sridharan M, Khosla A, et al. Recent development of nickel based chalcogenides for hydrogen generation. *Mater Today Proc*. 2023 Jan 1;73:316–22.

78 Liu T, Chen C, Liu S, Chen Z, Pu Z, Huang Q, et al. Transition metal phosphides as noble-metal-alternative co-catalysts for solar hydrogen production. *Coord Chem Rev*. 2024 Dec 15;521:216145.

79 Elreedy A, Ibrahim E, Hassan N, El-Dissouky A, Fujii M, Yoshimura C, et al. Nickel-graphene nanocomposite as a novel supplement for enhancement of

biohydrogen production from industrial wastewater containing mono-ethylene glycol. *Energy Convers Manag*. 2017 May 15;140:133–44.

80 Shi L, Chen Z, Jian Z, Guo F, Gao C. Carbon nanotubes-promoted Co–B catalysts for rapid hydrogen generation via NaBH4 hydrolysis. *Int J Hydrogen Energy* 2019 Jul 26;44(36):19868–77.

81 Rafique F, Ameen S, Tahir A, Haq T ul, Khan A, Ahmed SA, et al. Designing Nio-Fe_2O_3/Feooh Heterojunction Electrocatalysts for Enhanced Water Splitting: Synergistic Interfaces and Enhance Stability. 2025 [cited 2025 Feb 3]; Available from: https://papers.ssrn.com/abstract=5106170.

82 Salehmin MNI, Tiong SK, Mohamed H, Umar DA, Yu KL, Ong HC, et al. Navigating challenges and opportunities of machine learning in hydrogen catalysis and production processes: Beyond algorithm development. *J Energy Chem*. 2024 Dec 1;99:223–52.

83 Osman AI, Nasr M, Eltaweil AS, Hosny M, Farghali M, Al-Fatesh AS, et al. Advances in hydrogen storage materials: Harnessing innovative technology, from machine learning to computational chemistry, for energy storage solutions. *Int J Hydrogen Energy* 2024 May 20;67:1270–94.

84 Kim J, Jang HW. Can artificial intelligence boost developing electrocatalysts for efficient water splitting to produce green hydrogen? *Korean J Mater Res*. 2023;33(5):175–88.

85 Boretti A. Artificial intelligence driven evolution of perovskite-based solar thermochemical systems for hydrogen production: A narrative review. *Emerg Mater* [Internet]. 2024 Aug 16 [cited 2025 Feb 3];34:1–15. Available from: https://link.springer.com/article/10.1007/s42247-024-00823-w.

86 Mohamed NEDM, El Zoghby HM, Abdelhakam MM, Elmesalawy MM. AI-enabled smart hybrid energy optimization management system for green hydrogen-based islanded microgrids. *NILES 2024 - 6th Novel Intelligent and Leading Emerging Sciences Conference, Proceedings*. IEEE Xplore. 2024;539–44.

87 Slattery A, Wen Z, Tenblad P, Sanjosé-Orduna J, Pintossi D, Den Hartog T, et al. Automated self-optimization, intensification, and scale-up of photocatalysis in flow. *Science* (1979) [Internet]. 2024 Jan 1 [cited 2025 Feb 3];383(6681):1–12. Available from: https://www.science.org/doi/10.1126/science.adj1817.

4

Hydrogen Storage – Challenges and Breakthroughs

4.1 Introduction to Hydrogen Storage

Hydrogen is a promising energy carrier that has the potential to drive the transition toward a sustainable and carbon-neutral energy system. However, its widespread adoption is heavily dependent on the development of efficient, safe, and cost-effective storage technologies. Unlike conventional fossil fuels, hydrogen has a very low volumetric energy density under ambient conditions, making its storage a significant challenge. The need for high-density storage methods that ensure safety, economic viability, and compatibility with existing infrastructure is paramount for the successful integration of hydrogen into the global energy landscape. Several storage technologies have been explored, broadly categorized into physical storage, chemical storage, and solid-state storage. Each of these methods has unique advantages and limitations, requiring further research and technological improvements to achieve commercial scalability [1].

Hydrogen storage plays a crucial role in enabling its practical use in various sectors, including transportation, power generation, and industrial applications. While hydrogen can be produced efficiently using methods such as electrolysis, thermochemical cycles, and biological processes, its effective storage and transportation remain key barriers to widespread implementation [2]. The major challenges associated with hydrogen storage include low energy density, high costs, safety concerns, and the need for robust infrastructure. Addressing these challenges requires advancements in material science, engineering, and system integration to develop storage methods that optimize hydrogen's potential as a clean energy carrier [3]. One of the most used methods for hydrogen storage is physical storage, which involves compressing hydrogen gas or liquefying it at cryogenic temperatures to achieve higher energy densities. Compressed gas storage is a well-established technology, widely used in fuel cell vehicles and industrial applications. Hydrogen is typically stored at pressures of 350–700 bar in high-strength tanks made of advanced composite materials [4]. This method offers the advantage of fast refueling and a relatively simple storage mechanism. However, high-pressure storage poses challenges related to material durability, energy-intensive compression processes, and potential safety risks due to gas leakage or explosions.

Next-Generation Hydrogen Economy: Production, Storage, and Fuel Cell Technologies, First Edition.
Peeyush Phogat, Shreya Sharma and Satyam Rawat.

Cryogenic liquid hydrogen storage involves cooling hydrogen to 20 K (−253 °C) to increase its volumetric density significantly. This method is commonly used in aerospace applications and large-scale hydrogen transportation. Despite its high-energy density, liquid hydrogen storage suffers from boil-off losses, as maintaining such low temperatures requires continuous energy input. Additionally, the liquefaction process is highly energy-intensive, consuming approximately 30–40% of the stored hydrogen's energy content. This makes it less suitable for long-term storage and commercial applications where energy efficiency is a priority. In addition to physical storage methods, chemical hydrogen storage offers an alternative approach by binding hydrogen within chemical compounds and releasing it when needed. This method enables higher hydrogen densities and safer handling compared to compressed or liquid hydrogen. One of the most promising chemical hydrogen carriers is ammonia (NH_3), which contains 17.6 wt% hydrogen and can be easily liquefied under moderate pressure. Ammonia can be transported using existing fuel infrastructure and later decomposed to release hydrogen. However, ammonia storage and transport present challenges related to toxicity, catalyst development for decomposition, and energy-intensive cracking processes [5]. Another emerging chemical storage method involves liquid organic hydrogen carriers (LOHCs), such as toluene/methylcyclohexane and benzyl toluene, which can reversibly store and release hydrogen through catalytic hydrogenation and dehydrogenation reactions. LOHCs offer safe, high-density hydrogen storage, making them ideal for large-scale applications. Unlike ammonia, LOHCs are non-toxic and can be integrated into existing fuel supply chains. However, the main limitation of LOHCs is the high energy requirement for hydrogen release, which reduces overall efficiency [6].

Beyond chemical storage, solid-state hydrogen storage offers innovative solutions using materials that can absorb and release hydrogen at moderate pressures and temperatures. Metal hydrides, such as magnesium hydride (MgH_2), titanium hydride (TiH_2), and lanthanum nickel hydride ($LaNi_5H_6$), have been extensively studied for their ability to store hydrogen compactly and safely. Metal hydrides can achieve higher volumetric hydrogen densities compared to compressed gas and liquid hydrogen storage. However, they suffer from slow hydrogen absorption/desorption kinetics, high operating temperatures, and long-term degradation. Further research is required to develop metal hydrides with improved hydrogen storage capacities, faster reaction kinetics, and lower energy requirements [7].

Metal-organic frameworks (MOFs) and carbon-based nanostructures (such as graphene, carbon nanotubes (CNTs), and fullerenes) have gained attention for their potential in hydrogen storage. These materials provide high surface areas and tunable porosities, allowing for efficient hydrogen adsorption at moderate pressures. MOFs, in particular, have demonstrated high hydrogen uptake capacities, but their practical application is still limited by thermal and chemical stability issues. Carbon nanostructures, on the other hand, offer lightweight and scalable solutions, but their hydrogen storage capacities remain relatively low. Continued advancements in nanomaterials and hybrid composites may lead to breakthrough storage technologies that combine high efficiency, low cost, and enhanced durability. Despite the progress in hydrogen storage technologies, several challenges still need to be addressed to make hydrogen storage commercially viable on a large scale. One of the most significant challenges is hydrogen's low energy density, which affects storage and transportation efficiency. Even at 700 bar, compressed hydrogen has an energy density of 5.6 MJ/L, which is significantly lower than that of gasoline (32 MJ/L). This necessitates the development of storage solutions with higher volumetric efficiencies [8, 9].

Another critical challenge is safety. Hydrogen is highly flammable, and its small molecular size makes it prone to leakage through storage materials. Advanced safety mechanisms, such as high-strength containment materials, leak detection sensors, and explosion-resistant storage systems, are required to minimize risks associated with hydrogen storage. Material durability and system integrity also play a crucial role in ensuring the long-term stability of hydrogen storage technologies. Cost remains another major barrier to large-scale hydrogen storage adoption. The production and maintenance of high-pressure storage tanks, cryogenic systems, and chemical storage units involve high capital and operational costs. Developing low-cost materials and improving the efficiency of hydrogen storage cycles are essential for reducing costs and making hydrogen more competitive with conventional fuels [10].

The future of hydrogen storage lies in the development of hybrid and integrated storage systems that combine the benefits of different storage technologies. For example, combining physical storage with chemical storage could optimize energy density while improving safety and economic feasibility. Researchers are also exploring the use of artificial intelligence (AI) and machine learning to design novel hydrogen storage materials with enhanced performance. AI-driven computational models can help predict material stability, hydrogen absorption capacity, and reaction kinetics, accelerating the discovery of next-generation hydrogen storage solutions.

Looking ahead, hydrogen storage technologies must be designed to support the growing hydrogen infrastructure, enabling efficient transportation and distribution of hydrogen across global markets. The integration of hydrogen storage with renewable energy sources, such as solar and wind power, will be crucial for achieving a carbon-neutral hydrogen economy. Developing modular, decentralized storage solutions for small-scale applications, such as hydrogen-powered vehicles and residential fuel cells, will also contribute to a more flexible and sustainable energy system. Hydrogen storage is a critical component of the hydrogen economy, with multiple technologies being explored to improve efficiency, safety, and cost-effectiveness. While compressed gas and liquid hydrogen storage are widely used today, chemical and solid-state storage methods offer promising alternatives with higher storage densities and safer handling. The main challenges that need to be overcome include low energy density, safety concerns, cost constraints, and material limitations. Future research efforts should focus on developing advanced materials, hybrid storage systems, and AI-driven material discovery to enable large-scale hydrogen storage for diverse applications. With continued technological advancements, hydrogen has the potential to become a mainstream energy carrier, driving the world toward a cleaner and more sustainable future [11, 12]. Figure 4.1 provides a detailed representation of different hydrogen storage technologies given below.

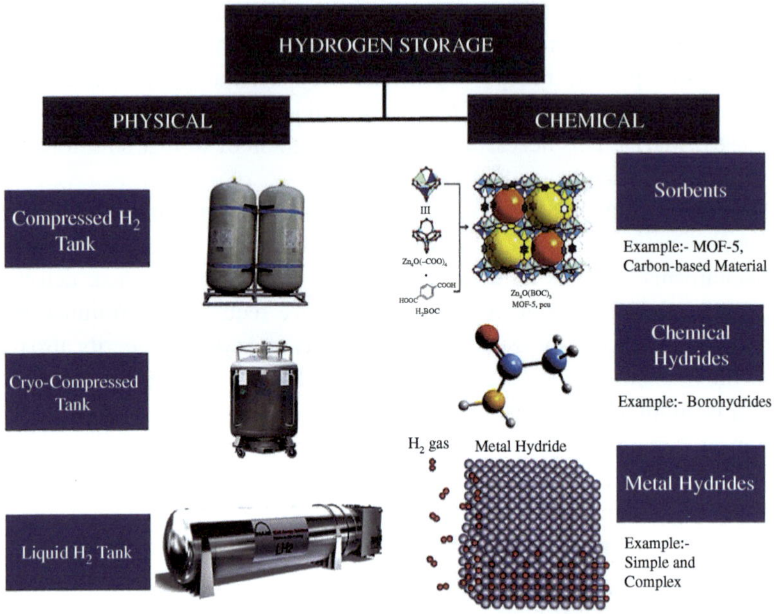

Figure 4.1 Schematic representation of different hydrogen storage technologies. Reproduced with permission from [13]/Elsevier/CC-BY-NC-ND.

4.2 Physical Hydrogen Storage

4.2.1 Compressed Gas Storage

Hydrogen, being the lightest element, has a very low volumetric energy density under ambient conditions, necessitating high-pressure storage to achieve practical energy densities. Compressed gas storage is one of the most established and widely used methods for storing hydrogen, particularly in fuel cell vehicles, industrial applications, and stationary energy systems. This method involves storing hydrogen in specially designed high-pressure tanks at pressures typically ranging from 350 to 700 bar (5,000 to 10,000 psi) to enhance its energy density while ensuring safe containment. The use of compressed hydrogen storage is driven by its relatively simple design, rapid refueling capabilities, and compatibility with existing hydrogen infrastructure. However, challenges related to material selection, energy consumption for compression, and safety concerns must be addressed for widespread adoption [14].

The working principle of compressed gas storage relies on increasing hydrogen's density by compressing it to high pressures while maintaining it within a robust storage vessel. Since hydrogen's volumetric energy density is low, increasing pressure allows for greater hydrogen storage within a fixed volume. For instance, at 700 bar, hydrogen has an energy density of 5.6 MJ/L, which, although still lower than conventional fuels such as gasoline (~32 MJ/L), is sufficient for practical applications like hydrogen-powered vehicles and backup power systems. Compressed gas storage provides a straightforward and mature solution but requires significant energy input for compression, which can consume up to 15% of the stored hydrogen's energy content. A critical component of compressed hydrogen storage systems is the high-pressure storage tank, which must be engineered to withstand extreme pressures while ensuring long-term durability and safety [15]. Various materials are used in the construction of these tanks, categorized into four main types based on their composition and structural reinforcement:

- Type I Tanks – These are fully metallic tanks made of steel or aluminum, offering high strength but being heavy and less efficient for mobility applications.
- Type II Tanks – Composed of a metallic liner reinforced with a composite overwrap, typically using glass or carbon fibers, improving weight efficiency while maintaining high strength.
- Type III Tanks – Feature a full composite overwrap with a metal liner, providing better strength-to-weight ratio and higher hydrogen storage efficiency.
- Type IV Tanks – Made of a full composite material with a polymer liner, offering lightweight, high-pressure tolerance, and corrosion resistance, making them the preferred choice for automotive and aerospace applications.

Among these, Type III and Type IV tanks are the most widely used for hydrogen fuel cell vehicles (FCVs) and mobile applications due to their superior weight reduction and enhanced pressure handling capabilities. Carbon fiber composites are commonly used in high-pressure tanks due to their exceptional strength-to-weight ratio, fatigue resistance, and non-corrosive properties. These materials ensure safety while maximizing hydrogen storage capacity. However, manufacturing carbon fiber tanks is costly, and further advancements are needed to develop low-cost alternatives with similar performance characteristics [16].

The safety of compressed hydrogen storage is a significant concern, as hydrogen's flammability, high diffusivity, and potential for leakage present unique challenges. The design of storage tanks must adhere to strict safety regulations and testing protocols to mitigate risks associated with overpressure, mechanical failure, and external impact damage. Some key safety considerations include:

a. **Leak Prevention:** Hydrogen's small molecular size makes it highly prone to leakage, which can lead to accidental ignition or explosion if exposed to an ignition source. To counter this, storage tanks are designed with multi-layer sealing mechanisms and high-integrity fittings [17].
b. **Overpressure Protection:** High-pressure storage necessitates the use of pressure relief devices (PRDs) that activate in case of excess pressure buildup, preventing catastrophic failure [18].
c. **Material Fatigue and Durability:** Repeated filling and discharging cycles can lead to material fatigue and microcracks, potentially compromising the tank's integrity. Advanced materials such as carbon composites and metal alloys are designed to withstand long-term mechanical stress and temperature variations [19].
d. **Impact Resistance:** Hydrogen storage tanks must undergo crash and puncture tests to ensure that they can withstand accidents or external impacts without rupture [20].

Despite these safety measures, compressed hydrogen storage still faces several limitations. One of the primary challenges is the energy-intensive nature of hydrogen compression. Compressing hydrogen to 700 bar consumes a significant portion of its stored energy, reducing the overall efficiency of the hydrogen supply chain. Moreover, high-pressure storage tanks add extra weight to vehicles, reducing their overall efficiency and range [21]. Another limitation is the large physical volume required for storage. Even at 700 bar, compressed hydrogen has a lower volumetric energy density than conventional liquid fuels, making it less practical for applications requiring compact energy storage. This is particularly challenging for aviation and maritime applications, where space and weight constraints are critical [22].

To improve the feasibility of compressed gas storage, ongoing research focuses on enhancing material properties, optimizing compression processes, and developing hybrid storage solutions. Innovations such as nano-engineered composite materials could lead to lighter and stronger storage tanks, reducing costs while improving performance. Additionally, integrating adsorbent materials within the tanks, such as MOFs or porous carbon materials, could enable low-pressure hydrogen storage at higher densities [23]. Future developments in compression technology aim to reduce the energy consumption associated with hydrogen pressurization. Electrically driven compressors and adiabatic compression methods are being explored to enhance efficiency. Additionally, hybrid storage solutions that combine compressed gas with other storage methods, such as metal hydrides or LOHCs, could provide a more versatile and efficient hydrogen storage ecosystem. Despite its challenges, compressed gas storage remains a crucial component of the hydrogen economy, particularly for applications where rapid refueling and high hydrogen purity are essential. Hydrogen FCVs, refueling stations, and industrial hydrogen supply chains continue to rely on compressed storage due to its maturity and widespread availability. However, further advancements in materials science, compression technology, and storage system design will be necessary to make compressed hydrogen storage more cost-effective, efficient, and scalable for future energy applications [24]. Figure 4.2 depicts the scheme of compressed hydrogen storage tanks.

4.2.2 Cryogenic Liquid Hydrogen Storage

Hydrogen can be stored in various forms, with cryogenic liquid hydrogen (LH_2) storage being one of the most effective methods for achieving high-energy density. This approach involves cooling hydrogen to extremely low temperatures (20 K or −253 °C), where it transitions into a liquid state. Liquid hydrogen has a significantly higher volumetric energy density than compressed gas storage, making it an attractive option for applications requiring large-scale hydrogen transport, aerospace propulsion, and high-energy-density storage systems. However, the process of liquefying hydrogen is energy-intensive, and maintaining cryogenic conditions presents challenges related to insulation, boil-off losses, and material compatibility [25].

The fundamental principle of cryogenic liquid hydrogen storage is based on achieving and maintaining hydrogen in its liquid state. Hydrogen gas must be cooled down through a multi-stage liquefaction process that involves compression, expansion,

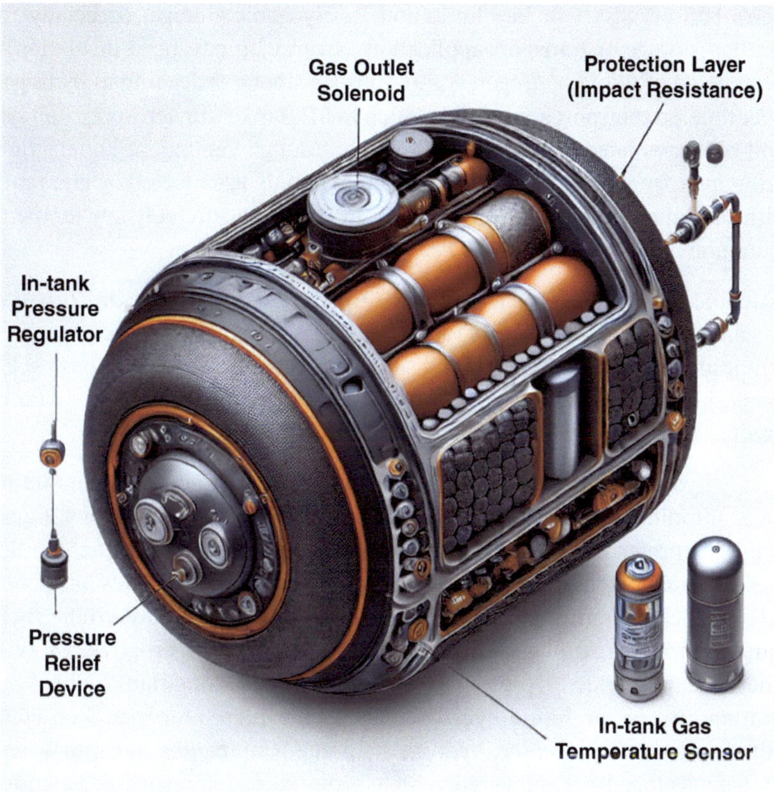

Figure 4.2 Schematic of compressed hydrogen storage tanks. Generated with AI using ChatGPT – OpenAI.

and heat exchange cycles. The Claude or Joule-Thomson expansion cycle is commonly used for hydrogen liquefaction, where high-pressure hydrogen undergoes successive cooling and expansion stages until it reaches its liquid state at 20 K. This process requires approximately 10–13 kWh/kg of energy, which accounts for nearly 30–40% of the energy content of the stored hydrogen. Despite this high energy cost, liquid hydrogen storage remains one of the most viable options for high-capacity storage and transport. A major challenge associated with cryogenic storage is minimizing heat transfer to prevent hydrogen from boiling off and causing evaporation losses. Hydrogen has a low heat capacity and high thermal conductivity, making it prone to heat absorption from the surroundings [26]. To counteract this, advanced insulation techniques are employed, including:

a. **Multi-layer Insulation (MLI):** Composed of reflective foil layers with vacuum gaps, reducing radiative heat transfer [27].
b. **Vacuum-insulated Double-walled Tanks:** These tanks contain an outer vacuum layer, significantly reducing conduction and convection heat losses [28].
c. **Active Cooling Systems:** Some large-scale liquid hydrogen storage facilities use cryocoolers or vented refrigeration systems to maintain stable cryogenic conditions [29].

Hydrogen's boil-off loss is a key limitation in cryogenic storage, especially for long-duration storage or transport applications. Even with advanced insulation, a small percentage of liquid hydrogen continuously evaporates, leading to hydrogen losses over time. In transport applications, such as LH_2 tankers or aerospace fuel systems, boil-off losses must be managed efficiently through venting, re-liquefaction, or utilizing hydrogen for auxiliary power. The materials used for cryogenic tanks must withstand thermal contraction and embrittlement at extremely low temperatures. Common materials include:

- Austenitic stainless steel (e.g., 304L, 316L) – Offers high strength, corrosion resistance, and thermal stability [30].
- Aluminum alloys – Lightweight and suitable for aerospace applications [31].
- Composite materials with polymer insulation – Used in advanced storage tanks to reduce weight and enhance thermal resistance [32].

Cryogenic hydrogen storage is widely utilized in various industries, including aerospace, automotive, and industrial sectors. In aerospace applications, liquid hydrogen is the preferred fuel for space propulsion due to its high specific impulse and clean combustion characteristics. NASA and other space agencies have been using LH_2 in rocket engines such as the Space Launch System and Ariane rockets. Liquid hydrogen also plays a crucial role in future hydrogen-powered aviation, where it can provide zero-emission fuel solutions for long-haul flights [33]. In the automotive sector, liquid hydrogen has been explored for hydrogen FCVs as an alternative to compressed hydrogen storage. Companies like BMW and Toyota have developed LH_2-powered vehicle prototypes, demonstrated extended driving range and fast refueled advantages [34]. However, challenges related to infrastructure, insulation, and refueling station requirements limit widespread adoption. In industrial applications, cryogenic liquid hydrogen is essential for hydrogen refueling stations, metallurgy, and energy storage systems. Large-scale hydrogen transport via LH_2 tankers and pipelines provides an efficient means of hydrogen distribution, especially in regions with limited access to hydrogen production sites. Comparing compressed gas storage and cryogenic liquid storage highlights key differences in pressure, temperature, energy efficiency, and safety considerations, as detailed in Table 4.1.

LH_2 storage offers higher energy density and efficient large-scale transport, but at the cost of higher energy input for liquefaction and the challenge of boil-off losses. On the other hand, compressed gas storage is easier to implement but requires high-pressure tanks with lower volumetric efficiency. To address the challenges of liquefaction energy consumption and storage losses, research is focused on improving cryogenic insulation materials, developing advanced liquefaction technologies, and integrating hybrid storage solutions. Some potential advancements include:

- **Magnetic Refrigeration:** Utilizing magnetocaloric effects to achieve more energy-efficient hydrogen liquefaction [43].
- **Cryo-adsorption Storage:** Incorporating porous materials (e.g., MOFs, aerogels) to enhance hydrogen density at cryogenic temperatures [44].

- **Active Thermal Management:** Employing integrated heat exchangers and smart insulation systems to minimize boil-off losses [45].

Despite these challenges, cryogenic liquid hydrogen storage remains a critical component of the hydrogen economy, especially for aviation, space exploration, and large-scale hydrogen transport. Its high-energy density and efficient distribution potential make it an essential part of future hydrogen infrastructure. However, further advancements in liquefaction efficiency, tank materials, and hybrid storage approaches will be needed to make it a cost-effective and practical solution for widespread adoption. Figure 4.3 depicts the scheme of cryogenic hydrogen storage tanks.

Table 4.1 Comparison of compressed gas and cryogenic liquid storage based on pressure, temperature, energy efficiency, and safety.

Parameter	Compressed gas storage	Cryogenic liquid storage
Pressure	350–700 bar [35]	~1 bar [35]
Temperature	Ambient (~298 K) [36]	20 K (−253 °C) [36]
Energy efficiency	Compression losses (10–15% of stored energy) [37]	Liquefaction losses (30–40% of stored energy) [38]
Volumetric energy density	Lower (5.6 MJ/L at 700 bar) [39]	Higher (~8.5 MJ/L) [40]
Storage tank material	Carbon fiber composites, metal alloys [41]	Stainless steel, aluminum alloys [42]
Safety considerations	Risk of high-pressure leaks and tank rupture	Boil-off losses, cryogenic burns
Common applications	FCVs, industrial storage	Aerospace, long-distance hydrogen transport

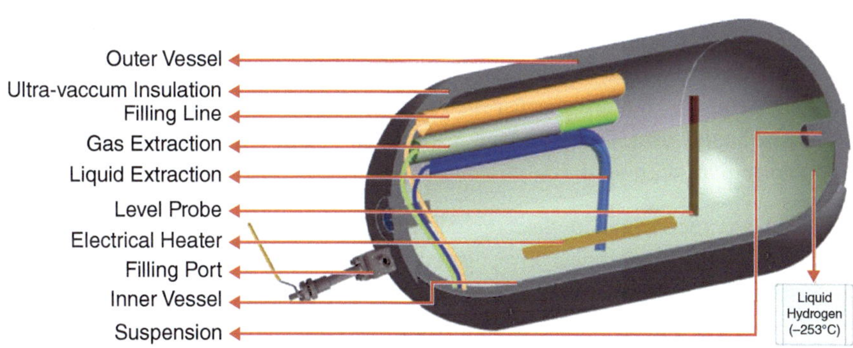

Figure 4.3 Schematic of cryogenic liquid storage tanks. Reproduced with permission from [46]/Elsevier.

4.3 Chemical Hydrogen Storage

4.3.1 Ammonia as a Hydrogen Carrier

Ammonia (NH_3) has emerged as a promising hydrogen carrier due to its high hydrogen content, ease of liquefaction, and well-established infrastructure. As a chemical compound, ammonia contains 17.6% hydrogen by weight, making it a highly efficient medium for hydrogen storage and transport. Unlike compressed or cryogenic hydrogen, ammonia can be liquefied at relatively mild conditions (under 8.6 bar at room temperature or at −33 °C under atmospheric pressure), which makes it easier to handle compared to cryogenic liquid hydrogen. Additionally, ammonia production is already an established industrial process, with large-scale manufacturing and distribution networks in place. One of the biggest advantages of ammonia as a hydrogen carrier is its high-energy density and cost-effective transportability. Compared to compressed hydrogen storage, which requires high-pressure tanks, and cryogenic liquid hydrogen storage, which demands extremely low temperatures, ammonia can be stored and transported efficiently using existing chemical storage infrastructure, such as pressurized tanks and pipelines. Moreover, ammonia is already widely used in the fertilizer, chemical, and refrigeration industries, meaning global supply chains and safety protocols are well developed [47].

To utilize ammonia as a hydrogen carrier, it must be decomposed (cracked) into hydrogen and nitrogen. Ammonia decomposition, also known as ammonia cracking, occurs through catalytic thermal decomposition, where ammonia molecules are broken down into hydrogen (H_2) and nitrogen (N_2) gases at high temperatures (600–900 °C). The reaction is represented as:

$$2NH_3 \rightarrow 3H_2 + N_2 \tag{4.1}$$

This reaction requires high temperatures and efficient catalysts to achieve high hydrogen yields. Various catalytic systems have been developed to optimize this process, including:

- **Nickel-based Catalysts:** Cost-effective but require high temperatures (>800 °C) for efficient conversion [48].
- **Ruthenium-based Catalysts:** Highly active even at lower temperatures (~400–600 °C) but are expensive and scarce [49].
- **Iron-based and Cobalt-based Catalysts:** Being explored as cost-effective alternatives with moderate efficiency [50].

Recent advancements in plasma-assisted, electrochemical, and photocatalytic ammonia cracking are being investigated to lower the energy input required for hydrogen release, improving the feasibility of ammonia as a clean hydrogen carrier. One of the major concerns with using ammonia for hydrogen storage is its toxicity and safety risks. Ammonia is a toxic and corrosive gas with a pungent odor, and exposure to high concentrations can cause respiratory issues and skin irritation. In addition, ammonia is highly soluble in water, which poses environmental risks if leaks occur. Handling ammonia safely requires strict safety regulations, leak detection systems, and proper ventilation [51].

Storage and transportation of ammonia also present logistical and infrastructure challenges. While ammonia can be stored in pressurized or refrigerated tanks, it requires specialized materials to prevent corrosion and leakage. Moreover, current hydrogen refueling stations and fuel cell systems are designed for pure hydrogen gas, meaning ammonia-to-hydrogen conversion technologies must be integrated for seamless adoption. Despite these challenges, ammonia has significant potential for large-scale hydrogen transport and storage, particularly in international hydrogen supply chains. Several countries, including Japan, Australia, and Germany, are investing in green ammonia production from renewable energy sources, ensuring a carbon-neutral hydrogen economy [52]. As shown in Figure 4.4(a), the release of hydrogen from ammonia involves catalytic decomposition, where ammonia molecules break down into hydrogen and nitrogen gases. This hydrogen can then be purified and used in fuel cells or other hydrogen-based applications. Ammonia presents a viable alternative for hydrogen storage and transport, with benefits such as high

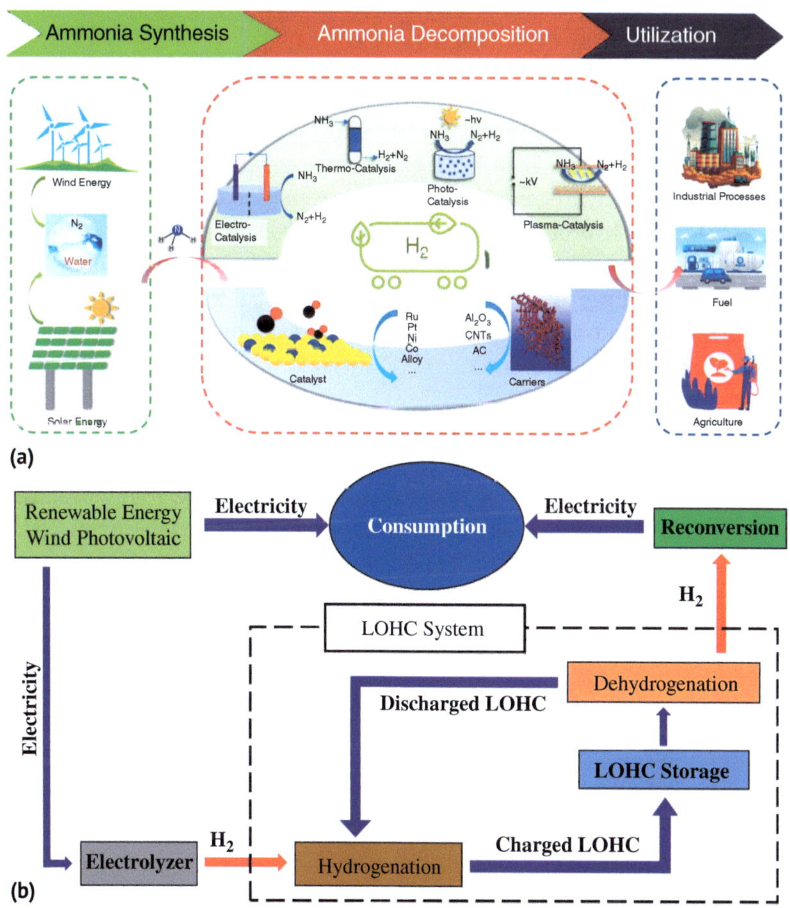

Figure 4.4 Hydrogen release from chemical hydrogen carriers. (a) Release of hydrogen from ammonia through catalytic decomposition. [58]/MDPI/CC BY 4.0 (b) Release of hydrogen from LOHCs via catalytic dehydrogenation.

hydrogen content, existing infrastructure, and cost-effective handling. However, challenges related to toxicity, decomposition efficiency, and compatibility with hydrogen fuel cell technology must be addressed to enable widespread adoption. Continued research into advanced ammonia cracking catalysts, low-temperature decomposition methods, and safe handling techniques will play a crucial role in establishing ammonia as a key player in the global hydrogen economy.

4.3.2 LOHCs

LOHCs have emerged as a promising hydrogen storage solution, offering a safe, efficient, and scalable method for transporting and storing hydrogen. LOHCs are liquid-phase organic compounds that can chemically absorb and release hydrogen through hydrogenation and dehydrogenation reactions. These molecules function as reversible hydrogen carriers, enabling hydrogen storage at ambient conditions without the need for high-pressure tanks or cryogenic cooling. This makes LOHCs a highly attractive option for large-scale hydrogen transport and distribution [53].

One of the key advantages of LOHCs is their ability to store hydrogen in a chemically bound form, reducing the risks associated with hydrogen leakage, flammability, and high-pressure containment. The most commonly studied LOHC systems include toluene/methylcyclohexane (MCH) and benzyl toluene/perhydrobenzyl toluene, both of which offer high hydrogen storage capacities and good chemical stability over multiple hydrogenation and dehydrogenation cycles. The hydrogenation process (loading hydrogen into the LOHC) involves the reaction of hydrogen with an unsaturated organic compound in the presence of a metal catalyst (typically platinum, palladium, or ruthenium-based catalysts) at moderate temperatures (100–200 °C) and pressures (20–50 bar). The general reaction for toluene to methylcyclohexane conversion is as follows:

$$C_7H_8 + 3H_2 \rightarrow C_7H_{14} \tag{4.2}$$

where toluene (C_7H_8) absorbs hydrogen to form methylcyclohexane (MCH, C_7H_{14}). This hydrogenated form can then be stored and transported safely. To retrieve hydrogen from the LOHC, the dehydrogenation process is carried out, where hydrogen is released by applying heat (250–350 °C) and using catalytic dehydrogenation systems. This reaction is the reverse of hydrogenation:

$$C_7H_{14} \rightarrow C_7H_8 + 3H_2 \tag{4.3}$$

As shown in Figure 4.4(b), the release of hydrogen from LOHCs occurs through a catalytic dehydrogenation process, where hydrogen gas is liberated from the liquid carrier and can be used in fuel cells or industrial applications. This reversible nature of LOHCs makes them highly suitable for hydrogen storage and long-distance transport.

Compared to ammonia-based hydrogen storage, LOHCs offer several advantages in terms of safety, ease of handling, and compatibility with existing fuel infrastructure. However, ammonia has a higher hydrogen storage capacity (17.6 wt%) compared to LOHCs (typically 6–7 wt%), meaning that larger volumes of LOHCs are required to store the same amount of hydrogen. Additionally, LOHC systems require efficient catalysts to minimize energy losses during hydrogenation and dehydrogenation. The key differences between ammonia and LOHC-based hydrogen storage are summarized in Table 4.2, highlighting hydrogen capacity, efficiency, and feasibility for large-scale deployment. One of the biggest advantages of LOHC technology is its compatibility with existing liquid fuel infrastructure. Unlike compressed hydrogen, which requires specialized high-pressure tanks, LOHCs can be stored and transported using conventional fuel transport systems, such as oil pipelines, tankers, and refueling stations. This makes LOHCs a highly practical solution for integrating hydrogen storage into current energy supply chains [54, 55].

However, there are still challenges to widespread commercialization of LOHC technology. The energy-intensive dehydrogenation process remains a significant hurdle, as it requires high temperatures (250–350 °C) and efficient catalysts to ensure a high rate of hydrogen release. Additionally, the long-term stability of LOHC molecules must be maintained to prevent chemical degradation and loss of hydrogen storage capacity over multiple cycles. The development of more efficient, low-cost catalysts and alternative LOHC molecules with higher hydrogen storage densities is crucial for improving the feasibility and economic viability of LOHC technology. Despite these challenges, research and industrial investments in LOHCs are increasing, with several pilot projects demonstrating their real-world applicability. Countries like Germany, Japan, and the Netherlands are actively exploring LOHC-based hydrogen transport solutions, with projects aimed at integrating LOHCs into hydrogen refueling stations and large-scale energy storage systems.

Table 4.2 Comparison of ammonia and LOHC-based hydrogen storage.

Parameter	Ammonia (NH_3)	LOHCs
Hydrogen storage capacity (wt%)	17.6%	6–7% [59]
Storage conditions	Stored as liquid under mild pressure (−33 °C at 1 bar)	Stored at ambient temperature and pressure [60]
Hydrogen release process	Catalytic cracking at high temperature (600–800 °C)	Catalytic dehydrogenation at moderate temperature (250–350 °C) [61]
Energy efficiency	Moderate (energy required for cracking)	Moderate (energy required for dehydrogenation) [62]
Safety considerations	Toxic and corrosive, requires strict handling regulations	Safer, non-toxic, and compatible with existing fuel infrastructure [63]
Infrastructure requirements	Requires specialized storage and transport infrastructure	Can be transported using conventional liquid fuel infrastructure [64]
Commercial readiness	Well developed but needs further optimization for efficiency	Emerging technology with ongoing research for efficiency improvements

As catalyst technology improves and renewable hydrogen production expands, LOHCs are expected to play a crucial role in the global hydrogen economy, providing a safe, efficient, and scalable solution for hydrogen transport and storage [56, 57].

4.4 Solid-state Hydrogen Storage

4.4.1 Metal Hydrides for Hydrogen Storage

Metal hydrides have emerged as a promising solution for hydrogen storage due to their ability to reversibly absorb and desorb hydrogen under controlled conditions. These materials consist of metal atoms that chemically bond with hydrogen, forming hydrides with varying stability. Unlike compressed gas or cryogenic liquid storage, metal hydrides offer higher volumetric hydrogen density and safer storage conditions, making them attractive for various applications, including fuel cells and portable energy storage systems. The working principle of metal hydrides relies on the reversible reaction between metal alloys and hydrogen gas, where hydrogen is absorbed into the metal lattice at moderate pressures and released when heat is applied. This process allows for efficient hydrogen storage at lower pressures compared to gaseous hydrogen storage methods [65].

Among the various metal hydrides studied for hydrogen storage, magnesium hydride (MgH_2) is one of the most researched materials due to its high hydrogen storage capacity (up to 7.6 wt%). However, MgH_2 suffers from slow hydrogen absorption/desorption kinetics and requires high temperatures (above 300 °C) for hydrogen release, limiting its practical applications. To address these issues, researchers have explored doping MgH_2 with transition metals such as Ti, Fe, and Nb to improve hydrogen diffusion rates and reduce operating temperatures [66]. Another commonly studied hydride is titanium hydride (TiH_2), which exhibits good hydrogen storage properties but is primarily used in industrial applications rather than as a direct hydrogen carrier due to its relatively low storage capacity [67].

Lanthanum-nickel-based hydrides, such as $LaNi_5H_6$, have also been widely used in hydrogen storage systems. These intermetallic compounds demonstrate fast hydrogen absorption/desorption kinetics at room temperature, making them suitable for low-pressure hydrogen storage applications. However, their hydrogen storage capacity is lower compared to MgH_2, which limits their use in large-scale hydrogen storage solutions [68]. Similarly, sodium alanate ($NaAlH_4$), a complex metal hydride, has attracted attention due to its moderate operating temperatures and improved hydrogen storage characteristics when catalyzed with titanium-based additives. $NaAlH_4$ can reversibly store and release hydrogen under relatively mild conditions, making it a promising candidate for on-board hydrogen storage in FCVs [69].

Despite their advantages, metal hydrides face several challenges that hinder widespread commercialization. One major limitation is their high weight, which reduces the overall gravimetric hydrogen storage capacity of a system. Since many metal hydrides have high atomic weights, they add significant mass to hydrogen storage systems, making them less practical for transportation applications. Additionally, hydrogen uptake and release kinetics remain a critical challenge, as many hydrides require high temperatures and extended reaction times for efficient hydrogen desorption. Researchers are actively investigating ways to enhance reaction kinetics by introducing nanoscale hydrides, catalysts, and composite materials that lower activation energy and improve hydrogen diffusion [70].

Another challenge associated with metal hydrides is their operating temperature range. While some hydrides release hydrogen at ambient temperatures, others, such as MgH_2, require elevated temperatures, necessitating additional energy input. This makes them less suitable for energy-efficient hydrogen storage in mobile applications like FCVs. To address these issues, recent research has focused on developing lightweight, high-capacity metal hydrides and composite hydride systems that combine the advantages of multiple materials to optimize performance. For example, incorporating nanostructured carbon materials like graphene or carbon nanotubes into metal hydrides can enhance hydrogen diffusion, improve thermal conductivity, and reduce operating temperatures. Metal hydrides offer a viable alternative to traditional hydrogen storage methods due to their high volumetric hydrogen density, safety, and reversibility. However, challenges related to weight, hydrogen absorption/desorption kinetics, and operating temperature requirements must be addressed for their widespread adoption. Advances in material engineering, nanotechnology, and catalytic doping are expected to enhance the performance of metal hydrides, bringing them closer to real-world applications in hydrogen-powered vehicles and stationary energy storage systems [71, 72].

4.4.2 MOFs and Carbon-based Materials

MOFs and carbon-based materials have emerged as promising candidates for hydrogen storage due to their high surface areas, tunable porosity, and excellent hydrogen adsorption properties. These materials offer an alternative to conventional hydrogen storage methods such as compressed gas, cryogenic liquid storage, and metal hydrides. MOFs and carbon nanostructures, including graphene, CNTs, and fullerenes, are being extensively studied for their potential to store hydrogen efficiently under moderate conditions, making them attractive for applications in hydrogen-powered vehicles, stationary energy storage, and portable fuel cells [73].

MOFs are crystalline materials composed of metal ions or clusters coordinated to organic linkers, forming highly porous structures with an exceptionally high surface area. This unique structure allows MOFs to adsorb significant amounts of hydrogen through physisorption, which is a process driven by weak van der Waals forces rather than strong chemical bonds. One of the key advantages of MOFs for hydrogen storage is their high tunability, meaning their pore size, surface chemistry, and structural stability can be adjusted by selecting appropriate metal centers and organic linkers. Several MOFs, such as MOF-5, HKUST-1, and MIL-101, have demonstrated excellent hydrogen adsorption capabilities at cryogenic temperatures. However, their hydrogen uptake at room temperature remains a challenge due to weak adsorption energies. To improve their performance, researchers are exploring strategies such as metal doping, functionalization of organic linkers, and hybridizing MOFs with other materials [74].

Carbon-based materials have also gained attention for hydrogen storage due to their lightweight nature, high surface area, and excellent adsorption properties. Graphene, a single layer of carbon atoms arranged in a hexagonal lattice, has been widely studied for hydrogen adsorption. Its unique two-dimensional structure allows for the efficient interaction of hydrogen molecules with its surface. However, like MOFs, graphene primarily relies on physisorption, which limits its hydrogen storage capacity at ambient temperatures. Efforts to enhance hydrogen adsorption on graphene include metal doping, defect engineering, and chemical functionalization to create active sites that improve binding energy. Similarly, CNTs, which are cylindrical structures composed of rolled-up graphene sheets, have shown promising hydrogen adsorption behavior. The adsorption capacity of CNTs depends on factors such as diameter, chirality, and surface modifications. Studies have shown that single-walled carbon nanotubes (SWCNTs) exhibit higher hydrogen storage potential compared to multi-walled carbon nanotubes (MWCNTs) due to their larger internal surface area. Additionally, hydrogen spillover techniques, where hydrogen molecules dissociate on a metal catalyst and migrate onto the carbon surface, have been explored to improve the hydrogen uptake of CNTs. Fullerenes, another class of carbon nanostructures, have also been investigated for hydrogen storage applications. These spherical carbon cages, composed of hexagonal and pentagonal carbon rings, provide a unique architecture for trapping hydrogen molecules. While fullerenes exhibit some degree of hydrogen adsorption, their storage capacity is limited compared to MOFs and CNTs. Researchers are working on chemical modifications, such as hydrogenation and metal doping, to enhance their hydrogen binding energy and improve their overall efficiency [75].

Despite the significant advantages of MOFs and carbon-based materials, several challenges must be addressed before they can be widely adopted for hydrogen storage. One major issue is thermal stability, as some MOFs and carbon structures may degrade under high temperatures or prolonged cycling conditions. Additionally, while these materials offer high hydrogen adsorption capacities at cryogenic temperatures, their storage performance at room temperature remains insufficient for practical applications. Efforts are ongoing to increase adsorption energies through material engineering and hybridization strategies. Another critical challenge is scalability. Although MOFs and carbon nanomaterials exhibit excellent hydrogen storage properties in laboratory settings, large-scale production and integration into commercial hydrogen storage systems remain complex. The synthesis of MOFs can be expensive and time-consuming, requiring precise control over reaction conditions. Carbon-based materials, while more abundant, often require complex processing techniques to achieve the desired structural and adsorption properties [76]. As shown in Figure 4.5, FESEM and TEM images illustrate the structural characteristics of metal hydrides and MOF-based hydrogen storage materials.

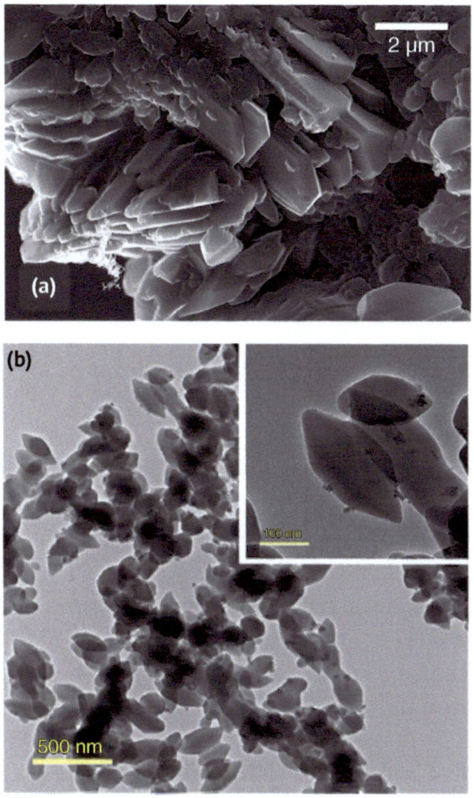

Figure 4.5 (a) FESEM image of MOF-based hydrogen material. Reproduced with permission from [77]/American Chemical Society. (b) TEM Image of MOF-based hydrogen material. Reproduced with permission from [78]/American Chemical Society.

4.5 Safety Considerations and Risk Assessment in Hydrogen Storage

4.5.1 Safety Protocols in Hydrogen Storage Systems

Hydrogen is a promising clean energy carrier, but its safe storage, handling, and transportation are crucial for its widespread adoption. Due to its high flammability, low molecular weight, and ability to permeate materials, strict safety protocols have been established to minimize risks associated with hydrogen storage. Various international regulatory bodies, including the International Organization for Standardization (ISO), the US Department of Energy (DOE), and the European Union (EU), have developed safety guidelines to ensure the secure production, storage, transport, and utilization of hydrogen. Compliance with these standards and best practices is essential to facilitate the safe deployment of hydrogen infrastructure worldwide [79].

4.5.1.1 Overview of Storage Safety Requirements for Different Hydrogen Carriers

Hydrogen can be stored in three primary forms: compressed gas, cryogenic liquid, and chemically bound storage (e.g., ammonia, metal hydrides, and LOHCs). Each method presents unique safety challenges that require specific containment strategies:

- Compressed Hydrogen Storage:
  - Typically stored at 350–700 bar (5,000–10,000 psi) in reinforced carbon fiber or metal tanks.
  - Key risks: High-pressure release, rupture due to structural failure, and hydrogen embrittlement in metallic tanks.
  - Safety measures: Leak detection systems, pressure relief valves, and impact-resistant composite tank materials [80].
- Cryogenic Liquid Hydrogen Storage:
  - Stored at extremely low temperatures (~20 K, or −253 °C) to maintain a liquid state.
  - Key risks: Boil-off losses due to heat leakage, frostbite hazards from liquid hydrogen spills, and rapid expansion upon evaporation.
  - Safety measures: Vacuum-insulated tanks, redundant venting systems, and temperature monitoring sensors [81].
- Chemical Hydrogen Storage (Ammonia, LOHCs, and Metal Hydrides):
  - Hydrogen is chemically bonded to a carrier material, such as NH_3, LOHCs, or metal hydrides.
  - Key risks: Toxicity (ammonia), flammability (LOHCs), and slow release rates (metal hydrides).
  - Safety measures: Chemical containment barriers, controlled catalytic release processes, and closed-loop refueling systems [82].

4.5.1.2 International Regulations and Standards for Hydrogen Handling

To ensure global safety compliance, several organizations and government agencies have developed detailed regulations and guidelines for hydrogen storage, handling, and distribution. These regulations cover design standards, operational safety, emergency response, and material compatibility. Some of the most important regulatory bodies include:

- ISO – Global Hydrogen Safety Standards:
  - ISO 14687: Defines purity standards for hydrogen used in fuel cells.
  - ISO 16110-1: Covers safety in hydrogen generation and storage systems.
  - ISO 19880-1: Specifies safety requirements for hydrogen refueling stations.
  - ISO 9809 & ISO 11114: Provide standards for high-pressure gas cylinders and material compatibility for hydrogen containment [83].
- DOE & National Fire Protection Association (NFPA):
  - DOE Hydrogen Safety Codes (CFR 10 Part 851): Covers safety regulations for hydrogen storage, handling, and infrastructure [84].
  - NFPA 2: Hydrogen Technologies Code: Establishes safety requirements for hydrogen refueling stations and pipeline transport [85].

  - Occupational Safety and Health Administration: Mandates hydrogen workplace safety measures, including protective equipment, ventilation, and spill containment [84].
- EU Hydrogen Safety Framework:
  - EC Directive 2014/94/EU: Defines safety standards for alternative fuel infrastructure, including hydrogen storage.
  - EN 17127:2019: Specifies hydrogen refueling station safety requirements in Europe.
  - ADR (European Agreement on Dangerous Goods Transport): Regulates the safe transportation of hydrogen and hydrogen carriers across EU member states [86].

Compliance with these global standards ensures hydrogen safety, promotes regulatory harmonization, and enables the scalability of hydrogen technologies while mitigating risks.

4.5.1.3 Best Practices for Safe Storage, Transportation, and Refueling Infrastructure

To ensure safe handling and storage of hydrogen, industry best practices have been developed to prevent leaks, explosions, and contamination. These practices apply to hydrogen production plants, fueling stations, transport vehicles, and stationary storage facilities.

1. Safe Storage Design and Maintenance
   - Hydrogen storage tanks and vessels should be made from high-strength materials resistant to embrittlement and stress cracking.
   - Use multiple containment layers (e.g., double-walled cryogenic tanks) to prevent leaks and rupture risks.
   - Employ hydrogen-compatible materials, such as stainless steel, aluminum alloys, and composite polymers, to reduce degradation.
   - Regular pressure testing and leak inspections must be conducted to ensure structural integrity [87].
2. Leak Detection and Monitoring Systems
   - Install hydrogen-specific gas sensors at storage facilities to detect even minor leaks, as hydrogen is odorless and colorless.
   - Use infrared thermal imaging and ultrasonic detectors to monitor storage tanks, pipelines, and transport vessels for potential leaks.
   - Implement automated emergency shut-off valves that isolate leaking storage systems [88].
3. Ventilation and Fire Protection Measures
   - Hydrogen storage sites must have well-ventilated areas to allow for rapid dispersal of leaked gas, reducing the risk of ignition.
   - Equip hydrogen storage rooms and fueling stations with spark-free electrical fixtures to prevent accidental ignition.
   - Install flame arrestors and explosion-proof barriers in hydrogen storage tanks [89].

4. Safe Transportation of Hydrogen
   - Hydrogen transport via pipelines requires high-pressure reinforced materials to prevent permeation and leakage.
   - For liquid hydrogen transport, cryogenic tankers must be insulated and pressure-regulated to minimize boil-off losses.
   - Ensure all transport vehicles comply with DOT (United States), ADR (EU), and ISO standards for hazardous gas transport [90].
5. Hydrogen Refueling Station Safety
   - Implement automated pressure regulation and flow control systems to prevent over-pressurization during refueling.
   - Hydrogen dispensers should be equipped with breakaway couplings to prevent accidental gas leaks.
   - Emergency venting systems and hydrogen flame detectors must be installed at refueling sites to prevent fire hazards [91].

The widespread adoption of hydrogen as a clean energy carrier depends on rigorous safety protocols in storage, transportation, and refueling. International regulations from ISO, DOE, and EU agencies ensure standardized safety measures, while best practices in hydrogen containment, leak detection, ventilation, and fire protection help minimize hazards. The implementation of next-generation safety systems, such as real-time gas monitoring, smart storage technologies, and advanced material engineering, will further improve hydrogen storage safety and enable its scalability in global energy systems.

4.5.2 Leakage Risks and Explosion Hazards

Hydrogen storage and transportation pose significant safety challenges due to the element's high flammability, small molecular size, and tendency to cause material degradation over time. Hydrogen is a colorless, odorless, and highly diffusive gas, making leak detection difficult and increasing the risk of ignition and explosion. These risks are particularly concerning in high-pressure storage systems, pipelines, and metal containment vessels, where permeability and structural weaknesses can lead to gas leaks, embrittlement, and catastrophic failures. Addressing these hazards is critical for the safe expansion of hydrogen as a clean energy carrier [92].

One of the primary safety concerns in hydrogen storage is its flammability and explosion risks, particularly in high-pressure tanks and refueling stations. Hydrogen has a wide flammability range (4–75% in air) and a low ignition energy (~0.02 mJ), meaning even a small leak can create an explosive mixture in the presence of an ignition source. Unlike liquid fuels, hydrogen flames are invisible and difficult to detect, increasing the risk of uncontrolled fires. High-pressure hydrogen storage tanks (350–700 bar) are particularly vulnerable, as ruptures or rapid pressure releases can result in ignition due to electrostatic discharge or contact with hot surfaces. To mitigate these risks, storage facilities use pressure relief valves, explosion-proof materials, and flame arrestors to minimize ignition hazards [93]. Additionally, advanced hydrogen sensors and infrared cameras are implemented in fueling stations and storage sites to detect invisible leaks before they become dangerous. Another major concern in hydrogen storage and distribution is permeability and leakage risks in pipelines and tanks. Hydrogen's small molecular size allows it to diffuse through metal and polymer materials, leading to gradual pressure loss and potential hazards in confined spaces. This phenomenon, known as hydrogen permeation, is particularly problematic in long-distance pipelines and high-pressure containment systems, where even minor leaks can accumulate over time. Unlike conventional fuels, hydrogen has a low density, meaning that leaks disperse quickly in open-air environments but pose significant risks in enclosed spaces such as tunnels, underground storage tanks, and transport vehicles. To reduce permeability risks, advanced multi-layer composite tanks and hydrogen-impermeable coatings (e.g., aluminum, glass, or polymer barriers) are being developed. Additionally, continuous leak monitoring systems and automated shut-off valves are installed in hydrogen infrastructure to detect and prevent gas escape [94].

One of the most serious long-term risks associated with hydrogen storage is hydrogen embrittlement, a phenomenon where hydrogen atoms penetrate metal surfaces, causing structural weakening, cracking, and eventual failure. This issue is particularly problematic in steel pipelines, pressure vessels, and fuel cell components, where exposure to hydrogen over time leads to microstructural changes that reduce mechanical integrity. Hydrogen embrittlement occurs through three primary mechanisms: hydrogen-induced cracking (HIC) [95], hydrogen-enhanced localized plasticity (HELP) [20], and decohesion of grain boundaries. Over time, these effects lead to material brittleness and increased susceptibility to fractures under stress. To mitigate embrittlement risks, researchers are developing hydrogen-resistant alloys such as nickel-based superalloys, high-entropy alloys, and coated steels that prevent hydrogen absorption. Additionally, thermal treatments and surface passivation techniques are employed to reduce metal susceptibility to embrittlement. Leakage risks and explosion hazards remain significant challenges in hydrogen storage and transport, requiring robust safety measures and advanced materials. High-pressure storage, permeation issues, and hydrogen embrittlement must be addressed through continuous monitoring, improved containment materials, and innovative engineering solutions. By integrating real-time leak detection systems, explosion prevention technologies, and hydrogen-resistant materials, the safe deployment of hydrogen as a global energy solution can be ensured [96].

4.5.3 Mitigation Strategies for Hydrogen Storage Hazards

Ensuring the safe storage and handling of hydrogen requires a combination of advanced detection systems, engineered safety mechanisms, and innovative materials. Given the high flammability, leakage risks, and explosion potential of hydrogen, various mitigation strategies have been developed to minimize hazards and improve long-term safety. One of the most effective approaches is the use of advanced hydrogen sensors and leak detection technologies, which provide real-time monitoring of hydrogen levels in storage and refueling systems. These sensors, based on electrochemical, thermal conductivity, and optical detection principles, can detect even trace amounts of hydrogen leaks and trigger automated safety responses before reaching flammable concentrations. As shown in Figure 4.6, modern hydrogen

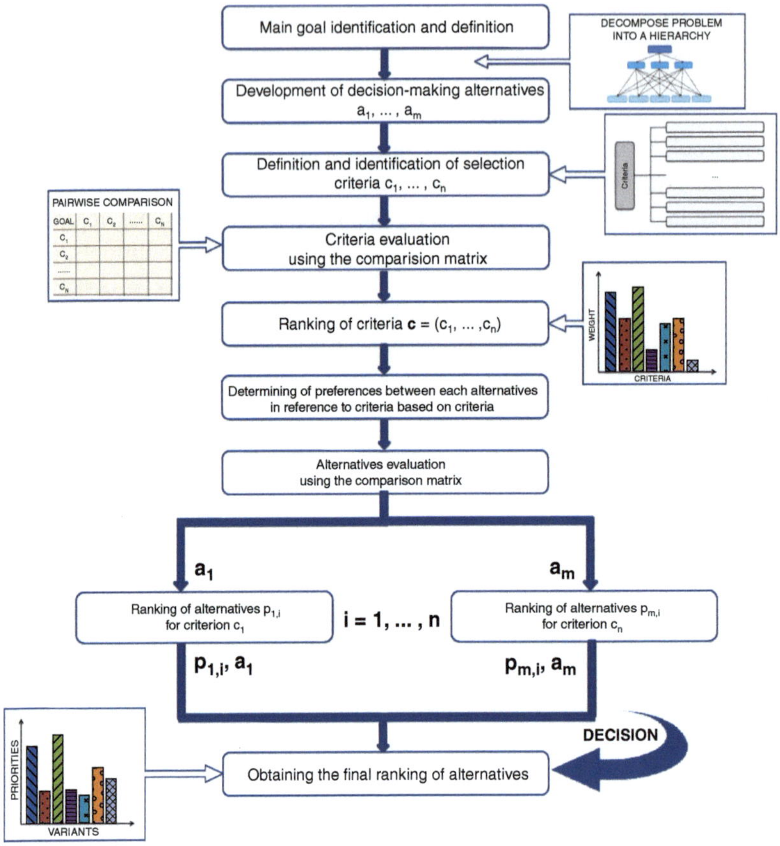

Figure 4.6 Hydrogen storage safety mechanisms and risk mitigation strategies. [99]/MDPI/CC BY 4.0.

infrastructure integrates infrared imaging and ultrasonic gas detection to enhance early leak identification and response. In addition to detection systems, passive and active safety mechanisms play a crucial role in mitigating risks. Passive safety measures include venting systems, PRDs, and explosion-resistant barriers, which help prevent overpressure buildup in hydrogen tanks. Meanwhile, active safety mechanisms such as emergency shut-off valves and automated containment systems are designed to immediately isolate leaks and prevent catastrophic failures. Many hydrogen storage sites now feature integrated suppression systems that deploy inert gas flooding or water misting to quench hydrogen flames in the event of ignition [97].

Material innovations have also significantly enhanced hydrogen storage safety. Researchers have developed composite tanks with reinforced carbon fiber layers, which offer high structural integrity and resistance to hydrogen embrittlement. Additionally, nano-coatings and self-sealing materials are being explored to reduce hydrogen permeability and prevent microscopic leaks in storage vessels. As summarized in Table 4.3, these advanced materials outperform traditional storage methods in terms of durability, leakage resistance, and operational lifespan, making them essential for next-generation hydrogen infrastructure. By

Table 4.3 Comparative safety assessment of compressed, cryogenic, and chemical hydrogen storage methods.

Storage method	Primary safety risks	Mitigation strategies	Overall safety rating
Compressed gas storage (350–700 bar)	High-pressure rupture, hydrogen embrittlement, slow leaks due to material permeability	Carbon fiber composite tanks, automated shut-off valves, real-time pressure monitoring [100]	★★★★☆ (High)
Cryogenic liquid hydrogen storage (~20K)	Boil-off losses, frostbite hazards, rapid expansion upon leaks	Vacuum-insulated tanks, redundant venting systems, temperature sensors [101]	★★★☆☆ (Moderate)
Ammonia-based hydrogen storage	Toxicity, risk of ammonia leaks, energy-intensive cracking process	Chemical-resistant containment, controlled catalytic release, ammonia sensors [102]	★★★★☆ (High)
LOHCs	Flammability, slow hydrogen release kinetics, high operating temperatures	Thermal management systems, flame-resistant storage, catalytic hydrogenation optimization [103]	★★★☆☆ (Moderate)
Metal hydride storage	Slow hydrogen absorption/desorption, weight limitations	Nano-engineered hydrides, heat-assisted desorption, smart thermal regulation [104]	★★★★☆ (High)

integrating sensor-based detection, robust safety mechanisms, and innovative materials, the risks associated with hydrogen storage and transportation can be significantly reduced, paving the way for safe and large-scale hydrogen adoption in the energy sector [98].

4.6 Future Directions in Hydrogen Storage

4.6.1 High-density Storage Solutions for Transportation and Stationary Applications

The need for high-density hydrogen storage is paramount for the success of hydrogen-powered vehicles and stationary power systems. Current storage technologies, such as compressed gas, cryogenic liquid hydrogen, and solid-state materials, still face limitations in terms of their energy density, efficiency, and cost-effectiveness. To make hydrogen storage more viable, the focus is shifting towards developing high-density solutions that can store greater amounts of hydrogen in smaller volumes [105].

- Development of advanced storage materials: Materials such as advanced metal hydrides, complex MOFs, and nanostructured carbon materials are being explored to significantly increase storage capacity. These materials aim to enhance hydrogen storage by improving their gravimetric and volumetric energy density.
- Improved pressure tanks: Innovations in carbon fiber composites and lightweight metals are expected to yield tanks with higher pressure tolerances, increasing storage capacity while reducing weight for hydrogen transportation.
- Cryogenic storage: As a potential solution for large-scale storage, cryogenic liquid hydrogen systems are being refined to optimize the energy needed for liquefaction and to enhance storage efficiency.

4.6.2 Integration of Hydrogen Storage with Fuel Cells and Renewable Energy Systems

As hydrogen emerges as a key energy carrier for future energy systems, its integration with fuel cells and renewable energy sources (solar, wind, etc.) is seen as essential for a sustainable hydrogen economy. Hydrogen can be produced through electrolysis powered by renewable energy, stored, and then used in fuel cells to generate electricity, offering a clean, sustainable energy cycle. The challenge lies in developing efficient storage systems that facilitate smooth integration with these technologies [106, 107].

- Decentralized hydrogen production and storage: Combining hydrogen storage with distributed renewable energy systems will enable localized energy storage and consumption, reducing the need for extensive transmission infrastructure.
- Hydrogen as a buffer: In renewable energy systems with variable output, such as solar and wind, hydrogen can be stored when there is excess production and then used when demand or generation is low. Efficient storage technologies that can handle this cyclical nature are crucial to the success of this model.
- Fuel cell integration: Advancements in low-temperature fuel cells (such as proton exchange membrane fuel cells) that use hydrogen directly as a fuel source require highly efficient storage solutions that ensure easy, safe, and effective transport and distribution of hydrogen.

4.6.3 Emerging Research in Hybrid Hydrogen Storage Combining Physical, Chemical, and Solid-state Approaches

Hybrid hydrogen storage solutions, which combine the benefits of physical, chemical, and solid-state storage methods, are gaining attention for their potential to overcome the limitations of individual technologies. These approaches aim to combine high-density storage with easy release and regeneration of hydrogen. By merging the advantages of different methods, it is hoped that storage systems can be developed that are both high-capacity and efficient across various operating conditions [108].

- Physical-chemical hybrid systems: Combining the high-energy density of physical storage (e.g., compressed hydrogen) with the chemical storage potential of ammonia or LOHCs could provide a more versatile approach. This could enable the storage of hydrogen under moderate pressure conditions while taking advantage of the higher volumetric energy densities of chemical storage.
- Solid-state hybrid systems: Solid-state materials like metal hydrides could be paired with chemical carriers or MOFs to create hybrid systems with higher storage capacities and improved thermodynamic properties for hydrogen uptake and release.
- Multi-mode systems: New systems may integrate different types of storage (e.g., liquid hydrogen storage with solid-state hybrid systems) to create multi-mode storage units, where hydrogen can be efficiently stored, transported, and released across various phases.

4.6.4 Economic and Policy Considerations for Large-scale Hydrogen Storage Deployment

While technological advancements are critical, economic and policy frameworks will play a significant role in the widespread adoption of hydrogen storage solutions. For hydrogen storage to become a key part of the global hydrogen economy, it is necessary to address the cost-effectiveness of storage systems, incentivize innovation, and implement policies that support large-scale deployment [109, 110].

- **Cost Reduction:** Large-scale adoption of hydrogen storage technologies will require substantial cost reductions across materials, manufacturing processes, and infrastructure. Policies that promote economies of scale for storage systems and manufacturing incentives for innovative storage technologies will be important in making these systems more affordable and commercially viable.
- **Infrastructure Development:** Building a comprehensive hydrogen storage infrastructure that is both scalable and adaptable to different applications (e.g., transportation, power generation, and industrial use) is crucial. Investment in storage-related infrastructure, such as distribution pipelines and refueling stations, will enable seamless hydrogen adoption.
- **Regulatory and Safety Standards:** The safety of hydrogen storage and the technologies used in hydrogen systems is a key concern. Regulatory bodies must establish safety standards and guidelines for storage and transportation, ensuring public confidence in hydrogen as a viable energy source.
- **Policy Incentives:** Governments will need to subsidize or incentivize research and development in hydrogen storage technologies. Policy measures such as tax breaks for companies investing in hydrogen infrastructure, grants for research into storage technologies, and carbon pricing will help accelerate hydrogen storage advancements.

As hydrogen storage continues to evolve, the integration of advanced materials and hybrid technologies will be key to achieving efficient, cost-effective, and scalable hydrogen storage solutions. A strong focus on safety, economic feasibility, and policy development will ensure that hydrogen storage technologies are prepared for large-scale deployment in transportation, stationary power, and other sectors, contributing to a sustainable and carbon-neutral hydrogen economy.

4.7 Summary and Future Perspectives

Hydrogen storage is a fundamental aspect of realizing a hydrogen economy, enabling hydrogen to be safely stored, transported, and utilized across various sectors. The main hydrogen storage methods include physical, chemical, and solid-state storage, each with unique advantages and challenges.

Physical Storage: This method involves storing hydrogen as a gas at high pressures or as a liquid at cryogenic temperatures. Compressed gas storage involves pressurizing hydrogen to 350–700 bar, with materials like carbon fiber composites and metal alloys used for tank construction. Cryogenic liquid hydrogen storage involves cooling hydrogen to extremely low temperatures (around 20K) to achieve higher density, often used in aerospace and industrial sectors. Both approaches face energy efficiency challenges and infrastructure limitations but remain essential for large-scale storage and transportation.

Chemical Storage: Chemical methods, such as ammonia and LOHCs, allow hydrogen to be stored in chemical compounds and released through catalytic processes. Ammonia offers a high volumetric energy density and a well-established infrastructure for storage and transportation. However, ammonia's toxicity and storage challenges need to be addressed. LOHCs offer an alternative for safe, reversible hydrogen storage with minimal energy loss. However, the catalytic processes involved still need optimization for commercial viability.

Solid-state Storage: This includes the use of materials like metal hydrides and MOFs, which store hydrogen in solid forms. Metal hydrides, such as MgH_2 and $LaNi_5H_6$, absorb hydrogen gas and release it under specific conditions, but they face challenges related to hydrogen uptake and release kinetics and operating temperatures. MOFs and carbon-based nanostructures provide high surface areas and efficient hydrogen adsorption but still face scalability and cost challenges.

Technological advancements, such as new materials with higher energy density, better stability, and reduced costs, are being actively researched. However, challenges remain in improving the efficiency of hydrogen uptake and release, addressing safety concerns, and scaling these technologies for mass deployment.

The future of hydrogen storage relies heavily on developing new materials, technologies, and infrastructure to support a sustainable hydrogen economy. Several promising avenues are being explored, including:

Breakthroughs in Hydrogen Storage Materials: Future research will likely focus on developing advanced materials that combine the benefits of physical, chemical, and solid-state storage methods. Key materials being researched include MOFs, nanostructured carbons, and novel metal hydrides. These materials aim to achieve high hydrogen storage densities, rapid hydrogen uptake and release kinetics, and improved thermal stability for practical applications. Hybrid systems combining the strengths of different storage methods could provide a versatile solution for different applications (e.g., transportation, stationary energy storage).

Optimization of Catalytic Processes: In chemical storage methods like LOHCs and ammonia storage, improving the efficiency of hydrogen release through catalytic hydrogenation and dehydrogenation processes is critical. Advanced catalysts, especially those based on non-precious metals or hybrid materials, could offer more cost-effective and sustainable solutions compared to conventional precious metal-based catalysts. Additionally, ammonia decomposition technologies that enable efficient release of hydrogen will likely play a significant role in advancing this storage method.

Storage for Specific Applications: Different storage technologies will have unique applications. For transportation, high-density, low-weight storage systems like compressed gas and cryogenic liquid storage will continue to be important, while solid-state storage solutions are more suitable for stationary applications such as energy storage and grid stabilization. Future research may focus on tailoring storage technologies to fit specific use cases and improving their integration with hydrogen production and fuel cell technologies.

Commercialization and Scalability: For large-scale deployment, hydrogen storage technologies need to be cost-effective, scalable, and integrated into existing infrastructure. The economics of hydrogen storage will depend on material costs, energy efficiency, and the ability to build and maintain storage infrastructure such as refueling stations, pipelines, and storage facilities. Public and private investments, along with policy incentives and government support, will play a key role in accelerating the commercial adoption of hydrogen storage solutions.

Role in the Future Hydrogen Economy: Hydrogen storage will be a critical enabler of the future hydrogen economy, which will rely on hydrogen as a clean, sustainable energy carrier. Efficient and cost-effective storage solutions will ensure that hydrogen can be produced when renewable energy sources (solar, wind) generate excess electricity, stored for future use, and transported over long distances to meet global energy demand. Hydrogen storage will also facilitate the integration of hydrogen with fuel cell technologies for zero-emission transportation and industrial processes. The development of a global hydrogen infrastructure is crucial to the widespread adoption of hydrogen technologies.

In conclusion, the continued development of hydrogen storage technologies, alongside advancements in materials science, catalysis, and system integration, will pave the way for a cleaner and more sustainable energy future. Ongoing research efforts and innovations are likely to address the current limitations and unlock the full potential of hydrogen as a cornerstone of the global energy transition.

References

1 Ma N, Zhao W, Wang W, Li X, Zhou H. Large scale of green hydrogen storage: Opportunities and challenges. *Int J Hydrogen Energy*. 2024 Jan 2;50:379–96.

2 Zhang YH, Jia ZC, Yuan ZM, Yang T, Qi Y, Zhao DL. Development and application of hydrogen storage. *J Iron Steel Res Int* [Internet]. 2015 Sep 1 [cited 2025 Mar 12]; 22(9):757–70. Available from: https://link.springer.com/article/10.1016/S1006-706X(15)30069-8.

3 Schüth F. Challenges in hydrogen storage. *Eur Phys J Spec Top* [Internet]. 2009 Sep 15 [cited 2025 Mar 12];176(1):155–66. Available from: https://link.springer.com/article/10.1140/epjst/e2009-01155-x.

4 Usman MR. Hydrogen storage methods: Review and current status. *Renew Sustain Energy Rev*. 2022 Oct 1;167:112743.

5 Zohuri B. Cryogenics and liquid hydrogen storage. *Hydrogen Energy* [Internet]. 2019 [cited 2025 Mar 12];121–39. Available from: https://link.springer.com/chapter/10.1007/978-3-319-93461-7_4.

6 Chu C, Wu K, Luo B, Cao Q, Zhang H. Hydrogen storage by liquid organic hydrogen carriers: Catalyst, renewable carrier, and technology—A review. *Carbon Resour Conv*. 2023 Dec 1;6(4):334–51.

7 Xu Y, Zhou Y, Li Y, Ding Z. Research progress and application prospects of solid-state hydrogen storage technology. *Molecules* [Internet]. 2024 Apr 12 [cited 2025 Mar 12];29(8):1767. Available from: https://www.mdpi.com/1420-3049/29/8/1767/htm.

8 Murray LJ, Dinc M, Long JR. Hydrogen storage in metal–organic frameworks. *Chem Soc Rev* [Internet]. 2009 Apr 21 [cited 2025 Mar 12];38(5):1294–314. Available from: https://pubs.rsc.org/en/content/articlehtml/2009/cs/b802256a.

9 Biris AR, Lupu D, Dervishi E, Li Z, Saini V, Saini D. Hydrogen storage in carbon-based nanostructured materials. *Particul Sci Technol* [Internet]. 2008 Jul [cited 2025 Mar 12];26(4):297–305. Available from: https://www.tandfonline.com/doi/abs/10.1080/02726350802084051.

10 Felderhoff M, Weidenthaler C, Von Helmolt R, Eberle U. Hydrogen storage: The remaining scientific and technological challenges. *Phys Chem Chem Phys* [Internet]. 2007 May 23 [cited 2025 Mar 12];9(21):2643–53. Available from: https://pubs.rsc.org/en/content/articlehtml/2007/cp/b701563c.

11 Epelle EI, Obande W, Udourioh GA, Afolabi IC, Desongu KS, Orivri U. Perspectives and prospects of underground hydrogen storage and natural hydrogen. *Sustain Energy Fuels* [Internet]. 2022 Jul 12 [cited 2025 Mar 12]; 6(14):3324–43. Available from: https://pubs.rsc.org/en/content/articlehtml/2022/se/d2se00618a.

12 Zhao L, Xu F, Zhang C, Wang Z, Ju H, Gao X. Enhanced hydrogen storage of alanates: Recent progress and future perspectives. *Prog Nat Sci Mater Int*. 2021 Apr 1;31(2):165–79.

13 Mulky L, Srivastava S, Lakshmi T, Sandadi ER, Gour S, Thomas NA. An overview of hydrogen storage technologies—Key challenges and opportunities. *Mater Chem Phys*. 2024 Oct 1;325:129710.

14 Boretti A. Technology readiness level of hydrogen storage technologies for transport. *Energy Storage* [Internet]. 2024 Feb 1 [cited 2025 Mar 12];6(1):e546. Available from: https://onlinelibrary.wiley.com/doi/full/10.1002/est2.546.

15 Orlova S, Mezeckis N, Vasudev VPK. Compression of hydrogen gas for energy storage: A review. *Latv J Phys Tech Sci*. 2023;60(2):4–16.

16 Ho Nguyen D, Hoon Kim J, To Nguyen Vo T, Kim N, Seon Ahn H. Design of portable hydrogen tank using adsorption material as storage media: An alternative to Type IV compressed tank. *Appl Energy*. 2022 Mar 15;310:118552.

17 Alssalehin E, Holborn P, Pilidis P. Assessment of hydrogen storage and pipelines for hydrogen farm. *Energies* [Internet]. 2025 Feb 27 [cited 2025 Mar 13]; 18(5):1167. Available from: https://www.mdpi.com/1996-1073/18/5/1167/htm.

18 Zheng J, Liu X, Xu P, Liu P, Zhao Y, Yang J. Development of high pressure gaseous hydrogen storage technologies. *Int J Hydrogen Energy*. 2012 Jan 1;37(1):1048–57.

19 Moradi R, Groth KM. Hydrogen storage and delivery: Review of the state of the art technologies and risk and reliability analysis. *Int J Hydrogen Energy*. 2019 May 3;44(23):12254–69.

20 Li H, Cao X, Liu Y, Shao Y, Nan Z, Teng L. Safety of hydrogen storage and transportation: An overview on mechanisms, techniques, and challenges. *Energy Rep*. 2022 Nov 1;8:6258–69.

21 Huot J, Lu ZH, Franco A, Giovannini C. Hydrogen gas compression for efficient storage: Balancing energy and increasing density. *Hydrogen* [Internet]. 2024 May 25 [cited 2025 Mar 13];5(2):293–311. Available from: https://www.mdpi.com/2673-4141/5/2/17/htm.

22 Bosu S, Rajamohan N. Recent advancements in hydrogen storage - Comparative review on methods, operating conditions and challenges. *Int J Hydrogen Energy*. 2024 Jan 2;52:352–70.

23 Makridis S. Hydrogen storage and compression. *Methane Hydrogen Energy Storage* [Internet]. 2017 Feb 15 [cited 2025 Mar 13];1(1):1–28. Available from: http://arxiv.org/abs/1702.06015.

24 Sadeq AM, Homod RZ, Hussein AK, Togun H, Mahmoodi A, Isleem HF. Hydrogen energy systems: Technologies, trends, and future prospects. *Sci Total Environ*. 2024 Aug 20;939:173622.

25 Qiu Y, Yang H, Tong L, Wang L, Arrigoni M, Amirkhiz S. Research progress of cryogenic materials for storage and transportation of liquid hydrogen. *Metals* [Internet]. 2021 Jul 10 [cited 2025 Mar 13];11(7):1101. Available from: https://www.mdpi.com/2075-4701/11/7/1101/htm.

26 Van Sciver SW. Cryogenic principles and applications. In: *Helium Cryogenics* [Internet]. New York, NY: Springer. 2012 [cited 2025 Mar 13];1–16. Available from: https://link.springer.com/chapter/10.1007/978-1-4419-9979-5_1.

27 Camplese D, Chianese C, Scarponi GE, Eberwein R, Otremba F, Cozzani V. Analysis of high temperature degradation of multi-layer insulation (MLI) systems for liquid hydrogen storage tanks. *Chem Eng Trans* [Internet]. 2024 [cited 2025 Mar 13];99:415–20. Available from: https://cris.unibo.it/handle/11585/960952.

28 Mzad H, Haouam A. Optimization approach of insulation thickness of non-vacuum cryogenic storage tank. *Prog Supercond Cryog* [Internet]. 2020 Mar 1 [cited 2025 Mar 13];22(1):17–23. Available from: https://doi.org/10.9714/psac.2020.22.1.017.

29 Wan C, Zhu S, Shi C, Bao S, Zhi X, Qiu L. Numerical simulation on pressure evolution process of liquid hydrogen storage tank with active cryogenic cooling. *Int J Refrig*. 2023 Jun 1;150:47–58.

30 Kim MS, Lee T, Son Y, Park J, Kim M, Eun H. Metallic material evaluation of liquid hydrogen storage tank for marine application using a tensile cryostat for 20 K and electrochemical cell. *Processes* [Internet]. 2022 Nov 15 [cited 2025 Mar 13];10(11): 2401. Available from: https://www.mdpi.com/2227-9717/10/11/2401/htm.

31 Salahi F, Zarei-Jelyani F, Rahimpour MR. Cryogenic liquid tankers for hydrogen transportation. *Hydrogen Transportation and Storage* [Internet]. CRC Press. 2024 Jan 1 [cited 2025 Mar 13];200–12. Available from: https://www.taylorfrancis.com/chapters/edit/10.1201/9781003382553-12/cryogenic-liquid-tankers-hydrogen-transportation-fatemeh-salahi-fatemeh-zarei-jelyani-mohammad-reza-rahimpour.

32 Yin L, Yang H, Ju Y. Review on the key technologies and future development of insulation structure for liquid hydrogen storage tanks. *Int J Hydrogen Energy*. 2024 Feb 29;57:1302–15.

33 Fesmire J, Swanger A, Jacobson J, Notardonato W. Energy efficient large-scale storage of liquid hydrogen. *IOP Conf Ser Mater Sci Eng* [Internet]. 2022 May 1 [cited 2025 Mar 13];1240(1):012088. Available from: https://iopscience.iop.org/article/10.1088/1757-899X/1240/1/012088.

34 Arnold G, Wolf J. Liquid hydrogen for automotive application next generation fuel for FC and ICE vehicles. *Teion Kogaku*. 2005;40(6):221–30.

35 Petitpas G, Aceves S. Hydrogen storage in pressure vessels: Liquid, cryogenic, and compressed gas. In: *Hydrogen Storage Technology: Materials and Applications* [Internet]. CRC Press. 2016 Apr 21 [cited 2025 Mar 13];91–108. Available from: https://www.taylorfrancis.com/chapters/edit/10.1201/b13685-10/hydrogen-storage-pressure-vessels-liquid-cryogenic-compressed-gas-guillaume-petitpas-salvador-aceves.

36 Yanxing Z, Maoqiong G, Yuan Z, Xueqiang D, Jun S. Thermodynamics analysis of hydrogen storage based on compressed gaseous hydrogen, liquid hydrogen and cryo-compressed hydrogen. *Int J Hydrogen Energy*. 2019 Jun 21;44(31):16833–40.

37 Dindorf R. Study of the energy efficiency of compressed air storage tanks. *Sustainability* [Internet]. 2024 Feb 17 [cited 2025 Mar 13];16(4):1664. Available from: https://www.mdpi.com/2071-1050/16/4/1664/htm.

38 Aziz M. Liquid hydrogen: A review on liquefaction, storage, transportation, and safety. *Energies* [Internet]. 2021 Sep 17 [cited 2025 Mar 13];14(18):5917. Available from: https://www.mdpi.com/1996-1073/14/18/5917/htm.

39 Odukomaiya A, Abu-Heiba A, Gluesenkamp KR, Abdelaziz O, Jackson RK, Daniel C. Thermal analysis of near-isothermal compressed gas energy storage system. *Appl Energy*. 2016 Oct 1;179:948–60.

40 Aceves SM, Espinosa-Loza F, Ledesma-Orozco E, Ross TO, Weisberg AH, Brunner TC. High-density automotive hydrogen storage with cryogenic capable pressure vessels. *Int J Hydrogen Energy*. 2010 Feb 1;35(3):1219–26.

41 Compressed Gas Association Inc. *Handbook of Compressed Gases*. Singapore: Springer Nature. 2012.

42 Senthil Kumar S, Bibin C, Ramachandran M. Design and analysis of hydrogen storage tank with different materials by ansys. *IOP Conf Ser Mater Sci Eng* [Internet]. 2020 Mar 1 [cited 2025 Mar 13];810(1):012016. Available from: https://iopscience.iop.org/article/10.1088/1757-899X/810/1/012016.

43 Numazawa T, Kamiya K, Utaki T, Matsumoto K. Magnetic refrigerator for hydrogen liquefaction. *Cryogenics* (Guildf). 2014 Jul 1;62:185–92.

44 Petitpas G, Bénard P, Klebanoff LE, Xiao J, Aceves S. A comparative analysis of the cryo-compression and cryo-adsorption hydrogen storage methods. *Int J Hydrogen Energy*. 2014 Jul 3;39(20):10564–84.

45 Shu QS, Demko JA, Fesmire JE. Heat switch technology for cryogenic thermal management. *IOP Conf Ser Mater Sci Eng* [Internet]. 2017 Dec 1 [cited 2025 Mar 13];278(1):012133. Available from: https://iopscience.iop.org/article/10.1088/1757-899X/278/1/012133.

46 Sharma P, Burolia AK, Mandal A, Neogi S. Commercially available resources for physical hydrogen storage and distribution. In: *Towards Hydrogen Infrastructure: Advances and Challenges in Preparing for the Hydrogen Economy*. Amsterdam: Elsevier. 2024 Jan 1;225–56.

47 Klerke A, Christensen CH, Nørskov JK, Vegge T. Ammonia for hydrogen storage: Challenges and opportunities. *J Mater Chem* [Internet]. 2008 May 7 [cited 2025 Mar 13];18(20):2304–10. Available from: https://pubs.rsc.org/en/content/articlehtml/2008/jm/b720020j.

48 Lee YJ, Cha J, Kwak Y, Park Y, Jo YS, Jeong H. Top-down syntheses of nickel-based structured catalysts for hydrogen production from ammonia. *ACS Appl Mater Interfaces* [Internet]. 2021 Jan 13 [cited 2025 Mar 13];13(1):597–607. Available from: https://pubs.acs.org/doi/abs/10.1021/acsami.0c18454.

49 Su Z, Guan J, Liu Y, Shi D, Wu Q, Chen K. Research progress of ruthenium-based catalysts for hydrogen production from ammonia decomposition. *Int J Hydrogen Energy*. 2024 Jan 2;51:1019–43.

50 Zheng C, Guan B, Guo J, Su T, Zhou J, Chen J. Research status, optimization strategies, and future prospects of ammonia decomposition catalysts for COx-free hydrogen. *Ind Eng Chem Res* [Internet]. 2023 Jul 26 [cited 2025 Mar 13];62(29):11305–36. Available from: https://pubs.acs.org/doi/abs/10.1021/acs.iecr.3c01261.

51 Fang W, Ding C, Chen L, Zhou W, Wang J, Huang K. Review of hydrogen storage technologies and the crucial role of environmentally friendly carriers. *Energy Fuels* [Internet]. 2024 Aug 1 [cited 2025 Mar 13];38(15):13539–64. Available from: https://pubs.acs.org/doi/abs/10.1021/acs.energyfuels.4c01781.

52 Negro V, Noussan M, Chiaramonti D. The potential role of ammonia for hydrogen storage and transport: A critical review of challenges and opportunities. *Energies* [Internet]. 2023 Aug 25 [cited 2025 Mar 13];16(17):6192. Available from: https://www.mdpi.com/1996-1073/16/17/6192/htm.

53 Preuster P, Papp C, Wasserscheid P. Liquid organic hydrogen carriers (LOHCs): Toward a hydrogen-free hydrogen economy. *Acc Chem Res* [Internet]. 2017 Jan 17 [cited 2025 Mar 13];50(1):74–85. Available from: https://pubs.acs.org/doi/abs/10.1021/acs.accounts.6b00474.

54 Niermann M, Beckendorff A, Kaltschmitt M, Bonhoff K. Liquid organic hydrogen carrier (LOHC)—Assessment based on chemical and economic properties. *Int J Hydrogen Energy*. 2019 Mar 8;44(13):6631–54.

55 Rao PC, Yoon M. Potential liquid-organic hydrogen carrier (LOHC) systems: A review on recent progress. *Energies* [Internet]. 2020 Nov 19 [cited 2025 Mar 13];13(22):6040. Available from: https://www.mdpi.com/1996-1073/13/22/6040/htm.

56 Lin A, Bagnato G. Revolutionising energy storage: The latest breakthrough in liquid organic hydrogen carriers. *Int J Hydrogen Energy*. 2024 Apr 18;63:315–29.

57 Markiewicz M, Zhang YQ, Bösmann A, Brückner N, Thöming J, Wasserscheid P. Environmental and health impact assessment of liquid organic hydrogen carrier (LOHC) systems—Challenges and preliminary results. *Energy Environ Sci* [Internet]. 2015 Mar 5 [cited 2025 Mar 13];8(3):1035–45. Available from: https://pubs.rsc.org/en/content/articlehtml/2015/ee/c4ee03528c.

58 Huang X, Lei K, Mi Y, Fang W, Li X. Recent Progress on hydrogen production from ammonia decomposition: Technical roadmap and catalytic mechanism. *Molecules* [Internet]. 2023 Jul 6 [cited 2025 Mar 12];28(13):5245. Available from: https://www.mdpi.com/1420-3049/28/13/5245/htm.

59 Byun M, Lee A, Cheon S, Kim H, Lim H. Preliminary feasibility study for hydrogen storage using several promising liquid organic hydrogen carriers: Technical, economic, and environmental perspectives. *Energy Convers Manag*. 2022 Sep 15;268:116001.

60 Teichmann D, Arlt W, Wasserscheid P, Freymann R. A future energy supply based on liquid organic hydrogen carriers (LOHC). *Energy Environ Sci* [Internet]. 2011 Aug 2 [cited 2025 Mar 13];4(8):2767–73. Available from: https://pubs.rsc.org/en/content/articlehtml/2011/ee/c1ee01454d.

61 Niermann M, Timmerberg S, Drünert S, Kaltschmitt M. Liquid organic hydrogen carriers and alternatives for international transport of renewable hydrogen. *Renew Sustain Energy Rev*. 2021 Jan 1;135:110171.

62 Noh H, Kang K, Seo Y. Environmental and energy efficiency assessments of offshore hydrogen supply chains utilizing compressed gaseous hydrogen, liquefied hydrogen, liquid organic hydrogen carriers and ammonia. *Int J Hydrogen Energy*. 2023 Mar 5;48(20):7515–32.

63 Singh R, Singh M, Gautam S. Hydrogen economy, energy, and liquid organic carriers for its mobility. *Mater Today Proc*. 2021 Jan 1;46:5420–7.

64 Li H, Zhang X, Zhang C, Ding Z, Jin X. Application and analysis of liquid organic hydrogen carrier (LOHC) technology in practical projects. *Energies* [Internet]. 2024 Apr 19 [cited 2025 Mar 13];17(8):1940. Available from: https://www.mdpi.com/1996-1073/17/8/1940/htm.

65 Rusman NAA, Dahari M. A review on the current progress of metal hydrides material for solid-state hydrogen storage applications. *Int J Hydrogen Energy*. 2016 Jul 27;41(28):12108–26.

66 Jinzhe L, Lider AM, Kudiiarov VN, Huang LJ, Shi ST, Cui J. Optimizing hydrogen storage in magnesium hydride using carbon-based catalysts. *J Phys Conf Ser* [Internet]. 2024 Nov 1 [cited 2025 Mar 13];2893(1):012077. Available from: https://iopscience.iop.org/article/10.1088/1742-6596/2893/1/012077.

67 Qureshi T, Khan MM, Pali HS. The future of hydrogen economy: Role of high entropy alloys in hydrogen storage. *J Alloys Compd*. 2024 Nov 5;1004:175668.

68 Larpruenrudee P, Bennett NS, Luo Z, Hossain MJ, Haque N, Sauret E. A review on the overall performance of metal hydride-based hydrogen storage systems. *Energies* [Internet]. 2025 Mar 6 [cited 2025 Mar 13];18(5):1291. Available from: https://www.mdpi.com/1996-1073/18/5/1291/htm.

69 Mustafa NS, Noor NAM, Omar NAMY, Omar Z, Ismail M. Influence of CeCl3 incorporation on the dehydrogenation characteristics of NaAlH4 for solid-state hydrogen storage. *Int J Hydrogen Energy*. 2025 Mar 4;105:278–85.

70 Lim KL, Kazemian H, Yaakob Z, Daud WRW. Solid-state Materials and methods for hydrogen storage: A critical review. *Chem Eng Technol* [Internet]. 2010 Feb 1 [cited 2025 Mar 13];33(2):213–26. Available from: https://onlinelibrary.wiley.com/doi/full/10.1002/ceat.200900376.

71 Nivedhitha KS, Beena T, Banapurmath NR, Umarfarooq MA, Ramasamy V, Soudagar MEM. Advances in hydrogen storage with metal hydrides: Mechanisms, materials, and challenges. *Int J Hydrogen Energy*. 2024 Apr 3;61:1259–73.

72 Rajesh Jesudoss Hynes N, Sankaranarayanan R, Senthamarai Kannan P, Khan A, Khan AAP, Asiri AM. Solid-state hydrides for hydrogen storage. In: *Nanomaterials for Hydrogen Storage Applications*. 2021 Jan 1;249–64.

73 Salehabadi A, Umar MF, Ahmad A, Ahmad MI, Ismail N, Rafatullah M. Carbon-based nanocomposites in solid-state hydrogen storage technology: An overview. *Int J Energy Res* [Internet]. 2020 Nov 1 [cited 2025 Mar 13];44(14):11044–58. Available from: https://onlinelibrary.wiley.com/doi/full/10.1002/er.5674.

74 Xia Y, Yang Z, Zhu Y. Porous carbon-based materials for hydrogen storage: Advancement and challenges. *J Mater Chem A Mater* [Internet]. 2013 Jul 31 [cited 2025 Mar 13];1(33):9365–81. Available from: https://pubs.rsc.org/en/content/articlehtml/2013/ta/c3ta10583k.

75 Carbon Nanostructures: Fundamentals to Applications. Carbon Nanostructures [Internet]. 2021 Oct 28 [cited 2025 Mar 13]; Available from: https://pubs.aip.org/books/monograph/44/Carbon-NanostructuresFundamentals-to-Applications. https://doi.org/10.1063/9780735423114.

76 Du YL, Sun ZY, Fu B, Huang Q. Unleashing the power of hydrogen: Challenges and solutions in solid-state storage. *Int J Hydrogen Energy*. 2025 Mar 5;136:1112–23.

77 Dhawa T, Chattopadhyay S, De G, Mahanty S. In situ Mg/MgO-embedded mesoporous carbon derived from magnesium 1,4-benzenedicarboxylate metal organic framework as sustainable Li-S battery cathode support. *ACS Omega* [Internet]. 2017 Oct 31 [cited 2025 Mar 12];2(10):6481–91. Available from: https://pubs.acs.org/doi/full/10.1021/acsomega.7b01156.

78 Lee J, Kwak SY. Mn-Doped maghemite (γ-Fe2O3) from metal-organic framework accompanying redox reaction in a bimetallic system: The structural phase transitions and catalytic activity toward NOx removal. *ACS Omega* [Internet]. 2018 Mar 31 [cited 2025 Mar 12];3(3):2634–40. Available from: https://pubs.acs.org/doi/full/10.1021/acsomega.7b01865.

79 Davies E, Ehrmann A, Schwenzfeier-Hellkamp E. Safety of hydrogen storage technologies. *Processes* [Internet]. 2024 Oct 8 [cited 2025 Mar 19];12(10):2182. Available from: https://www.mdpi.com/2227-9717/12/10/2182/htm.

80 Elberry AM, Thakur J, Santasalo-Aarnio A, Larmi M. Large-scale compressed hydrogen storage as part of renewable electricity storage systems. *Int J Hydrogen Energy*. 2021 Apr 26;46(29):15671–90.

81 Simonetto M, Pascoe JA, Sharpanskykh A. Preliminary safety assessment of a liquid hydrogen storage system for commercial aviation. *Safety* 27 [Internet].

2025 Mar 11 [cited 2025 Mar 19];11(1):27. Available from: https://www.mdpi.com/2313-576X/11/1/27/htm.

82 Sattar MA, Rasul MG, Jahirul MI, Hasan MM. An up-to-date review on the progress and challenges of hydrogen storage, and its safety and economic analysis. *Sustain Energy Fuels* [Internet]. 2024 Aug 6 [cited 2025 Mar 19];8(16):3545–73. Available from: https://pubs.rsc.org/en/content/articlehtml/2024/se/d4se00281d.

83 Kumar S, Nanan-Surujbally A, Sharma DP, Pathak D. Hydrogen safety/standards (national and international document standards on hydrogen energy and fuel cell). In: *Towards Hydrogen Infrastructure: Advances and Challenges in Preparing for the Hydrogen Economy*. Amsterdam: Elsevier. 2024 Jan 1;315–46.

84 Touhiduzzaman M, Louie MS hui, Veeramany A, Ehrhart BD. *Large-Scale Hydrogen Storage Cyber Risk Assessment*. USA: US Department of Energy Office of Scientific and Technical Information. 2024 Oct 1 [cited 2025 Mar 19]; Available from: https://www.osti.gov/servlets/purl/2480226/.

85 Louie MS, Ehrhart BD. *Regulations, Codes, and Standards Review for Underground Hydrogen Storage*. Albuquerque: Sandia National Laboratories, U.S Department of Energy SHASTA Project. 2024 Apr 1 [cited 2025 Mar 19]; Available from: https://www.osti.gov/servlets/purl/2369636/.

86 Moretto P, Quong S. Legal requirements, technical regulations, codes, and standards for hydrogen safety. In: *Hydrogen Safety for Energy Applications: Engineering Design, Risk Assessment, and Codes and Standards*. 2022 Jan 1;345–96.

87 Magliano A, Perez Carrera C, Pappalardo CM, Guida D, Berardi VP. A comprehensive literature review on hydrogen tanks: Storage, safety, and structural integrity. *Appl Sci* [Internet]. 2024 Oct 14 [cited 2025 Mar 19];14(20):9348. Available from: https://www.mdpi.com/2076-3417/14/20/9348/htm.

88 Qanbar MW, Hong Z. A review of hydrogen leak detection regulations and technologies. *Energies* [Internet]. 2024 Aug 15 [cited 2025 Mar 19];17(16):4059. Available from: https://www.mdpi.com/1996-1073/17/16/4059/htm.

89 Fuster B, Houssin-Agbomson D, Jallais S, Vyazmina E, Dang-Nhu G, Bernard-Michel G. Guidelines and recommendations for indoor use of fuel cells and hydrogen systems. *Int J Hydrogen Energy*. 2017 Mar 16;42(11):7600–7.

90 Yang M, Hunger R, Berrettoni S, Sprecher B, Wang B. A review of hydrogen storage and transport technologies. *Clean Energy* [Internet]. 2023 Feb 1 [cited 2025 Mar 19];7(1):190–216. Available from: https://dx.doi.org/10.1093/ce/zkad021.

91 Genovese M, Blekhman D, Fragiacomo P. An exploration of safety measures in hydrogen refueling stations: Delving into hydrogen equipment and technical performance. *Hydrogen* [Internet]. 2024 Feb 17 [cited 2025 Mar 19];5(1):102–22. Available from: https://www.mdpi.com/2673-4141/5/1/7/htm.

92 Le ST, Nguyen TN, Linforth S, Ngo TD. Safety investigation of hydrogen energy storage systems using quantitative risk assessment. *Int J Hydrogen Energy*. 2023 Jan 22;48(7):2861–75.

93 Rigas F, Sklavounos S. Evaluation of hazards associated with hydrogen storage facilities. *Int J Hydrogen Energy*. 2005 Oct 1;30(13–14):1501–10.

94 Amaral PCS, Oh CB, Do KH, Choi B Il. Risk assessment of hydrogen leakage and explosion in a liquid hydrogen facility using computational analysis. *Int J Hydrogen Energy*. 2024 Nov 19;91:950–64.

95 Cavaliere P. Safety Issues and Regulations. In: *Water Electrolysis for Hydrogen Production* [Internet]. Singapore: Spriger Nature. 2023 [cited 2025 Mar 19];729–91. Available from: https://link.springer.com/chapter/10.1007/978-3-031-37780-8_19.

96 Ma Q, He Y, You J, Chen J, Zhang Z. Probabilistic risk assessment of fire and explosion of onboard high-pressure hydrogen system. *Int J Hydrogen Energy*. 2024 Jan 2;50:1261–73.

97 Qasim M, Fatima A, Akhtar T, Fizza S, Batool E, Abdullah K. Underground hydrogen storage: A critical review in the context of climate change mitigation. *Sch Acad J Biosci*. 2024;(7):220–31.

98 Zhang C, Cao X, Bujlo P, Chen B, Zhang X, Sheng X. Review on the safety analysis and protection strategies of fast filling hydrogen storage system for fuel cell vehicle application. *J Energy Storage*. 2022 Jan 1;45:103451.

99 Uliasz-Misiak B, Lewandowska-śmierzchalska J, Matuła R. Selection of underground hydrogen storage risk assessment techniques. *Energies* [Internet]. 2021 Dec 1 [cited 2025 Mar 19];14(23):8049. Available from: https://www.mdpi.com/1996-1073/14/23/8049/htm.

100 Muthukumar P, Kumar A, Afzal M, Bhogilla S, Sharma P, Parida A. Review on large-scale hydrogen storage systems for better sustainability. *Int J Hydrogen Energy*. 2023 Oct 15;48(85):33223–59.

101 Correa-Jullian C, Groth KM. Data requirements for improving the quantitative risk assessment of liquid hydrogen storage systems. *Int J Hydrogen Energy*. 2022 Jan 19;47(6):4222–35.

102 Aziz M, TriWijayanta A, Nandiyanto ABD. Ammonia as effective hydrogen storage: A review on production, storage and utilization. *Energies* [Internet]. 2020 Jun 12 [cited 2025 Mar 19];13(12):3062. Available from: https://www.mdpi.com/1996-1073/13/12/3062/htm.

103 Rüde T, Bösmann A, Preuster P, Wasserscheid P, Arlt W, Müller K. Resilience of liquid organic hydrogen carrier based energy-storage systems. *Energy Technol* [Internet]. 2018 Mar 1 [cited 2025 Mar 19];6(3):529–39. Available from: https://onlinelibrary.wiley.com/doi/full/10.1002/ente.201700446.

104 Sreeraj R, Aadhithiyan AK, Anbarasu S. Integration of thermal augmentation methods in hydride beds for metal hydride based hydrogen storage systems: Review and recommendation. *J Energy Storage*. 2022 Aug 25;52:105039.

105 Rasul MG, Hazrat MA, Sattar MA, Jahirul MI, Shearer MJ. The future of hydrogen: Challenges on production, storage and applications. *Energy Convers Manag*. 2022 Nov 15;272:116326.

106 Ehteshami SMM, Chan SH. The role of hydrogen and fuel cells to store renewable energy in the future energy network—Potentials and challenges. *Energy Policy*. 2014 Oct 1;73:103–9.

107 Arsad AZ, Hannan MA, Al-Shetwi AQ, Mansur M, Muttaqi KM, Dong ZY. Hydrogen energy storage integrated hybrid renewable energy systems: A review analysis for future research directions. *Int J Hydrogen Energy*. 2022 May 5; 47(39):17285–312.

108 Mandal TK, Gregory DH. Hydrogen storage materials: Present scenarios and future directions. *Annu Rep Sect A* [Internet]. 2009 Jun 2 [cited 2025 Mar 13]; 105(0):21–54. Available from: https://pubs.rsc.org/en/content/articlehtml/2009/ic/b818951j.

109 Kumar N, Lee SY, Park SJ. Advancements in hydrogen storage technologies: A comprehensive review of materials, methods, and economic policy. *Nano Today* 2024 Jun 1;56:102302.

110 McDowall W, Eames M. Towards a sustainable hydrogen economy: A multi-criteria sustainability appraisal of competing hydrogen futures. *Int J Hydrogen Energy*. 2007 Dec 1;32(18):4611–26.

5

Hydrogen Fuel Cells – Types and Mechanisms

5.1 Overview of Hydrogen Fuel Cells

Hydrogen fuel cells are a promising technology in the quest for clean and sustainable energy solutions. These devices generate electricity by converting chemical energy from hydrogen into electrical energy through an electrochemical reaction, without combustion, heat, or harmful emissions. The central appeal of hydrogen fuel cells lies in their ability to produce electricity while emitting only water vapor as a byproduct, making them environmentally friendly and aligned with global sustainability goals [1]. The fundamental process involves the reaction of hydrogen (H_2) with oxygen (O_2), which results in the generation of electricity, heat, and water. This reaction takes place in a fuel cell, which consists of an anode, a cathode, and an electrolyte. The fuel cell works in a manner similar to a battery but differs in that it continuously generates electricity as long as fuel (hydrogen) is supplied. The core components of a hydrogen fuel cell include the anode, cathode, and electrolyte. At the anode, hydrogen gas is introduced and split into protons (H^+) and electrons (e^-). This process, facilitated by a catalyst, occurs at a very high efficiency, allowing the protons to travel through the electrolyte to the cathode while the electrons flow through an external circuit, producing electricity. The protons and electrons meet again at the cathode, where they combine with oxygen from the air to form water (H_2O), which is expelled as the only emission from the process. This clean byproduct makes hydrogen fuel cells an attractive alternative to fossil fuel-based power generation, which releases harmful pollutants such as carbon dioxide (CO_2) into the atmosphere [2].

One of the key advantages of hydrogen fuel cells over traditional combustion engines and batteries is their high efficiency. Hydrogen fuel cells can achieve efficiencies between 40% and 60% in electricity generation, and up to 85% to 90% when the waste heat is utilized for cogeneration (combined heat and power). In contrast, internal combustion engines typically operate at lower efficiencies, often below 30%. Furthermore, fuel cells can operate continuously as long as hydrogen fuel is available, unlike batteries that need to be recharged after a certain period of use. This continuous operation, combined with high efficiency, makes hydrogen fuel cells

Next-Generation Hydrogen Economy: Production, Storage, and Fuel Cell Technologies, First Edition.
Peeyush Phogat, Shreya Sharma and Satyam Rawat.

particularly appealing for applications that require long operational hours, such as heavy-duty transport, industrial applications, and backup power systems [3].

The versatility of hydrogen fuel cells makes them suitable for a variety of applications. They are used in the automotive industry in hydrogen fuel cell vehicles (HFCVs), which include buses, trucks, and passenger cars. HFCVs offer a zero-emissions alternative to gasoline and diesel-powered vehicles, with the added benefit of shorter refueling times compared to electric vehicles, which require longer charging periods. In addition to transportation, hydrogen fuel cells are also used in stationary applications, such as backup power for critical infrastructure, remote energy supply in off-grid areas, and as a power source in industries where high energy density is needed. The fuel cell technology can be scaled up for large-scale industrial applications or down for use in portable devices, giving it a wide range of flexibility [4]. While hydrogen fuel cells have significant advantages, they are not without challenges. The production, storage, and distribution of hydrogen are the primary hurdles that limit the widespread adoption of fuel cells. Hydrogen, while abundant, is rarely found in its pure form in nature. As a result, it must be produced through energy-intensive methods such as electrolysis or steam methane reforming (SMR), both of which currently have their own economic and environmental challenges. Electrolysis, for example, requires large amounts of electricity, which can make the process expensive unless renewable energy sources are used. Steam methane reforming, which is cheaper, generates CO_2 as a byproduct, thus undermining the environmental benefits of hydrogen fuel cells unless carbon capture and storage (CCS) technologies are employed [5]. Hydrogen storage and transportation also present significant challenges. Hydrogen has a very low energy density by volume, meaning that it must be stored either as a compressed gas, a cryogenic liquid, or chemically bound to other materials (such as ammonia or metal hydrides) for efficient transport and storage. Each of these storage methods comes with its own set of challenges, including high compression costs, energy losses during liquefaction, and potential safety concerns related to high-pressure hydrogen storage. Moreover, establishing the infrastructure necessary to support hydrogen fuel cells – such as hydrogen production plants, storage facilities, and refueling stations – requires substantial investment and collaboration across industries and governments [6].

Another key barrier to the large-scale adoption of hydrogen fuel cells is their cost. Fuel cell systems and hydrogen storage solutions are currently expensive compared to conventional internal combustion engine systems and battery-electric vehicles. While costs have been steadily decreasing due to advancements in materials, manufacturing processes, and economies of scale, they are still a significant factor holding back the mainstream adoption of hydrogen-powered solutions [7]. Furthermore, the lack of widespread refueling infrastructure makes it difficult to achieve the network density needed to make hydrogen fuel cell vehicles practical for everyday use. Despite these challenges, the potential of hydrogen fuel cells remains significant [8]. They offer a path toward decarbonizing industries that are difficult to electrify directly, such as heavy-duty transport, steel and cement manufacturing, and other energy-intensive sectors. In the coming years, continued research and development efforts are expected to improve the performance, efficiency, and affordability of hydrogen fuel cells, making them more competitive with other low-emission technologies like electric batteries [9]. Additionally, the expansion of hydrogen production and distribution infrastructure, combined with supportive government policies and incentives, will be crucial in unlocking the full potential of hydrogen fuel cells as a clean energy solution. In summary, hydrogen fuel cells are a highly promising technology that offers several advantages, including high efficiency, zero emissions, and versatility across different sectors. Despite the challenges related to hydrogen production, storage, and cost, ongoing advancements in fuel cell technology and infrastructure development are expected to pave the way for a wider adoption of hydrogen-powered systems. These systems have the potential to play a pivotal role in global efforts to reduce carbon emissions and transition to a more sustainable, clean energy future [10].

5.1.1 Basic Operating Principles of Fuel Cells

A hydrogen fuel cell operates on a simple but highly efficient electrochemical principle, converting chemical energy from hydrogen and oxygen into electrical energy, water, and heat. Unlike combustion-based power generation methods, fuel cells do not burn hydrogen to produce energy, which means that they operate at higher efficiencies and emit only water vapor as a byproduct. The basic operation of a fuel cell is driven by the electrochemical reaction between hydrogen and oxygen, which occurs in the presence of a catalyst, typically platinum [11].

The central components of a hydrogen fuel cell include the anode, cathode, and electrolyte. The fuel cell's structure is designed to facilitate the flow of protons (hydrogen ions) and electrons, ensuring efficient energy production [12]. Here's a detailed breakdown of how a fuel cell works:

a. **Hydrogen Input (Anode Side):** The fuel cell's operation begins at the anode, where hydrogen gas (H_2) is introduced. The hydrogen molecules are split into protons (H^+) and electrons (e^-) by a catalyst (usually platinum) placed at the anode. The catalyst helps break the bond between the hydrogen atoms and allows the release of electrons, which travel through an external circuit, generating electric current. The remaining protons (hydrogen ions) pass through the electrolyte towards the cathode side.
b. **Electron Flow (External Circuit):** The freed electrons, after being separated from the hydrogen atoms, travel along an external circuit, creating an electric current that can be used to power electrical devices or motors. The flow of electrons through the circuit is essentially the electrical energy generated by the fuel cell. This electricity can be used to power everything from electric vehicles to stationary power systems, depending on the fuel cell's size and application.
c. **Electrolyte (Proton Conduction):** The electrolyte is a crucial component of the fuel cell. It allows only protons (H^+) to pass through while blocking the flow of electrons, forcing the electrons to travel through the external circuit. The electrolyte also serves to maintain the separation between the hydrogen and oxygen sides of the cell, which is necessary for the electrochemical reaction to proceed.

d. **Oxygen Input (Cathode Side):** On the cathode side, oxygen gas (O_2) is supplied from the air. The oxygen molecules combine with the protons (H^+) that have traveled through the electrolyte and the electrons that have flowed through the external circuit. This reaction results in the formation of water (H_2O), which is the only byproduct of the fuel cell's operation. The oxygen is typically sourced from ambient air, and the water produced is expelled as steam or liquid.

The chemical reaction in a hydrogen fuel cell can be summarized as follows:

$$2H_2 + O_2 \rightarrow 2H_2O + \text{electricity} + \text{heat}$$

This reaction demonstrates that hydrogen and oxygen react to form water and electricity, without any harmful emissions like CO_2 or nitrogen oxides (NOx), making hydrogen fuel cells a clean and sustainable energy source. The basic working mechanism of hydrogen fuel cells is illustrated in Figure 5.1.

5.1.2 Advantages of Hydrogen Fuel Cells

Hydrogen fuel cells present several compelling advantages, particularly in the context of clean energy transition and efforts to decarbonize industries and transportation.

a. **Environmental Benefits:** One of the most significant advantages of hydrogen fuel cells is their environmental impact. The primary byproduct of the fuel cell's operation is water (H_2O), which makes hydrogen fuel cells a zero-emission technology. This stands in stark contrast to conventional internal combustion engines, which emit CO_2, NOx, and particulate matter – key contributors to air pollution and climate change. Hydrogen fuel cells produce no greenhouse gas emissions, making them an ideal solution for sectors looking to reduce their carbon footprint [14].

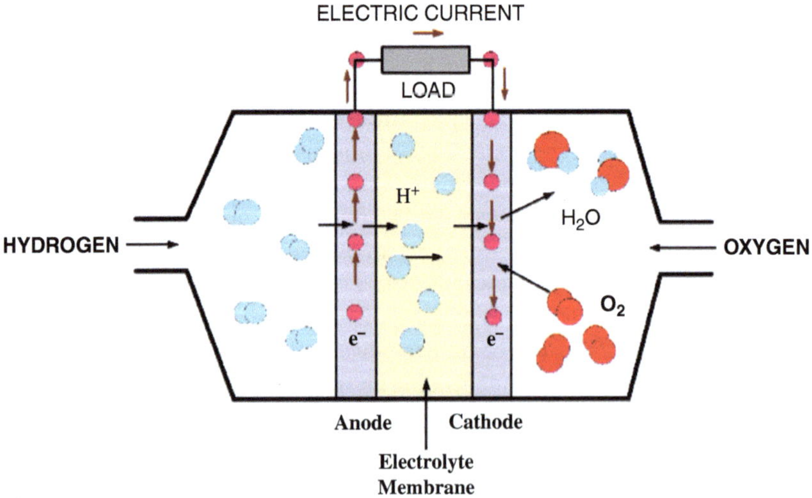

Figure 5.1 Schematic representation of the basic operating mechanism of a hydrogen fuel cell, showing hydrogen input, proton and electron flow, and water formation. [13]/IEEE/CC BY 4.0.

b. **High Efficiency:** Fuel cells are more efficient than traditional combustion engines because they convert chemical energy directly into electrical energy, bypassing the intermediate step of combustion. The efficiency of hydrogen fuel cells is typically in the range of 40–60% for stationary applications, and even higher in combined heat and power (CHP) systems. In electric vehicles, hydrogen fuel cells achieve an energy efficiency of about 60%, compared to around 20–30% for gasoline engines. This greater efficiency leads to less energy waste and improved overall performance [15].
c. **Energy Density:** Hydrogen fuel cells have a higher energy density than many conventional batteries, making them particularly well-suited for applications where weight and space are important factors, such as in heavy-duty transport and aviation. Hydrogen's high energy content per kilogram allows fuel cells to deliver longer ranges in vehicles without the need for large, heavy batteries. This advantage is especially crucial for long-distance travel and for industries that require large amounts of power [16].
d. **Scalability:** Hydrogen fuel cells are highly versatile and can be used in a wide range of applications. From small portable devices to large-scale power generation, fuel cells can be scaled up or down to meet specific energy needs. In the transportation sector, hydrogen fuel cells can power passenger cars, buses, trucks, and trains. In the industrial sector, fuel cells can be used for backup power, stationary energy generation, and even as part of industrial processes like hydrogenation [17].
e. **Energy Independence:** Hydrogen fuel cells offer the potential for greater energy independence, particularly in regions that have access to renewable energy sources but lack fossil fuel reserves. Hydrogen can be produced locally from renewable sources such as solar, wind, and hydroelectric power, reducing reliance on imported oil and gas. This is especially relevant for countries seeking to diversify their energy sources and achieve energy security [18].
f. **Quiet Operation:** Unlike internal combustion engines, which produce significant noise, hydrogen fuel cells operate quietly. This feature makes fuel cells particularly attractive for applications in urban environments or in industries where noise pollution is a concern, such as in public transportation, commercial vehicles, or residential backup power systems [19].

5.2 Types of Hydrogen Fuel Cells

Hydrogen fuel cells represent a promising technology for clean energy production, with the potential to decarbonize various sectors, including transportation, industrial applications, and stationary power generation. The different types of hydrogen fuel cells have unique operating principles, materials, and design characteristics, all of which contribute to their efficiency, performance, and cost-effectiveness. Each type of fuel cell operates by converting chemical energy from hydrogen into electrical energy through an electrochemical process, but the specific technologies used vary. This variety allows hydrogen fuel cells to be tailored for different applications, depending on factors such as power output, operating temperature, efficiency, and the specific challenges presented by the environment in which they are deployed. Understanding the various types of hydrogen fuel cells is essential for evaluating their role in the broader energy landscape. From high-temperature fuel cells that are ideal for industrial applications to low-temperature fuel cells that are more suitable for transportation, each type offers distinct advantages that make it suitable for particular uses. Among the most widely studied and utilized fuel cell types are proton exchange membrane fuel cells (PEMFC) [20], solid oxide fuel cells (SOFC) [21], alkaline fuel cells (AFC) [22], phosphoric acid fuel cells (PAFC) [23], molten carbonate fuel cells (MCFC) [24], and direct methanol fuel cells (DMFC) [25]. These different technologies are at varying stages of commercialization and development, and their unique characteristics make them suitable for specific applications in a variety of settings. Some fuel cells, for instance, are highly efficient and compact, making them ideal for use in vehicles, while others are better suited for stationary power generation or industrial processes. In this section, we will delve deeper into each of these fuel cell types, exploring their basic operating principles, advantages, challenges, and applications. By understanding the different fuel cell technologies, we can assess their potential in helping meet the growing demand for clean, renewable energy and their role in the global shift toward sustainable energy systems. A comparative overview of various fuel cell types is shown in Figure 5.2.

5.2.1 PEMFCs

PEMFCs are one of the most widely used and studied types of hydrogen fuel cells. They have gained significant attention due to their relatively low operating temperature, high efficiency, and scalability, making them ideal for applications in the transportation sector and portable power generation. PEMFCs are known for their ability to provide clean, efficient, and flexible energy solutions, and they have already been successfully incorporated into various commercial applications,

Fuel Cell Type	Operating Temperature	Electrical Efficiency	Applications
Polymer Electrolyte Membrane (PEM)	50 °C–100 °C	53–58% (Transportation) 25–35% (Stationary)	Backup Power, Portable Power, Small Distributed Generation, Transportation, Specialty Vehicles
Alkaline (AFC)	90 °C–100 °C	60%	Military, Space
Phosphoric Acid (PAFC)	150 °C–200 °C	>40%	Distributed Generation
Molten Carbonate (MCFC)	600 °C–700 °C	45–47%	Electric Utility, Large Distributed Generation
Solid Oxide (SOFC)	600 °C–1000 °C	35–43%	Auxiliary Power, Electric Utility, Large Distributed Generation

Figure 5.2 Comparative overview of different types of hydrogen fuel cells in terms of operating temperature, efficiency, and applications. Adapted from [26].

including hydrogen-powered vehicles and backup power systems. The technology behind PEMFCs is based on the use of a solid polymer electrolyte membrane that facilitates the transport of protons while preventing the flow of electrons, thus enabling the electrochemical conversion of hydrogen into electricity [27].

The basic operating principle of a PEMFC involves the reaction of H_2 with O_2 to produce electricity, water, and heat. This is achieved through a process called electrolysis, which takes place across the membrane. At the anode side of the fuel cell, hydrogen gas is supplied, where it is split into H^+ and e^- by a catalyst, typically made of platinum. The protons pass through the proton exchange membrane, which is a key component of the fuel cell, while the electrons travel along an external circuit, generating an electrical current that can be used to power devices or vehicles. At the cathode, oxygen from the air combines with the protons and electrons, forming H_2O and releasing heat as a byproduct [28]. The internal structure and key components of a PEMFC are depicted in Figure 5.3.

One of the key advantages of PEMFCs is their low operating temperature, typically around 80 °C, which allows for quick start-up times and high power density. This makes them particularly suitable for applications that require rapid response and high performance, such as in fuel cell vehicles. The low temperature also enables the use of commercially available materials, such as platinum catalysts, which are effective at the lower temperatures compared to other fuel cell types that require higher operating temperatures. PEMFCs also operate with high efficiency, often in the range of 40–60%, and can achieve even higher efficiencies when integrated with CHP systems. However, PEMFCs do have certain limitations and challenges [29]. One of the most significant is the cost of the platinum catalyst. Platinum is expensive and scarce, and its use in PEMFCs has raised concerns about the long-term sustainability of the technology. Researchers are exploring alternatives to platinum,

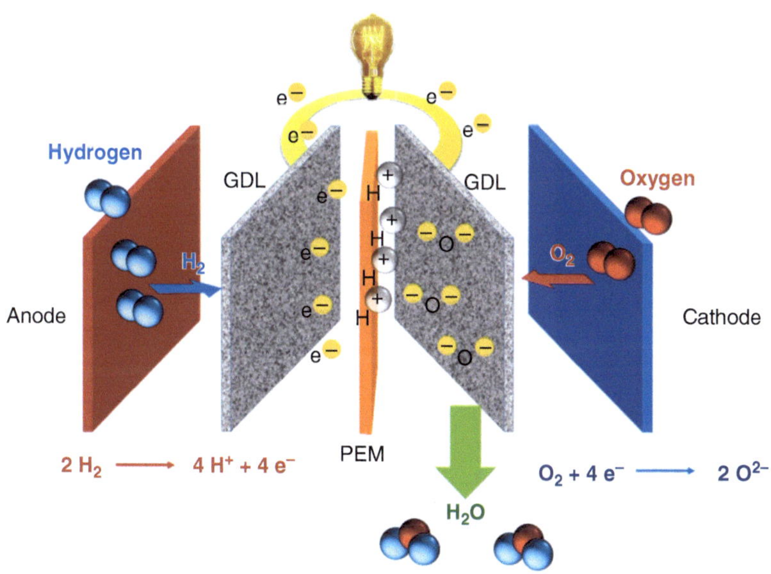

Figure 5.3 Structure of a PEMFC. [33]/MDPI/CC BY 4.0.

such as platinum-alloy catalysts and non-precious metal catalysts, to reduce costs and improve the commercialization prospects of PEMFCs. Another challenge is the durability of the proton exchange membrane itself, which can degrade over time, especially when exposed to high humidity and temperatures. Ongoing research is focused on improving the membrane's resistance to degradation, as well as enhancing the overall lifespan of the fuel cell. Despite these challenges, PEMFCs remain one of the most promising hydrogen fuel cell technologies, particularly for transportation applications. In the automotive industry, PEMFCs are being used to develop hydrogen-powered vehicles, which offer significant environmental benefits over conventional gasoline and diesel vehicles. These FCVs emit only water vapor as a byproduct, making them a zero-emission alternative to traditional internal combustion engine vehicles. Companies such as Toyota, Honda, and Hyundai have already introduced commercial PEMFC-powered vehicles, with infrastructure for hydrogen refueling stations gradually expanding in key markets [30].

In addition to transportation, PEMFCs are also being explored for use in stationary applications, such as residential backup power systems, remote area power generation, and uninterruptible power supplies (UPS). Their ability to deliver reliable and consistent power, coupled with their low environmental impact, makes them an attractive option for these applications. The scalability of PEMFCs, from small portable systems to larger-scale installations, further broadens their potential uses across different sectors. The development of PEMFC technology is also closely linked to advancements in hydrogen production, storage, and distribution systems. As hydrogen infrastructure improves and the cost of hydrogen fuel decreases, PEMFCs are expected to become increasingly competitive with other forms of energy generation [31]. Furthermore, ongoing research into materials science, catalyst development, and system integration is expected to enhance the performance and reduce the costs of PEMFCs in the coming years, bringing them closer to widespread commercialization. PEMFCs are a key technology in the pursuit of clean, sustainable energy solutions. Their efficiency, scalability, and low operating temperature make them well-suited for a range of applications, particularly in the transportation and stationary power sectors. Although challenges such as high catalyst costs and membrane durability remain, continued research and technological advancements are likely to address these issues, paving the way for broader adoption of PEMFCs as a cornerstone of the hydrogen economy. As the global demand for clean energy solutions grows, PEMFCs will play an increasingly important role in driving the transition to a sustainable and decarbonized future [32].

5.2.2 SOFCs

SOFCs are a type of high-temperature fuel cell that use a solid ceramic electrolyte to conduct ions from the anode to the cathode. SOFCs operate at much higher temperatures compared to other types of hydrogen fuel cells, typically ranging from 600 °C to 1,000 °C, which gives them a distinct set of advantages and challenges. These high temperatures allow SOFCs to achieve very high efficiency levels and make them particularly suitable for stationary power generation, including both residential and industrial applications. Unlike PEMFCs that operate at lower temperatures,

SOFCs can use a wider variety of fuels, including hydrogen, natural gas, biogas, and even carbon monoxide, providing greater fuel flexibility. The operation of an SOFC involves the electrochemical conversion of hydrogen (or other fuels) into electricity, water, and heat through a process of ion conduction within the solid electrolyte. The cell consists of three main components: the anode, the electrolyte, and the cathode. The anode is typically made from a porous metal or ceramic material that allows the hydrogen fuel to diffuse through and react. At the anode, H_2 is split into H^+ and e^-. The protons pass through the solid electrolyte (which is typically made from zirconia doped with yttria), while the electrons are forced to travel through an external circuit, generating electric current. At the cathode, oxygen from the air combines with the electrons to form oxygen ions (O^{2-}), which then travel through the electrolyte to meet the protons at the anode to form water [34].

A key feature of SOFCs is their high efficiency, which is primarily due to their high operating temperatures. The high temperature enables the direct electrochemical oxidation of fuels, which minimizes losses associated with other energy conversion processes, such as combustion. This results in overall electrical efficiencies of up to 60%, and when used in CHP configurations, SOFCs can achieve efficiencies greater than 80%, making them highly attractive for stationary applications. The ability to use various fuels also makes SOFCs versatile, offering opportunities for integration into diverse energy systems, including those based on renewable or low-carbon fuels [35]. However, the high operating temperature of SOFCs also presents several challenges. One of the main concerns is the material degradation over time due to thermal cycling and the potential for the formation of undesirable compounds at the anode and cathode interfaces. These issues can lead to a decrease in the lifespan and performance of the fuel cell. Additionally, the need for high-temperature operation requires complex and expensive materials, particularly for the solid electrolyte and electrode components, which adds to the initial cost of SOFC systems [36]. Furthermore, the high-temperature nature of SOFCs means that they have relatively slow start-up times, which can be a limitation in applications that require rapid response. Despite these challenges, significant research efforts are being directed towards improving the durability, efficiency, and cost-effectiveness of SOFCs. One promising development is the use of new materials and coatings that can withstand the harsh conditions of high-temperature operation, improving the fuel cell's longevity and performance. Additionally, efforts are underway to reduce the costs of manufacturing and materials, which could help lower the capital costs of SOFC systems and make them more competitive with other energy generation technologies [37].

SOFCs are particularly suited for large-scale, stationary applications such as power plants, industrial backup power, and distributed generation systems. They are well-suited to provide steady, high-efficiency power for both residential and commercial users, particularly in areas where access to the electric grid may be limited. Their ability to operate on multiple fuels also makes them an excellent option for transitioning to cleaner energy sources, as they can easily be adapted to use biogas, methane, or hydrogen, which can significantly reduce their carbon emissions. In addition to stationary applications, SOFCs are also being explored for use in integrated systems for industrial processes, where their waste heat can be captured and utilized for further processes, such as heating or desalination [38]. This makes them ideal for use in industries with high energy demands, including chemical processing, steel manufacturing, and desalination plants. The growing interest in SOFC technology is also linked to advancements in hydrogen infrastructure, as hydrogen can serve as an ideal fuel for SOFCs, offering a clean and sustainable alternative to traditional fossil fuels. As hydrogen production becomes more widespread and affordable, SOFCs could play a key role in the global transition to a low-carbon economy [39]. SOFCs represent a highly efficient and flexible technology for converting hydrogen and other fuels into electricity. Their high operating temperatures offer significant efficiency advantages, particularly when used in CHP applications. While challenges related to material durability, high cost, and slow start-up times remain, ongoing advancements in materials science and system design are likely to overcome these barriers, expanding the commercial viability of SOFCs. With their ability to operate on a variety of fuels and provide steady, reliable power, SOFCs have the potential to play an important role in the future of hydrogen energy and the global energy transition [40].

5.2.3 AFCs

AFCs are one of the earliest types of fuel cells, first developed in the 1960s for space missions by NASA. They operate using an alkaline electrolyte solution, typically potassium hydroxide (KOH), which facilitates the conduction of hydroxide ions (OH^-) between the anode and cathode. AFCs are characterized by their relatively simple construction, efficiency, and suitability for a variety of applications, ranging from portable devices to larger-scale power generation. Though their use has been somewhat limited in comparison to other fuel cell types, AFCs have a strong historical presence in specialized sectors like space exploration and military applications. The fundamental operation of AFCs is based on the electrochemical reaction between hydrogen and oxygen to produce electricity, water, and heat. At the anode, H_2 is split into H^+ and e^-. The protons then combine with OH^- from the alkaline electrolyte, forming water. Meanwhile, the electrons travel through an external circuit, creating an electrical current. At the cathode, oxygen (from the air) combines with electrons that have traveled through the circuit to form hydroxide ions, which then move toward the anode to complete the cycle. The electrolyte, typically in liquid form, ensures the efficient movement of hydroxide ions and plays a crucial role in maintaining the cell's overall functionality [41].

One of the primary advantages of AFCs is their high efficiency. They offer significantly higher performance in terms of efficiency compared to traditional combustion engines, with an electrical efficiency of around 50–60%. When used in CHP applications, AFCs can reach even higher efficiencies. The low operating temperature, typically between 60 °C and 100 °C, allows AFCs to start quickly, making them ideal for applications that require rapid power availability. Additionally, AFCs can use pure hydrogen or hydrogen-rich fuels, such as natural gas, methanol, and even biogas, depending on the configuration. Another advantage of AFCs is their ability to generate electricity at a lower temperature than other high-temperature fuel cells, such as SOFCs. This makes AFCs more suitable for portable applications and situations where heat management is critical [42]. For example, AFCs have been used in military applications, such as powering communication systems, where their lightweight design and quick start-up time offer distinct operational benefits. AFCs are also used in some commercial applications, including backup power for telecommunications and small-scale power generation, where space, weight, and efficiency are important factors. However, AFCs face several challenges that have limited their broader adoption. One of the main limitations is their sensitivity to CO_2. The presence of CO_2 in the air or the fuel stream can lead to the formation of carbonates in the electrolyte, which reduces the conductivity of the electrolyte and ultimately decreases the performance of the fuel cell. This sensitivity makes AFCs less suited for use in environments where CO_2 is abundant or in applications that require the use of impure hydrogen, such as from natural gas reformation. As a result, AFCs generally require a purer hydrogen feedstock, which can increase the cost of fuel supply and infrastructure, limiting their cost competitiveness with other types of fuel cells. Additionally, the electrolyte used in AFCs, typically a liquid, can present challenges in terms of system durability and stability. The alkaline solution must be carefully managed and replenished, which can add to the complexity of system maintenance. The use of liquid electrolytes also introduces issues related to leakage, corrosion, and the need for ongoing monitoring and management to ensure optimal performance. These factors can increase the cost of operating AFCs, particularly in larger, commercial applications [43].

Despite these challenges, AFCs remain a promising technology, particularly in niche applications where their unique advantages are most beneficial. In recent years, advances in membrane technology and the development of new materials for the anode and cathode electrodes have been explored to address some of the challenges associated with AFCs. Researchers are focused on enhancing the resistance of AFCs to CO_2 contamination and improving the longevity and durability of the electrolyte. Some of these advancements have already resulted in prototype systems that exhibit better performance, even in more demanding conditions [44]. Moreover, AFCs are being actively researched for their potential use in a variety of emerging energy systems, including renewable energy storage. When combined with renewable sources such as wind and solar, AFCs can provide a means of storing energy in the form of hydrogen, which can later be used to generate electricity when needed. This ability to store and convert renewable energy into a readily usable form makes AFCs an attractive option for the energy transition toward low-carbon economies. The future of AFCs in the hydrogen economy is tied to overcoming the challenges of CO_2 sensitivity and electrolyte management. With ongoing advancements in fuel processing, material science, and system integration, AFCs have the potential to play a more significant role in both stationary and portable energy systems. Their efficiency, quick start-up, and ability to operate at relatively low temperatures offer a unique combination of benefits, particularly in areas where size, weight, and rapid response are essential [45].

AFCs are an established but still niche technology in the world of hydrogen fuel cells. While they offer a range of benefits, such as high efficiency, quick start-up, and lower operating temperatures, their sensitivity to CO_2 and challenges with electrolyte management have limited their widespread use. However, ongoing technological advancements, particularly in materials and system design, are helping to address these issues, and AFCs may find greater application in sectors where their specific characteristics are highly valued. As the hydrogen economy continues to evolve, AFCs may play an increasingly important role in specific energy applications, particularly in areas where portability, efficiency, and quick power availability are key considerations [46].

5.2.4 PAFCs

PAFCs are a type of hydrogen fuel cell that utilizes phosphoric acid (H_3PO_4) as the electrolyte to facilitate the electrochemical reaction between hydrogen and oxygen. This technology has been in development since the 1960s and has seen significant use in both stationary power generation and industrial applications. PAFCs are particularly known for their ability to operate at moderate temperatures, typically between 150 °C and 220 °C, which allows them to offer a balance between efficiency and operational stability. While PAFCs are not as widely used as PEMFCs in mobile applications, they have established themselves as reliable sources of power for commercial and industrial purposes, including backup power for buildings, distributed energy generation, and cogeneration systems. PAFCs operate on the basic electrochemical principles of fuel cells, where hydrogen and oxygen react to produce electricity, water, and heat. In a PAFC, hydrogen gas is supplied to the anode, where it is split into H^+ and e^-. The protons pass through the electrolyte (phosphoric acid), while the electrons flow through an external circuit, generating an electric current. At the cathode, oxygen from the air combines with the electrons and protons to form water. The phosphoric acid electrolyte serves as a medium for proton conduction and plays a critical role in the fuel cell's efficiency and longevity. Since the electrolyte in PAFCs is a liquid, it can be effectively managed and easily replenished, making it simpler to maintain compared to some other fuel cell types that rely on solid-state electrolytes [47].

One of the primary advantages of PAFCs is their ability to generate power at relatively high efficiencies, especially when utilized in CHP applications. PAFCs are typically able to reach electrical efficiencies of around 40–50%, and when waste heat is captured and used for heating purposes, overall system efficiencies can exceed 80%. This makes them well-suited for buildings and industrial plants that require both electricity and thermal energy. The moderate operating temperature also contributes to the long-term durability and stability of PAFCs, as it allows for smoother thermal management and reduces the stress on materials [48]. Additionally, PAFCs are less sensitive to impurities in the hydrogen fuel than other fuel cell types, such as PEMFCs, making them more tolerant of feedstocks that may contain trace amounts of contaminants. The design of PAFCs enables relatively high-power output, making them suitable for medium to large-scale power generation applications. They are used in a range of industries and commercial sectors where continuous, reliable power is needed. For example, PAFCs have been employed in hospitals, data centers, and commercial buildings, providing backup power and supporting energy resilience. The ability of PAFCs to work in tandem with other energy systems in a cogeneration setup, where both electricity and useful heat are produced simultaneously, further increases their appeal. Additionally, PAFCs are suitable for grid stabilization applications, where they can provide reliable, distributed power generation in remote or decentralized areas [49].

However, despite these advantages, PAFCs are not without their limitations. The main challenge is their relatively high operating temperature compared to other fuel cell types such as PEMFCs. This higher temperature increases the wear on the fuel cell components, which can lead to a shorter operational lifespan. The need for high-temperature materials that can withstand the thermal stresses is a significant engineering challenge, and the development of cost-effective, durable materials is essential for improving the longevity and cost-effectiveness of PAFC systems. Furthermore, the electrolyte itself, phosphoric acid, while offering advantages in terms of stability and conductivity, is also corrosive, which can pose challenges for the materials used in the construction of the fuel cell. The overall cost of PAFCs is another limitation, particularly for small-scale applications. The relatively high cost of the electrolyte, along with the need for specialized materials to handle the operating temperature and the corrosive nature of the phosphoric acid, contributes to the higher capital costs of PAFC systems. While PAFCs offer high efficiency, the upfront investment may be prohibitive in some cases, especially when compared to other fuel cell types like PEMFCs, which are better suited for smaller-scale, portable applications. However, research and development in PAFC technology are focused on addressing these challenges. Efforts are being made to improve the efficiency and reduce the cost of PAFCs by exploring new materials for both the electrolyte and the fuel cell components. Innovations in catalyst design, for example, are aimed at improving the electrochemical efficiency and reducing the dependency on precious metals like platinum, which are used in many fuel cells. Additionally, advances in system integration are helping to reduce the overall system cost and improve the economic viability of PAFC-based power generation [50].

In terms of environmental impact, PAFCs are considered a clean energy technology since their only by-products are water and heat, with no harmful emissions produced during their operation. When used in cogeneration applications, the waste heat from the fuel cell can be utilized to provide heating or industrial process heat, further reducing overall energy consumption and improving efficiency. The ability of PAFCs to operate on hydrogen, which can be produced from renewable sources, such as wind or solar power, aligns well with the global shift toward cleaner energy systems and the decarbonization of various sectors. PAFCs are a proven and reliable technology for medium to large-scale power generation, particularly in commercial and industrial settings. Their moderate operating temperature and relatively high efficiency, especially in combined heat and power applications, make them an attractive option for stationary power generation. While PAFCs face challenges such as high operating temperatures, corrosion, and high upfront costs, ongoing research and development efforts are aimed at improving their performance and cost-effectiveness. As the demand for clean, efficient energy solutions continues to grow, PAFCs have the potential to play a key role in the global transition to sustainable energy systems [51].

5.2.5 MCFCs

MCFCs are a type of high-temperature fuel cell that use a molten carbonate salt mixture as the electrolyte to conduct ions between the anode and cathode. These fuel cells operate at temperatures between 600 °C and 700 °C, which is significantly higher than most other fuel cell types. MCFCs are known for their ability to efficiently utilize a variety of fuels, including hydrogen, natural gas, and even biogas, making them highly versatile in different applications. Their high operating temperature allows them to offer substantial efficiency benefits, especially in stationary power generation, where large-scale energy production is needed. MCFCs have gained attention for their potential in both CHP systems and large-scale industrial applications, due to their efficiency, fuel flexibility, and suitability for use in integrated energy systems. The operating principle of MCFCs is based on the electrochemical reaction that takes place between hydrogen (or other fuels) and oxygen (usually from air) to produce electricity, water, and heat. In MCFCs, hydrogen gas is supplied to the anode, where it is oxidized into H^+ and e^-. The electrons travel through an external circuit, generating electrical power, while the protons migrate through the molten carbonate electrolyte to the cathode. At the cathode, oxygen from the air combines with the electrons from the external circuit and the protons from the electrolyte to produce water and CO_2. The molten carbonate electrolyte serves not only as an ion conductor but also helps in the process of CO_2 absorption, which is unique to MCFC technology. This feature allows MCFCs to tolerate CO_2 in the anode feed gas, which is an important advantage when using natural gas or other carbon-based fuels [52].

One of the major advantages of MCFCs is their ability to achieve high electrical efficiencies, often in the range of 45% to 55%. When waste heat from the fuel cell is captured and used for cogeneration, the overall system efficiency can be further increased to 80% or higher. This makes MCFCs an excellent option for large-scale power generation applications, where both electricity and useful heat are required. The high operating temperature of MCFCs also means they are more efficient in the use of heat and can be integrated into CHP systems. The thermal energy produced by MCFCs can be used in industrial processes, heating buildings, or even for desalination purposes, adding significant value to the overall system. The flexibility of MCFCs is another important characteristic. These fuel cells can operate on a variety of fuels, including hydrogen, natural gas, biogas, and even landfill gas. The ability to use fuels other than pure hydrogen is a key advantage of MCFC technology, as it enables cost savings and greater fuel availability. By reforming natural gas or biogas into hydrogen and carbon monoxide before it enters the fuel cell, MCFCs are able to produce electricity from a range of feedstocks, which makes them attractive in regions where hydrogen infrastructure is still being developed. Furthermore, the ability to use carbon-based fuels like natural gas allows MCFCs to play a role in decarbonizing sectors that are difficult to electrify with renewable energy, such as heavy industry and high-demand industrial applications. In terms of durability and lifetime, MCFCs have shown considerable promise. The high operating temperature helps in maintaining a stable and efficient electrochemical reaction, and the materials used in MCFCs are designed to withstand the extreme conditions of the fuel cell's operation [53].

However, despite these advantages, there are challenges that must be overcome for wider adoption. One of the main challenges is the need for specialized materials that can endure the high temperatures and corrosive environment inside the fuel cell. The electrolyte, which is a molten carbonate salt, can be corrosive to the components of the cell if not properly managed, requiring high-quality materials that are resistant to corrosion and thermal stress. As a result, the cost of MCFC systems is typically higher than that of lower-temperature fuel cells like PEMFCs or PAFCs, though the overall system efficiency justifies the higher cost in many applications. Another challenge is the need for efficient CO_2 management. MCFCs inherently absorb CO_2 from the anode side, which is beneficial in some respects as it can help with fuel utilization and improve performance. However, this also requires the system to carefully manage CO_2 concentrations to avoid excessive accumulation that could reduce the efficiency of the cell over time. In addition, the handling of CO_2 and the by-products of the fuel cell process presents challenges in terms of emissions control and ensuring that the system operates at optimal efficiency. Despite these challenges, research into improving the performance, durability, and cost-effectiveness of MCFCs is ongoing. Efforts to develop advanced materials that are resistant to high temperatures and corrosion are key to reducing the long-term costs of MCFC technology. Advances in fuel reforming technologies are also helping to lower the cost of hydrogen production, making MCFCs more economically competitive. Additionally, MCFCs' ability to operate on diverse fuels makes them an attractive solution for a range of industries, particularly in areas where natural gas is abundant or where renewable fuels like biogas are available [54].

From an environmental perspective, MCFCs are considered a cleaner energy solution compared to traditional combustion-based power generation. Since their primary by-products are water, heat, and carbon dioxide, MCFCs produce fewer harmful pollutants such as NOx or particulate matter, which are typically associated with fossil fuel combustion. When operated on hydrogen or other renewable fuels, MCFCs offer a pathway toward low-emission power generation, contributing to efforts to decarbonize the energy sector. MCFCs represent a promising technology for large-scale power generation due to their high efficiency, fuel flexibility, and suitability for CHP systems. Although challenges remain, including the high cost and the need for specialized materials, ongoing advancements in material science and fuel reforming technology hold the potential to improve the performance and reduce the costs of MCFC systems. With their ability to operate on a variety of fuels and produce both electricity and useful heat, MCFCs have the potential to play a significant role in the global transition to cleaner, more efficient energy systems, particularly in industrial applications and large-scale power generation [55].

5.2.6 DMFCs

DMFCs are a type of fuel cell that directly uses methanol as a fuel, bypassing the need for a reforming process to convert methanol into hydrogen. This distinctive characteristic makes DMFCs particularly attractive for portable and small-scale applications, as they provide a convenient and easily transportable source of fuel. DMFCs operate similarly to other fuel cell technologies, using an electrochemical reaction to generate electricity, but with methanol being the direct fuel rather than hydrogen. The overall reaction in a DMFC involves the oxidation of methanol at the anode, which produces H^+, e^-, and CO_2, while the reduction of oxygen at the cathode combines with the protons and electrons to produce water. One of the primary advantages of DMFCs is their fuel source: methanol is a liquid at room temperature, making it easy to store and transport compared to gaseous hydrogen. This characteristic of methanol makes DMFCs particularly well-suited for applications where compactness, portability, and ease of refueling are crucial, such as in small electronic devices, laptops, smartphones, and even small vehicles. The liquid nature of methanol also allows for high energy density and easier handling compared to compressed hydrogen, which requires specialized infrastructure and storage conditions. These factors position DMFCs as a promising candidate for portable power generation, particularly in applications that require lightweight, compact, and efficient energy sources. In terms of operational principles, DMFCs function by utilizing a proton-conducting membrane, typically Nafion, which separates the anode and cathode. The anode is supplied with methanol and water, where methanol is oxidized into carbon dioxide and H^+. The protons pass through the electrolyte membrane to the cathode, while the electrons travel through an external circuit, generating electric current. At the cathode, oxygen from the air combines with the protons and electrons to form water. Unlike hydrogen fuel cells, DMFCs do not require a separate hydrogen production process, making them simpler to operate. The direct utilization of methanol allows for lower overall system complexity, which can lead to reduced costs in certain applications [56].

The main advantages of DMFCs are their high energy density, ease of fuel storage and transport, and the simplicity of their fuel delivery system. Methanol, being a liquid, provides a higher volumetric energy density than hydrogen gas, which is essential for applications requiring long-duration energy supply in a compact form. Additionally, methanol is widely available, and refueling infrastructure for methanol is already in place in many regions, reducing the need for the specialized infrastructure required for hydrogen fuel cells. This accessibility and convenience make DMFCs an ideal solution for powering portable electronic devices, military equipment, backup power systems, and other applications where weight and size constraints are significant concerns. Despite these advantages, DMFCs face several challenges that need to be addressed for broader adoption. One of the main challenges is the relatively low efficiency of DMFCs compared to other fuel cell types, such as PEMFCs. The efficiency of DMFCs can be reduced by methanol crossover, a phenomenon where methanol from the anode leaks through the membrane to the cathode side, leading to the consumption of methanol without contributing to the production of electricity. This results in the loss of fuel and reduced overall efficiency. Efforts are underway to develop better membranes that minimize methanol crossover and improve the efficiency of the fuel cells. Another issue that impacts the performance of DMFCs is the management of the reaction products, particularly the CO_2 generated during the oxidation of methanol. While CO_2 is a natural by-product of the fuel cell's operation, the accumulation of CO_2 can affect the long-term performance of the cell and the fuel cell stack, potentially requiring additional measures for CO_2 management. Some of the challenges in this area can be mitigated through the use of advanced fuel cell designs and better fuel processing techniques. Additionally, the energy conversion efficiency of DMFCs is still lower compared to other alternative energy sources like lithium-ion batteries, which are commonly used in portable electronics and vehicles [57].

The development of direct methanol fuel cells is actively progressing, with research focusing on improving efficiency, reducing costs, and enhancing durability. Several advancements have been made in the areas of catalyst development, membrane technology, and system integration. Research has led to the creation of more efficient catalysts for the methanol oxidation reaction, which has contributed to increased power output and better overall efficiency. Additionally, improvements in the membranes used in DMFCs are helping to reduce methanol crossover, increase proton conductivity, and enhance the overall performance of the fuel cell. As research continues, DMFCs are expected to become more viable for a wider range of applications, particularly in small-scale portable energy devices. In terms of environmental impact, DMFCs are considered a cleaner alternative to conventional combustion-based power generation. They produce minimal emissions, with the main by-products being water and carbon dioxide. When powered by renewable methanol, DMFCs offer the potential for a low-carbon solution in portable power generation. However, if conventional fossil fuel-based methanol is used, the environmental benefits are diminished due to the carbon emissions associated with methanol production. Nonetheless, when integrated with renewable energy sources such as biomass, DMFCs can play a role in decarbonizing the portable power sector. DMFCs offer a unique solution to portable and small-scale power generation by utilizing methanol as a direct fuel source. With advantages such as easy fuel storage, high energy density, and fuel availability, DMFCs are well-suited for applications in portable electronics and backup power systems. However, challenges such as low efficiency, methanol crossover, and CO_2 management remain significant barriers to widespread adoption. Ongoing research and technological advancements are focused on improving the efficiency, reducing the cost, and enhancing the overall performance of DMFCs. As these challenges are overcome, DMFCs hold promise for playing an essential role in the transition toward cleaner, more efficient energy systems for small-scale, portable applications [58, 59].

5.3 Mechanisms of Hydrogen Fuel Cells

Hydrogen fuel cells are electrochemical devices that convert chemical energy into electrical energy through the reaction of hydrogen with oxygen, without combustion. These fuel cells operate by facilitating an electrochemical reaction in which H_2 is supplied to the anode, and O_2 is provided to the cathode, leading to the generation of electricity, water, and heat. The mechanisms governing the operation of hydrogen fuel cells are fundamental to their performance and efficiency. Understanding the electrochemical reactions, the role of catalysts, hydrogen production, and efficiency is critical for advancing the technology and optimizing fuel cell systems for various applications. Below, we delve into the specific mechanisms that underpin the functioning of hydrogen fuel cells [60].

5.3.1 Electrochemical Reactions in Hydrogen Fuel Cells

The heart of a hydrogen fuel cell is the electrochemical reaction that occurs between hydrogen and oxygen. This reaction takes place at two electrodes, the anode and the cathode, separated by an electrolyte membrane. At the anode, hydrogen molecules (H_2) are introduced into the fuel cell, where they are split into H^+ and e^- by a catalyst. The protons pass through the electrolyte, which is typically a proton-conducting membrane, while the electrons flow through an external circuit, generating electricity [61].

The electrochemical process is divided into two half-reactions: the oxidation of hydrogen at the anode and the reduction of oxygen at the cathode. The anode reaction can be represented as:

$$H_2 \rightarrow 2H^+ + 2e^-$$

This reaction releases electrons, which travel through the external circuit, creating an electric current. Meanwhile, the hydrogen ions (protons) pass through the electrolyte towards the cathode. At the cathode, O_2 from the air reacts with the protons and electrons to form H_2O. The cathode reaction is as follows:

$$O_2 + 4H^+ + 4e^- \rightarrow 2H_2O$$

As the two half-reactions occur simultaneously, the hydrogen fuel cell produces electricity, water, and heat as by-products. This process is highly efficient, as it avoids the thermodynamic losses associated with traditional combustion methods, making fuel cells an environmentally friendly energy source. The direct conversion of chemical energy into electrical energy is central to the fuel cell's ability to deliver power efficiently with minimal waste.

5.3.2 Role of Catalysts in Fuel Cells

Catalysts are integral to the efficient functioning of hydrogen fuel cells, as they significantly accelerate the electrochemical reactions that occur within the cell, enabling the production of electricity from hydrogen. The fundamental role of a catalyst in a hydrogen fuel cell is to reduce the activation energy required for the chemical reactions at the electrodes (anode and cathode), thereby making the process more efficient and viable for practical applications. Without the use of catalysts, the reactions involved in hydrogen oxidation at the anode and oxygen reduction at the cathode would occur too slowly to be useful for large-scale energy production. The critical role of catalysts in hydrogen fuel cells is visually represented in Figure 5.4.

A. **Role of Catalysts at the Anode:** At the anode, H_2 is introduced to the fuel cell and is split into H^+ and e^- by the action of the catalyst. This process is crucial because it generates the electrons that flow through the external circuit to generate electricity. The anode reaction can be written as:

$$H_2 \rightarrow 2H^+ + 2e^-$$

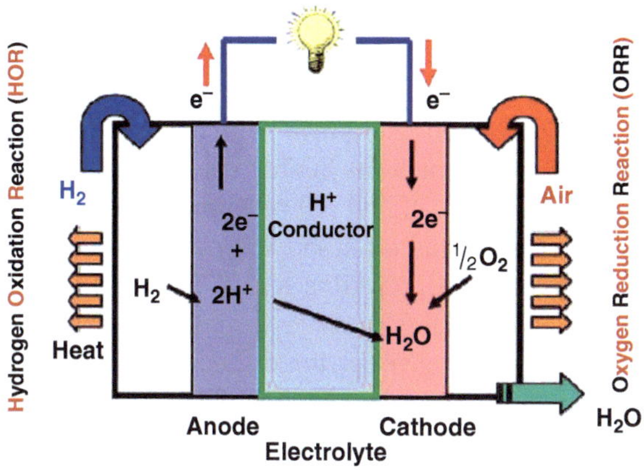

Figure 5.4 Diagram illustrating the role of platinum-based catalysts in hydrogen oxidation and ORRs. Reproduced with permission from [66]/Elsevier.

This reaction is highly dependent on the catalyst material, which helps facilitate the dissociation of hydrogen molecules into hydrogen atoms, which then combine with the catalyst surface to form hydrogen ions and release electrons. The catalyst used at the anode must be capable of efficiently adsorbing hydrogen molecules, breaking their bonds, and enabling the separation of electrons and protons with minimal energy input. The most commonly used catalyst material in hydrogen fuel cells is platinum due to its excellent electronic conductivity and chemical stability. Platinum has a unique ability to facilitate the breaking of the hydrogen-hydrogen bond, which is energetically difficult without a catalyst. However, platinum is an expensive and rare material, making it a costly choice for widespread commercial adoption. As a result, there is a strong push for research into alternative catalysts that could lower the cost of fuel cell technology without sacrificing efficiency. To enhance the efficiency of platinum-based catalysts, researchers are exploring several methods, such as nanostructuring platinum or combining it with other metals like palladium or ruthenium. This approach can reduce the amount of platinum needed while maintaining or even improving performance. For example, using platinum alloys or nanoparticles increases the available surface area for reactions, allowing for more effective hydrogen dissociation at a lower cost [62].

B. Role of Catalysts at the Cathode: At the cathode, O_2 from the air react with hydrogen ions (H^+) that have passed through the electrolyte and electrons from the external circuit to form H_2O as a byproduct. This reaction is called the oxygen reduction reaction (ORR), and it is significantly slower and more difficult than the hydrogen oxidation reaction at the anode. In the case of oxygen reduction, the catalyst is required not only to speed up the reaction but also to lower the energy barrier for oxygen molecules to accept electrons and combine with protons to form water [63].

The cathode reaction can be written as:

$$O_2 + 4H^+ + 4e^- \rightarrow 2H_2O$$

Platinum is also commonly used as a catalyst at the cathode for similar reasons as at the anode, namely its ability to facilitate electron transfer and bond breaking in oxygen molecules. However, the ORR is a slower process than hydrogen oxidation, requiring a more complex catalyst design to improve the rate of the reaction. Platinum's effectiveness at the cathode is somewhat limited due to the relatively slow kinetics of the ORR, which hinders overall fuel cell performance. To address this challenge, research into non-precious metal catalysts, such as iron, cobalt, and nitrogen-doped carbon materials, has shown promising results in enhancing the oxygen reduction reaction. These materials are much less expensive than platinum and may provide a cost-effective alternative for the cathode catalyst in fuel cells. For example, iron-based catalysts have been shown to exhibit enhanced activity for ORR, and combining them with carbon nanotubes or graphene can further improve conductivity and performance. A key area of focus in catalyst development is increasing the surface area of the catalyst material. Higher surface area means more active sites for the electrochemical reactions to occur, which in turn leads to higher efficiency. Nanostructured catalysts, such as nanoparticles, nanowires, and thin films, are being explored for both anode and cathode applications. These nanomaterials provide an enormous increase in surface area, which enhances the rate of both hydrogen oxidation and oxygen reduction reactions, improving the overall efficiency of the fuel cell [64].

While platinum remains the gold standard for fuel cell catalysts due to its proven efficiency, its high cost and scarcity are significant barriers to the widespread adoption of hydrogen fuel cells. The scarcity of platinum, coupled with the fact that it is a precious metal, makes it unsustainable for large-scale applications. As a result, a significant portion of current research in fuel cell technology is dedicated to finding alternatives to platinum-based catalysts. A major challenge is designing catalysts that can maintain high efficiency while being low-cost and durable. Catalyst durability is another critical concern. Fuel cell catalysts degrade over time due to the extreme conditions within the fuel cell, such as high temperatures, voltages, and the presence of impurities in the hydrogen fuel. This degradation can lead to a reduction in performance and the need for frequent replacement, which adds to the cost of fuel cell systems. To overcome these challenges, researchers are investigating a wide range of materials and designs. Non-precious metal catalysts, such as transition metals and metal oxides, are promising candidates for reducing the dependence on platinum. Another approach involves using biomimetic catalysts, which replicate the mechanisms found in natural processes, such as enzymes that catalyze hydrogen oxidation and oxygen reduction in living organisms. By mimicking these natural systems, researchers hope to develop highly efficient, cost-effective catalysts for fuel cell applications [65].

5.3.3 Fuel Cell Efficiency and Performance

Fuel cell efficiency is a critical factor in determining the viability and attractiveness of hydrogen fuel cells for large-scale applications. The efficiency of a fuel cell depends on various factors, including the type of fuel cell, the operating conditions, and the quality of the fuel used. A key measure of fuel cell performance is its electrical efficiency, which refers to the amount of energy converted from the chemical energy of hydrogen into usable electrical energy. Fuel cells can theoretically achieve efficiencies of up to 60%, and even higher when waste heat is utilized in CHP systems [67].

The electrical efficiency of a hydrogen fuel cell can be defined as the ratio of electrical output power to the energy input (in the form of hydrogen fuel). In ideal conditions, a hydrogen fuel cell can operate at high efficiency because the conversion of hydrogen to electricity is a relatively direct process, avoiding the energy losses associated with conventional combustion-based power generation. The efficiency of hydrogen fuel cells is usually higher compared to internal combustion engines and other energy conversion technologies. However, efficiency can vary significantly based on the type of fuel cell and the operating environment. PEMFCs, one of the most common types used in transportation applications, typically operate at efficiencies of 40–60% depending on the specific system design and operating conditions. For SOFCs, which are primarily used in stationary power generation, the efficiency can be higher, ranging from 45% to 60%. When SOFCs are integrated into combined heat and power systems, the overall efficiency can rise to as much as 85%, as the excess heat generated in the fuel cell is used for heating purposes [68].

Another important factor influencing the overall performance of hydrogen fuel cells is thermal efficiency. Thermal efficiency refers to the effectiveness of a fuel cell system in converting the energy released as heat into useful work. In traditional fuel cells, a significant portion of the energy is lost as heat, which reduces the overall system efficiency. However, by using this heat for other purposes, such as heating water or providing space heating, the overall energy efficiency of the system can be greatly improved. This is particularly the case with high-temperature fuel cells such as SOFCs and MCFCs. These cells generate considerable amounts of heat as a byproduct of the electrochemical reactions. When this heat is captured and used in CHP systems, the total efficiency of the system can reach levels much higher than that of a stand-alone power generation system. In this way, hydrogen fuel cells can become a more efficient and sustainable option, contributing to both electricity generation and thermal energy needs [69].

Several factors can influence the efficiency of hydrogen fuel cells. These include the type of fuel cell, its operating temperature, the quality of the hydrogen fuel, and the design of the fuel cell stack. One of the most important considerations is the electrochemical reaction kinetics. The rate at which hydrogen reacts at the anode and oxygen reacts at the cathode determines the efficiency of the cell. Higher reaction rates result in higher current densities, which contribute to increased power output and efficiency. Fuel cell temperature plays a significant role in determining efficiency as well. For instance, PEMFCs operate best at temperatures between 60–80 °C, where the electrochemical reactions proceed efficiently. In contrast, high-temperature fuel cells like SOFCs can operate at temperatures of 600–1,000 °C, which allows for greater energy conversion efficiency but may also lead to material degradation if the system is not carefully designed. Another factor affecting efficiency is the quality of hydrogen fuel. Hydrogen fuel must be of a high purity to avoid contamination of the catalyst and other components of the fuel cell. Impurities in the hydrogen, such as carbon monoxide, can poison the catalyst, leading to decreased performance and efficiency over time [70].

Beyond efficiency, the overall performance of a hydrogen fuel cell system is determined by a combination of various metrics, including power density, lifespan, and durability. Power density refers to the amount of power generated per unit of fuel cell volume or weight. High power density is particularly important for mobile applications like hydrogen-powered vehicles, where space and weight constraints are critical. The lifetime of a fuel cell is another important performance metric. Over time, fuel cells can degrade due to factors such as catalyst poisoning, material wear, and thermal cycling. A fuel cell's durability can be measured in terms of operating hours, and achieving longer lifetimes is critical for reducing the overall cost of fuel cell systems. Fuel cell manufacturers continuously work on improving materials and designs that can extend the lifespan and reduce the maintenance needs of fuel cell systems. Table 5.1 highlights the differences in efficiency, power density, and operating conditions for various fuel cell types. The selection of the appropriate fuel cell technology for a specific application depends on several factors, including power requirements, operating environment, and cost considerations. Through ongoing advancements in materials and design, the performance of hydrogen fuel cells is expected to improve, further enhancing their potential for widespread adoption across various industries [71]. Figure 5.5 compares the efficiencies and operating temperatures of various hydrogen fuel cells.

Future research in fuel cell technology is focused on improving both efficiency and performance. Key areas of development include reducing the operating costs, improving the fuel cell stack design, and enhancing the performance of catalysts

Table 5.1 A comparison of the efficiency, power density, and typical applications of various types of hydrogen fuel cells.

Fuel cell type	Efficiency (electrical) (%)	Power density (W/kg)	Operating temperature (°C)	Typical applications
PEMFC	40–60	500–1,000	60–80	Automotive, portable electronics, backup power systems [72]
SOFC	45–60	200–500	600–1,000	Stationary power generation, industrial CHP [73]
AFC	50–60	400–1,200	50–100	Space applications, backup power systems [74]
PAFC	40–45	200–500	150–220	Stationary power generation, industrial applications [75]
MCFC	50–60	250–600	600–700	Industrial power generation, large-scale CHP [76]
DMFC	20–30	100–200	50–90	Portable electronics, military applications [77]

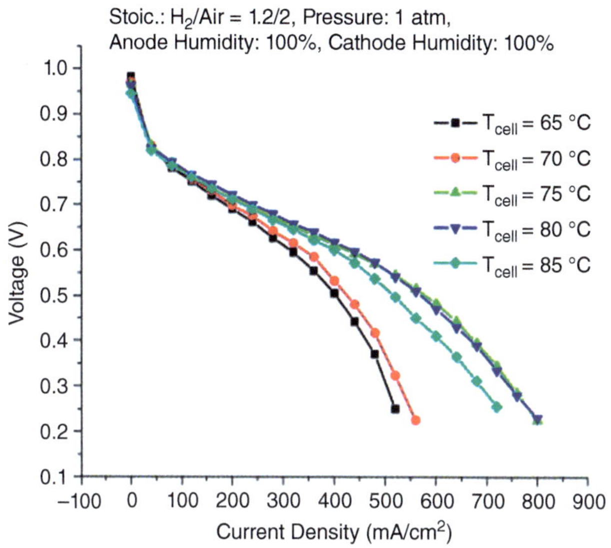

Figure 5.5 Graphical comparison of operating temperature ranges and efficiency levels for common hydrogen fuel cell types. Reproduced with permission from [78]/Elsevier.

and electrolytes. Innovations in materials science, nanotechnology, and computational modeling are likely to play a significant role in achieving these goals. By improving the efficiency and performance of fuel cells, researchers aim to make hydrogen fuel cells a more viable and cost-competitive alternative to conventional power generation and energy storage technologies.

5.4 Fuel Cell Components and Design

The design and functionality of hydrogen fuel cells rely on several key components that work together to convert chemical energy into electrical energy. These components are optimized for specific fuel cell types and applications, but they all follow a similar design principle based on the electrochemical conversion process. The performance, durability, and efficiency of a fuel cell are largely determined by the materials used and the configuration of these components. A deeper understanding of the individual fuel cell components, including the membrane electrolyte assembly (MEA), electrodes and catalysts, bipolar plates, and current collectors, is crucial for comprehending how fuel cells operate and how their performance can be improved. Fuel cell technology is made up of several critical components that each contribute to the overall electrochemical process. The MEA is the heart of the fuel cell, where the hydrogen and oxygen reactions occur, generating electricity. The electrodes facilitate the electrochemical reactions at both the anode and cathode, while the catalysts ensure that these reactions occur at a high rate. To facilitate the efficient flow of hydrogen and oxygen to the MEA, bipolar plates and current collectors are used to ensure proper distribution of reactants and to collect the electrical output. Finally, the fuel and oxidant supply systems provide the hydrogen and oxygen (or air) required for the electrochemical reactions to occur [79].

5.4.1 MEA

The MEA is the core component of a hydrogen fuel cell and plays a central role in its operation. It consists of a proton-conducting electrolyte membrane, which separates the anode and cathode and facilitates the transport of protons (hydrogen ions) from one side of the cell to the other. The MEA is composed of several layers, including the electrolyte membrane, gas diffusion layers, and catalyst layers, all of which work together to enable the electrochemical reactions that produce electricity. The electrolyte membrane is usually made of Nafion, a polymer membrane that has excellent proton conductivity, which is crucial for the electrochemical process. The catalyst layers on either side of the membrane typically consist of platinum or platinum alloys, which facilitate the oxidation of hydrogen at the anode and the reduction of oxygen at the cathode. These catalyst layers promote the reactions that generate the free electrons (which flow through the external circuit, generating electricity) and protons. The gas diffusion layers (GDLs) are placed on both sides of the membrane and serve as channels that help distribute the reactant gases (hydrogen and oxygen/air) evenly over the catalyst surfaces. The MEA's efficiency is strongly dependent on the performance of these materials and layers. Optimizing the thickness and porosity of the gas diffusion layers, the structure of the catalyst layers, and the ion conductivity of the electrolyte membrane can significantly enhance the performance of the fuel cell. Research into alternative, less expensive materials for the MEA – such as non-precious metal catalysts – continues to be a priority in reducing fuel cell costs and improving long-term viability [80]. Figure 5.6 illustrates the essential components of a hydrogen fuel cell stack.

5.4.2 Electrodes and Catalysts

The electrodes in a fuel cell are responsible for facilitating the electrochemical reactions that occur during the conversion of hydrogen into electricity. There are two electrodes in a typical fuel cell: the anode, where hydrogen is oxidized, and the cathode, where oxygen is reduced. These electrodes serve as sites for the chemical reactions that produce electricity and are essential for fuel cell performance. The anode typically consists of a porous material, often carbon-based, which allows hydrogen

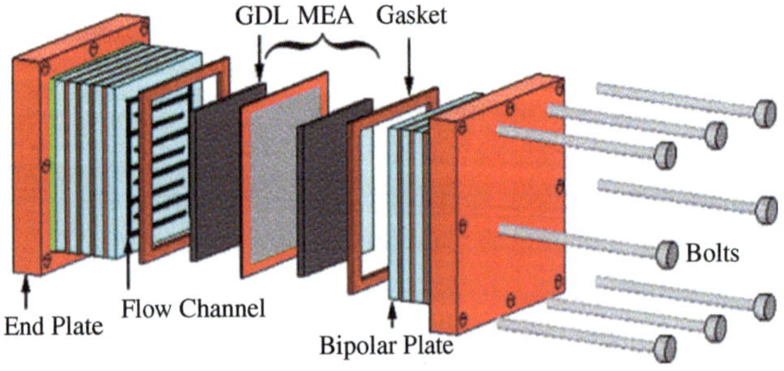

Figure 5.6 Exploded view of a fuel cell stack showing the MEA, bipolar plates, current collectors, and flow fields. Reproduced with permission from [81]/Elsevier.

gas to diffuse into the electrode surface where it is split into protons and electrons. The protons pass through the electrolyte membrane, while the electrons travel through the external circuit, creating the electrical current. The cathode is similarly constructed, but its function is to reduce oxygen from the air to form water when it combines with the protons and electrons coming from the external circuit. This reduction reaction is crucial for completing the electrochemical cycle. The catalysts in the electrodes, particularly at the anode, play a critical role in speeding up these electrochemical reactions. Platinum is commonly used as a catalyst because of its high efficiency in facilitating the oxidation of hydrogen and reduction of oxygen. However, platinum is expensive, and researchers are working to identify alternative, cheaper materials such as palladium, ruthenium, and non-precious metals that can perform similarly to platinum at a lower cost. The development of thin-film catalysts and nanocatalysts aims to increase efficiency while reducing the amount of platinum or other precious metals needed. The design of the electrodes must also ensure that the gas diffusion layers are appropriately structured to allow optimal gas flow to the catalyst sites. The pores must be sized to balance the distribution of gas with the ability of the electrode to conduct electrical current [82].

5.4.3 Bipolar Plates and Current Collectors

The bipolar plates are essential components in the fuel cell stack that distribute reactant gases (hydrogen and oxygen) to the anode and cathode, respectively, and also carry away the water produced as a byproduct of the reaction. Bipolar plates are placed between individual fuel cells in the stack and serve as both gas flow distributors and electrical conductors. These plates also help in the overall assembly of the fuel cell stack and in managing heat distribution. The materials used for bipolar plates must have high electrical conductivity, resistance to corrosion, and good mechanical properties to ensure longevity. Graphite has traditionally been used for bipolar plates due to its conductivity and stability, but its brittleness and tendency to degrade over time have led to the exploration of metallic materials such as stainless steel, titanium, and aluminum. These metals provide better mechanical strength and durability, though they come with challenges such as corrosion resistance and higher manufacturing costs. Another important component is the current collectors, which help collect the electrical current generated by the fuel cell reactions and guide it out of the fuel cell stack. These components must be designed to maximize electrical conductivity and minimize losses. They are often integrated with the bipolar plates to streamline the overall design [83].

5.4.4 Fuel and Oxidant Supply Systems

The fuel and oxidant supply systems are responsible for providing the necessary reactants (hydrogen and oxygen/air) to the fuel cell and ensuring that they are at the appropriate pressure, flow rate, and humidity for optimal performance. The hydrogen supply system typically consists of a hydrogen storage tank, a pressure regulator, and a series of valves and pipes to deliver hydrogen to the anode. The oxygen or air supply system includes an air compressor or blower to deliver the oxygen to the cathode, ensuring that the necessary gas pressure is maintained. The hydrogen storage method is a key consideration in fuel cell design. Hydrogen can be stored in various ways, including compressed hydrogen tanks, liquid hydrogen, or metal hydride systems. Each method has its own advantages and trade-offs in terms of energy density, weight, and storage conditions. The storage system must be designed to safely handle hydrogen under high pressure or at cryogenic temperatures. The air supply is typically pressurized to ensure a steady flow of oxygen to the cathode. The system also includes humidifiers to add moisture to the air and prevent dehydration of the electrolyte membrane. Too little moisture can cause the electrolyte to become non-conductive, leading to a decrease in performance, while excessive moisture can lead to flooding in the fuel cell.

Each of the components in a hydrogen fuel cell plays a critical role in ensuring efficient operation and long-term durability. The MEA is the heart of the cell, facilitating the crucial proton exchange process, while the electrodes and catalysts enable the electrochemical reactions that generate electricity. The bipolar plates and current collectors distribute gases and collect the electrical output, and the fuel and oxidant supply systems ensure that the fuel cell receives the necessary reactants to function effectively. Together, these components make hydrogen fuel cells a promising technology for a wide range of applications, from transportation to stationary power generation, contributing to a cleaner and more sustainable future [84].

5.5 Fuel Cell Durability and Lifetime

Fuel cell technology holds significant promise as a clean and efficient energy solution. However, one of the major challenges in bringing hydrogen fuel cells to large-scale commercial applications is their durability and lifetime. While fuel cells offer great efficiency and environmental benefits, ensuring that they can operate for extended periods without significant degradation is critical for their widespread adoption. The lifespan of a fuel cell depends on various factors such as operating conditions, materials used, and maintenance practices. Table 5.2 summarizes the factors affecting fuel cell durability. This section examines the factors affecting fuel cell lifespan, strategies to improve durability, and the importance of long-term performance in reducing costs. A long and reliable fuel cell lifetime is essential to make the technology competitive with other energy sources, particularly in sectors such as transportation and stationary power generation. The challenge lies in addressing degradation mechanisms that can affect performance over time, such as catalyst poisoning, membrane degradation, and material fatigue. Therefore, a comprehensive understanding of durability factors and the implementation of strategies to enhance longevity is crucial for optimizing fuel cell systems for both commercial viability and sustainable development [85].

5.5.1 Factors Affecting Fuel Cell Lifespan

Several factors impact the lifespan of a hydrogen fuel cell. These factors can be broadly categorized into materials degradation, operational stress, and environmental conditions.

Table 5.2 Factors affecting fuel cell durability.

Factor	Impact on durability	Strategies for mitigation
Catalyst degradation	Loss of catalytic activity, especially for platinum-based catalysts	Use of non-precious metals, catalyst encapsulation, and bimetallic catalysts [91]
Membrane degradation	Reduced proton conductivity due to chemical or physical damage	Development of more durable, high-temperature-resistant membranes [92]
Operating conditions	High temperature, fluctuating humidity, and load cycling accelerate degradation	Optimization of thermal management, load cycling, and humidity control [93]
Corrosion	Corrosion of bipolar plates and electrodes reduces efficiency	Use of corrosion-resistant materials, protective coatings [94]
Fuel impurities	Impurities in hydrogen or air can poison catalysts	Hydrogen purification and air filtration systems [95]

a. **Materials Degradation:** The most critical factor affecting fuel cell longevity is the degradation of materials, especially the MEA and catalysts. The PEM (typically made of Nafion® or similar materials) is susceptible to chemical degradation over time, especially under high temperatures, varying humidity, and oxidative environments. This degradation can lead to reduced proton conductivity and performance degradation. Catalysts, particularly platinum used in the anode and cathode, are also vulnerable to poisoning from impurities in the hydrogen or air supply, as well as flooding and drying issues that can disrupt the catalyst's effectiveness. The loss of catalytic activity reduces the efficiency of the fuel cell [86].

b. **Operational Stress:** Operating conditions such as temperature fluctuations, load cycling, and voltage fluctuations can significantly impact fuel cell performance and longevity. Inadequate thermal management can lead to thermal cycling, causing mechanical stress on components and accelerating material fatigue. Frequent load cycling, or rapid changes in power demand, can exacerbate these effects, accelerating the wear on internal components like the bipolar plates and the MEA. Additionally, high currents and voltages during operation can lead to electrical degradation [87].

c. **Environmental Conditions:** The environment in which the fuel cell operates can significantly affect its performance. Contaminants in the air supply (such as sulfur, carbon monoxide, or nitrogen compounds) can react with fuel cell materials, leading to catalyst poisoning and electrode degradation. Humidity levels are another crucial factor; too much humidity can cause flooding of the catalyst layer, while too little moisture can cause dehydration of the proton-conducting membrane, leading to reduced conductivity. Additionally, corrosive elements in the external environment can affect the structural integrity of components like the bipolar plates, which need to be resistant to corrosion [88].

5.5.2 Strategies for Improving Durability

To extend the lifespan of hydrogen fuel cells, several strategies have been developed to address the factors that contribute to material degradation and operational stress. These strategies focus on improving materials and system design as well as optimizing operational conditions.

a. **Improved Materials:** Developing more durable materials for the MEA, catalyst layers, and electrodes is one of the most important strategies for enhancing fuel cell lifespan. Researchers are exploring alternatives to platinum catalysts, such as non-precious metal catalysts, that can maintain catalytic efficiency for longer periods. Furthermore, novel membrane materials with better resistance to chemical degradation and enhanced proton conductivity at high temperatures are being investigated. Materials that are less susceptible to flooding and drying in the catalyst layers are also being developed.
b. **Advanced Catalyst Protection:** Catalyst durability can be improved through techniques such as catalyst encapsulation, where a protective layer is added around the catalyst particles to prevent degradation from impurities. Additionally, the development of bimetallic catalysts (which use a combination of two or more metals) can increase resistance to poisoning and improve performance over a longer lifespan. Platinum group metal (PGM)-free catalysts are also being explored to reduce the cost and improve the longevity of the catalyst. Figure 5.7 outlines the primary factors affecting fuel cell durability.
c. **Optimized Operating Conditions:** Optimizing the operating conditions of fuel cells can significantly extend their life. Maintaining steady-state operating temperatures and ensuring that the fuel cell operates within ideal humidity levels are critical for preventing excess moisture or dehydration. Additionally, systems that can monitor operational stress in real-time, adjusting parameters such as fuel flow rate and temperature, can prevent extreme conditions that lead to degradation. Load management systems that smooth out fluctuations in power demand can reduce wear on the system by avoiding frequent start-stop cycles.
d. **Protective Coatings and Corrosion Resistance:** Using protective coatings for bipolar plates and electrodes can enhance their corrosion resistance, which is especially important for components exposed to the harsh external environment. Materials such as carbon-based coatings or conductive polymers can help reduce the impact of corrosion, extending the life of these components. In addition, the integration of corrosion inhibitors into the fuel cell stack can help protect critical components from degradation due to exposure to environmental factors [89].

5.5.3 Long-term Performance and Cost Reduction

The long-term performance and cost reduction of fuel cells are closely linked. A fuel cell that deteriorates quickly or requires frequent maintenance will be more expensive to operate over time, limiting its commercial potential. Conversely, improving durability leads to longer operational life and lower total cost of ownership, making hydrogen fuel cells more competitive in various applications, particularly in transportation and stationary power generation. By addressing

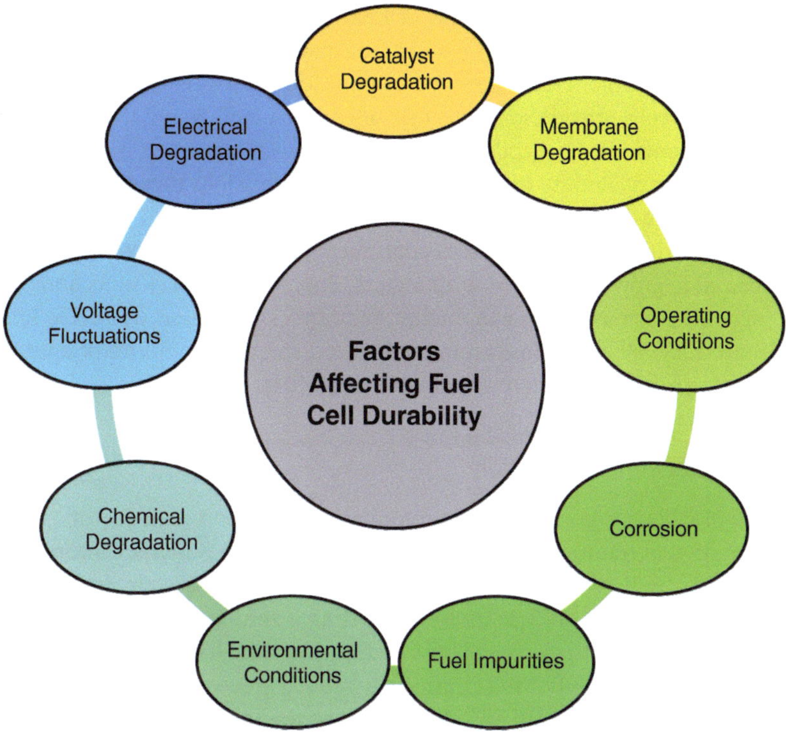

Figure 5.7 Factors affecting fuel cell durability.

material degradation, optimizing system design, and reducing operational stress, the lifetime of hydrogen fuel cells can be significantly extended. Ideally, fuel cells should be able to function for 5,000 to 10,000 hours or more before requiring significant maintenance or replacement of components. Fuel cell systems designed for high durability can maintain stable performance and efficiency over their operational life, which reduces the need for costly replacements and repairs. As the durability of fuel cells improves, the costs of fuel cell technology will likely decrease. One of the major barriers to the widespread adoption of hydrogen fuel cells has been their high initial cost, largely due to the use of precious metals like platinum. With the development of cheaper catalysts, durable membranes, and more efficient designs, the cost per kilowatt of fuel cell systems is expected to drop. Moreover, improvements in manufacturing techniques and economies of scale will further contribute to cost reduction. The longer a fuel cell lasts, the lower the overall cost per unit of energy becomes, making hydrogen fuel cells a more cost-effective option in the long term [90].

5.6 Applications of Hydrogen Fuel Cells

Hydrogen fuel cells are emerging as a promising energy technology, with applications across various sectors due to their efficiency, scalability, and environmental benefits. The diverse uses of hydrogen fuel cells make them an attractive alternative to traditional power generation and energy storage technologies. In this section, we will explore some of the key applications of hydrogen fuel cells, focusing on their role in transportation, stationary power generation, backup power systems, and portable devices and consumer electronics. Each application has unique requirements, but hydrogen fuel cells provide solutions that offer sustainability and innovation in meeting energy demands. Fuel cells offer significant advantages over conventional energy technologies, including higher efficiency, zero emissions (apart from water vapor), and the ability to generate electricity from renewable resources. As the global shift towards sustainable energy sources continues, the adoption of hydrogen fuel cells is expected to grow rapidly across all sectors, driven by advances in technology and the increasing demand for clean energy solutions. This section will delve into how hydrogen fuel cells are transforming industries, shaping the future of energy, and contributing to a cleaner, more sustainable world [96].

5.6.1 Fuel Cells in Transportation

Hydrogen fuel cells are poised to revolutionize the transportation sector, offering an alternative to traditional internal combustion engine vehicles powered by gasoline or diesel. In this context, fuel cells produce electricity from hydrogen and oxygen, which powers an electric motor, providing a zero-emission solution for various types of vehicles, from passenger cars to heavy-duty trucks. Fuel cell electric vehicles have the potential to significantly reduce greenhouse gas emissions and air pollutants, making them an integral part of the transition to clean transportation. Fuel cells have a distinct advantage over battery electric vehicles (BEVs), particularly in applications where long-range and quick refueling are crucial. For example, hydrogen fuel cells provide high energy density, which enables vehicles to cover longer distances compared to conventional electric vehicles, which require long charging times. Furthermore, hydrogen fuel cells can be refueled in a matter of minutes, unlike electric vehicles, which typically require hours to recharge. This makes hydrogen a promising solution for heavy-duty trucks, buses, and commercial vehicles that need to operate for extended hours and in demanding conditions. The transportation industry is already witnessing global adoption of hydrogen fuel cells. For example, Japan, South Korea, and several European countries have already launched commercial fleets of hydrogen-powered buses, trucks, and passenger vehicles. Hydrogen-powered train systems are also under development, particularly in regions where electrification of railway lines is not feasible due to geographical or logistical challenges. Additionally, major automotive manufacturers such as Toyota, Hyundai, and Honda have developed and commercialized hydrogen fuel cell passenger vehicles, including the Toyota Mirai and Hyundai Nexo. This growth in hydrogen-powered transportation reflects the industry's commitment to reducing carbon emissions and addressing the environmental challenges posed by conventional fuel-powered vehicles [97].

5.6.2 Fuel Cells in Stationary Power Generation

Hydrogen fuel cells also have significant potential in stationary power generation, providing a reliable and efficient way to generate electricity for both residential and industrial applications. Fuel cells can be used as primary power sources or as backup solutions in areas where grid reliability is critical. Unlike traditional power generation methods that rely on burning fossil fuels, hydrogen fuel cells produce electricity without emitting harmful pollutants, making them an attractive option for areas that prioritize clean energy solutions. Fuel cells used in stationary power generation offer several advantages, including high efficiency, modular scalability, and low maintenance requirements. These fuel cell systems can be deployed in a variety of settings, including remote areas, data centers, hospitals, and military installations. In addition, they can be combined with renewable energy sources, such as solar or wind power, to form hybrid systems that offer reliable and sustainable electricity generation. One promising application is the use of hydrogen fuel cells in distributed energy systems, where small-scale power generation units are installed near the point of use, reducing the need for long-distance transmission lines and minimizing energy loss. Fuel cell-based power systems can also be integrated with CHP applications, where both electricity and thermal energy (such as hot water or space heating) are generated simultaneously, improving the overall efficiency of the system [98].

5.6.3 Fuel Cells for Backup Power and Remote Applications

Hydrogen fuel cells are also an ideal solution for providing backup power in the event of grid failures or disruptions. Traditional backup power systems, such as diesel generators, have environmental drawbacks, emitting greenhouse gases and air pollutants. In contrast, hydrogen fuel cells offer a clean alternative, generating electricity with zero emissions while providing the necessary reliability and autonomy required for critical applications. Fuel cells are especially well-suited for use in remote locations where conventional power infrastructure is not available or feasible. These systems can operate off-grid, providing electricity for isolated communities, telecommunications towers, and remote industrial sites. The ability to store hydrogen and produce electricity on-demand makes fuel cells particularly effective in emergency power and disaster relief scenarios. Furthermore, the modular nature of fuel cell systems allows them to be scaled according to the specific needs of the application, ensuring flexibility in a variety of situations. In addition to providing backup power, hydrogen fuel cells can support energy security in areas with unreliable or aging electrical grids. For instance, fuel cells can be used to supply power to critical infrastructure such as hospitals, military bases, and research facilities, where continuous power supply is crucial [99].

5.6.4 Fuel Cells in Portable Devices and Consumer Electronics

Hydrogen fuel cells are also being explored for use in portable devices and consumer electronics, such as laptops, smartphones, and portable power banks. While battery technology currently dominates the portable device market, hydrogen fuel cells present a promising alternative that could offer significant advantages, such as longer run times and faster refueling times. Fuel cells are capable of providing higher energy density than traditional batteries, meaning they can potentially extend the operating time of devices without increasing their size. Additionally, fuel cells can be refueled quickly, unlike conventional batteries that require recharging.

These features could revolutionize the consumer electronics market, particularly for users who require continuous power in remote or outdoor environments, such as campers, travelers, and emergency responders. Researchers are already developing miniature fuel cell systems that are small enough to be integrated into portable devices, offering a clean and efficient alternative to rechargeable batteries. As fuel cell technology continues to advance and become more affordable, it is likely that portable fuel cells will become more common in consumer electronics and Internet of Things devices [100].

5.7 Recent Advances and Research in Fuel Cell Technology

The field of hydrogen fuel cell technology has seen significant progress in recent years, driven by advances in materials science, electrochemical engineering, and the integration of renewable energy systems. These developments are not only improving the performance and efficiency of fuel cells but also expanding their potential applications in diverse sectors. This section discusses some of the most recent advances in fuel cell technology, with a focus on next-generation fuel cell designs, innovations in catalyst materials, and fuel cell integration with renewable energy systems. These innovations are helping to overcome existing challenges and paving the way for widespread adoption of hydrogen fuel cells in both industrial and consumer applications [101].

5.7.1 Next-generation Fuel Cell Designs

One of the most exciting areas of research in fuel cell technology is the development of next-generation fuel cell designs. These designs aim to address the limitations of traditional fuel cell systems, such as high costs, limited durability, and inefficient performance at varying operating conditions. Researchers are exploring several new concepts and architectures to improve fuel cell functionality, making them more suitable for large-scale commercial use. One significant advancement is the development of solid-state fuel cells, which aim to replace the liquid or polymer electrolytes in conventional fuel cells with solid-state materials. Solid-state fuel cells offer numerous benefits, including enhanced thermal stability, reduced maintenance, and higher efficiency. These systems are particularly attractive for applications in automotive and stationary power generation due to their potential for operating at higher temperatures, which can increase energy output and reduce the need for complex cooling systems. Additionally, microfuel cells are being researched for use in portable devices and small-scale applications. These compact fuel cells are designed to provide reliable power for consumer electronics, remote sensors, and medical devices. Microfuel cells offer advantages in terms of size and weight, as well as the ability to operate for long periods without frequent recharging, making them a promising option for next-generation wearables and IoT devices. Researchers are also focusing on MEA advancements to improve fuel cell performance. New MEAs incorporate nanoscale materials and hybrid structures that enhance proton conductivity while reducing the need for expensive platinum catalysts. These innovations are expected to lower the overall cost of fuel cells while improving their efficiency and lifespan [102].

5.7.2 Innovations in Catalyst Materials

Catalysts are central to the functioning of fuel cells, as they facilitate the electrochemical reactions that convert hydrogen into electricity. The primary challenge with conventional catalysts, particularly platinum, is their high cost and susceptibility to degradation over time. To address these issues, researchers are exploring innovative catalyst materials that can enhance the performance of fuel cells while reducing costs. One promising area of research involves the development of non-platinum catalysts, which aim to replace platinum with more abundant and affordable materials, such as nickel, iron, and cobalt. These alternative catalysts can offer similar performance to platinum at a fraction of the cost, making fuel cells more economically viable for mass-market applications. Recent breakthroughs in nano-catalysis have enabled researchers to create nano-sized catalyst particles with improved surface area and activity, which can lead to better catalytic performance with lower material usage. Another innovation in catalyst technology is the use of biomimetic catalysts, which mimic the structure and function of natural enzymes. Enzymes found in nature, such as those involved in biological hydrogen production, are highly efficient catalysts. By mimicking these natural processes, researchers are developing bio-inspired catalysts that can operate more efficiently at lower temperatures, enhancing fuel cell performance while reducing the need for costly materials. Moreover, researchers are investigating composite catalysts that combine several different materials to optimize the performance of fuel cells. These composites can be tailored to improve the overall efficiency of fuel cells by enhancing the electrocatalytic activity while reducing degradation rates. This is particularly important for the longevity of fuel cells, as catalyst degradation is one of the primary factors limiting their lifespan [103].

5.7.3 Fuel Cell Integration with Renewable Energy Systems

The integration of fuel cells with renewable energy systems is a key area of research aimed at improving the sustainability and scalability of hydrogen as a clean energy source. Hydrogen fuel cells can be used to store energy produced by renewable sources such as solar and wind, which are intermittent in nature. By producing hydrogen from excess renewable electricity, fuel cells can serve as an effective energy storage solution, allowing surplus energy to be stored and used when renewable generation is low or demand is high. One example of this integration is the development of power-to-gas systems, where excess renewable electricity is used to produce hydrogen through electrolysis. The hydrogen can then be stored and used in fuel cells to generate electricity or heat. This system can help balance supply and demand by enabling the storage of renewable energy and providing a dispatchable power source when needed. In addition to storing energy for electricity generation, hydrogen can also be used in industrial applications, transportation, and home heating, providing a versatile solution for decarbonizing multiple sectors. Furthermore, fuel cells can complement renewable energy systems by providing grid stability and demand-side management. In regions where the grid is reliant on renewable power sources, fuel cells can be integrated with the grid to provide load balancing and frequency regulation. By using fuel cells to smooth out the fluctuations in renewable generation, the energy system becomes more stable, ensuring that there is always a reliable source of power. Another area of research is the development of hybrid systems, where hydrogen fuel cells are combined with battery storage or solar panels to create fully integrated energy solutions. These systems can optimize the strengths of each technology, providing a more resilient and efficient energy supply. Hybrid systems can be deployed in a variety of settings, from residential buildings to commercial enterprises and microgrids, ensuring that energy demand is met with minimal environmental impact [104].

5.8 Environmental Impact and Sustainability of Hydrogen Fuel Cells

Hydrogen fuel cells are increasingly recognized as a key technology in the transition to a more sustainable energy future. Their ability to generate electricity with zero emissions at the point of use, combined with the potential for producing hydrogen from renewable sources, makes them an attractive alternative to fossil fuel-based energy systems. However, to fully assess their environmental impact and long-term sustainability, it is important to consider the entire lifecycle of hydrogen fuel cells, from production to disposal. This section explores the environmental aspects of hydrogen fuel cells, focusing on lifecycle emissions and sustainability considerations, the role of hydrogen fuel cells in the circular economy, and their environmental benefits compared to other technologies [105].

5.8.1 Lifecycle Emissions and Sustainability Considerations

The environmental impact of hydrogen fuel cells extends beyond their operation. While fuel cells themselves produce only water vapor and heat as byproducts, the overall environmental footprint depends largely on how the hydrogen is produced and how fuel cell systems are manufactured, used, and disposed of. Understanding the full lifecycle emissions is essential to evaluate the true sustainability of hydrogen fuel cells. The most significant environmental consideration for hydrogen fuel cells is the source of hydrogen production. Currently, the majority of hydrogen is produced using natural gas through a process called SMR, which generates significant carbon emissions. This method, known as grey hydrogen production, results in a substantial carbon footprint, undermining the environmental benefits of hydrogen fuel cells. To mitigate this, blue hydrogen production, which incorporates CCS technologies to capture and store the carbon emissions from SMR, has been proposed as a more sustainable alternative. However, the scalability and efficiency of CCS are still under research, and its widespread implementation remains a challenge. A more sustainable option is green hydrogen, which is produced using renewable electricity to power electrolyzers that split water into hydrogen and oxygen. This process, known as water electrolysis, results in zero emissions when coupled with renewable energy sources such as wind, solar, or hydropower. The environmental impact of green hydrogen production is significantly lower, and as the share of renewable energy in global grids increases, green hydrogen is expected to become a key component of a sustainable hydrogen economy. However, the cost of renewable hydrogen production remains high compared to conventional methods, and further advances in electrolysis technology are needed to reduce costs and improve efficiency. In addition to hydrogen production, the manufacturing of fuel cell systems also has environmental implications. Fuel cells require rare materials like platinum for catalysts, and mining these materials can have adverse environmental impacts. Efforts to reduce the use of platinum by developing non-precious metal catalysts and improving catalyst efficiency are key to improving the overall sustainability of fuel cells. Finally, the disposal and recycling of fuel cells must be considered. The fuel cell stack and components, such as the MEA, electrodes, and catalyst materials, will eventually reach the end of their life and need to be disposed of or recycled. The ability to recycle fuel cell components is still in its early stages, and effective recycling processes will need to be developed to minimize environmental impacts and reduce waste [106].

5.8.2 Hydrogen Fuel Cells and Circular Economy

The concept of the circular economy is central to sustainability, focusing on minimizing waste and making the most of available resources. Hydrogen fuel cells, when integrated into a circular economy framework, offer the potential to close material and energy loops, reducing the need for virgin resources and enhancing the efficiency of energy systems. One key aspect of the circular economy for hydrogen fuel cells is the recycling of hydrogen. Hydrogen can be produced, stored, and used repeatedly within a closed-loop system. In this system, hydrogen is generated using renewable energy sources, stored for later use, and converted back into electricity via fuel cells. This circular process reduces dependence on fossil fuels, minimizes emissions, and enables the efficient use of resources. The ability to store hydrogen for use in times of peak demand or when renewable energy generation is low also helps stabilize energy grids and ensures a consistent supply of clean energy. Fuel cell systems can also be integrated into waste-to-energy strategies. For example, biomass or waste-derived hydrogen can be used in fuel cells, converting organic waste into valuable energy. This process further supports a circular economy by creating energy from otherwise discarded materials. Additionally, hydrogen fuel cells can be used in combination with renewable energy systems to provide power-to-gas solutions, where surplus renewable electricity is used to produce hydrogen, which is then stored and used when demand is high or when renewable generation is low. Moreover, the materials used in hydrogen fuel cells themselves can be recycled and reused at the end of their life, reducing the need for raw material extraction and minimizing environmental impact. For instance, platinum and other precious metals used in fuel cell catalysts can be reclaimed through recycling processes, which reduces the environmental burden of mining and resource extraction [107].

5.8.3 Environmental Benefits Compared to Other Technologies

When compared to other energy technologies, hydrogen fuel cells offer several distinct environmental benefits. One of the most significant advantages is their ability to produce zero-emission power at the point of use. Unlike conventional internal combustion engines that emit pollutants such as NOx, particulate matter, and CO_2, hydrogen fuel cells produce only water vapor as a byproduct, making them a much cleaner alternative for applications in transportation, stationary power generation, and industrial processes. In transportation, hydrogen FCVs have a clear environmental advantage over conventional gasoline and diesel-powered vehicles, as they contribute to a significant reduction in air pollution and greenhouse gas emissions. Hydrogen-powered vehicles do not emit harmful pollutants like CO, hydrocarbons, and NOx, which are commonly associated with internal combustion engines. Moreover, when green hydrogen is used, fuel cells offer a nearly carbon-neutral solution to the transportation sector. In stationary power generation, hydrogen fuel cells are advantageous for their low environmental footprint, particularly when paired with renewable hydrogen. Fuel cells offer a reliable and clean source of electricity, making them ideal for off-grid systems, remote locations, and backup power. In areas with intermittent renewable energy generation, hydrogen fuel cells can provide a continuous and clean source of power, helping to reduce reliance on fossil fuel-based power plants. Compared to battery electric technologies, hydrogen fuel cells offer greater energy density and faster refueling times. This makes them particularly useful for heavy-duty transport applications such as trucks, buses, and trains, where batteries are often not practical due to their weight and long charging times. Fuel cells can also be integrated into industrial processes to decarbonize sectors such as steel production, cement manufacturing, and refining, providing an alternative to the high emissions associated with these industries. Despite these advantages, hydrogen fuel cells still face challenges related to cost, hydrogen production, and infrastructure. However, their environmental benefits, particularly when powered by renewable energy, position them as a key technology in the effort to decarbonize multiple sectors and contribute to global sustainability goals [108].

5.9 Challenges and Future Prospects

Despite their tremendous potential to revolutionize energy systems and reduce greenhouse gas emissions, hydrogen fuel cells face a number of challenges that must be addressed to achieve widespread adoption and integration into the global energy landscape. These challenges span across technological, economic, and infrastructural domains, requiring a concerted effort from governments, industries, and researchers to overcome them.

a. **High Costs:** One of the most significant barriers to the widespread adoption of hydrogen fuel cells is their high cost. The production of fuel cells involves expensive materials, such as platinum, which is used as a catalyst in the electrochemical reaction. Although advancements in materials science and fuel cell design are gradually reducing costs, fuel cells remain prohibitively expensive for many potential applications, especially when compared to conventional internal combustion engines or battery electric vehicles [109].

b. **Hydrogen Production and Infrastructure:** While hydrogen is the most abundant element in the universe, it does not exist in its pure form and must be extracted from other substances, such as natural gas or water. Currently, most hydrogen is produced through a process called steam methane reforming (SMR), which is energy-intensive and emits carbon dioxide, undermining the environmental benefits of hydrogen. Although renewable hydrogen production methods, such as electrolysis powered by solar or wind energy, are being developed, they are still in the early stages and are costly. Additionally, the infrastructure needed to produce, store, and distribute hydrogen at scale remains underdeveloped, limiting the availability of hydrogen fuel in many regions [110].
c. **Storage and Transport:** Hydrogen has a low volumetric energy density, meaning it needs to be stored at high pressures or in liquid form to make it practical for transportation. These storage solutions, such as high-pressure tanks or cryogenic storage systems, are complex and costly to develop and maintain. Transporting hydrogen over long distances presents additional challenges, as pipelines must be constructed, or hydrogen must be transported by trucks, rail, or ships in liquid or compressed form. Both storage and transportation of hydrogen require specialized infrastructure and technology, which adds to the overall cost and complexity [111].
d. **Limited Fueling Infrastructure:** The limited availability of hydrogen refueling stations is another major challenge. In comparison to the widespread network of gasoline stations, the infrastructure for hydrogen refueling is minimal, particularly in regions outside of major metropolitan areas. This lack of refueling infrastructure presents a significant barrier for consumers and businesses considering the switch to hydrogen-powered vehicles or equipment. Expanding the hydrogen refueling network requires significant investment in infrastructure, which has been slow to materialize [112].
e. **Durability and Longevity:** Hydrogen fuel cells face durability and performance issues over time. The catalysts used in fuel cells, especially platinum, can degrade over time, leading to reduced efficiency and the need for costly replacements. Additionally, fuel cells are sensitive to impurities in the hydrogen fuel, which can lead to contamination and shorter lifespans. Ongoing research is focused on improving the durability of fuel cells by finding more robust materials and improving the overall design [113].
f. **Public Perception and Awareness:** Despite the clear environmental benefits of hydrogen fuel cells, public awareness and perception remain significant obstacles. Many consumers are unfamiliar with how hydrogen fuel cells work and may be hesitant to adopt this technology due to concerns about safety, fuel availability, and cost. There is also skepticism surrounding hydrogen's role in the broader energy transition, particularly given the current dominance of battery electric vehicles and the challenges related to hydrogen production [114].

Despite these challenges, the future of hydrogen fuel cells looks promising, driven by significant advancements in technology, economies of scale, and policy support. One of the most promising areas for growth is the advancement of hydrogen production technologies, particularly green hydrogen. As the global energy transition accelerates, there is increasing investment in the development of electrolyzers, which use renewable electricity to split water into hydrogen and oxygen. Innovations in electrolyzer efficiency, combined with falling costs for renewable energy, are expected to make green hydrogen more cost-competitive with traditional hydrogen production methods. This, in turn, will increase the availability of clean hydrogen for fuel cells and contribute to the decarbonization of the energy and transport sectors.

a. In the transportation sector, hydrogen fuel cells are expected to play a critical role in the decarbonization of heavy-duty vehicles such as trucks, buses, and trains. While BEVs are gaining traction in light-duty passenger cars, they are less suited for long-range, heavy-duty applications due to their limited energy density and long charging times. Hydrogen fuel cells, on the other hand, offer faster refueling times and greater energy density, making them ideal for long-distance haulage and other high-demand applications. Hydrogen-powered trucks, trains, and ships are already being tested in various regions, and as fuel cell technology improves and infrastructure expands, these vehicles are expected to become more commercially viable.

b. In the industrial sector, hydrogen fuel cells have the potential to decarbonize energy-intensive industries such as steel production, cement manufacturing, and chemical processing. By replacing traditional fossil fuels with clean hydrogen, these industries can significantly reduce their carbon footprints and contribute to global climate goals. As the cost of hydrogen production decreases and more industries adopt fuel cell technologies, hydrogen is expected to become an increasingly important fuel for industrial applications.

c. A key factor that will determine the success of hydrogen fuel cells is policy and regulatory support. Governments around the world are recognizing the potential of hydrogen as a key component of their energy transition strategies and are providing funding and incentives to support research, infrastructure development, and market adoption. The European Union, Japan, and South Korea are leading the way in hydrogen policies, with ambitious goals for expanding hydrogen infrastructure and accelerating the adoption of fuel cell technologies. If these policies are replicated on a global scale, the deployment of hydrogen fuel cells could be significantly accelerated.

d. Another area of future growth is the integration of hydrogen fuel cells with renewable energy systems. As the share of intermittent renewable energy sources such as wind and solar continues to increase, hydrogen can play a key role in energy storage and grid balancing. Surplus renewable electricity can be used to produce hydrogen via electrolysis, which can then be stored and used later for power generation. This power-to-gas technology offers a promising solution to the challenge of storing renewable energy, and its integration with hydrogen fuel cells could help stabilize energy grids and ensure a reliable supply of clean electricity [115].

5.10 Conclusion

Hydrogen fuel cells represent a transformative technology with the potential to revolutionize the global energy landscape, driving the transition toward a cleaner, more sustainable future. As an energy carrier, hydrogen offers numerous advantages, including zero-emissions operation, high efficiency, and versatile applications across transportation, stationary power generation, and industrial processes. Fuel cells, powered by hydrogen, have emerged as one of the most promising solutions for decarbonizing sectors traditionally reliant on fossil fuels, such as heavy-duty transportation, power generation, and manufacturing industries. While hydrogen fuel cell technology has made significant strides in recent years, several challenges remain. The high cost of fuel cell systems, particularly due to the use of precious metals in catalysts, remains a major hurdle to mass-market adoption. Similarly, the development of hydrogen production, storage, and refueling infrastructure is crucial to facilitate the widespread deployment of fuel cell technologies. Overcoming these technological and economic barriers will require continuous innovation, public-private partnerships, and strong policy support. Furthermore, ensuring the sustainability of hydrogen production methods, especially through the expansion of green hydrogen derived from renewable sources, is vital to unlocking the full potential of hydrogen as a clean energy solution. Looking forward, the future prospects for hydrogen fuel cells are promising. Ongoing research into alternative catalysts, electrolyzer technologies, and hydrogen storage solutions is expected to drive down costs and enhance performance. The growth of renewable energy sources, along with advancements in hydrogen infrastructure, will make hydrogen a competitive and viable alternative to fossil fuels in the coming decades. With continued global investment and collaboration, hydrogen fuel cells could play a pivotal role in achieving global climate goals, reducing carbon emissions, and accelerating the transition to a sustainable energy future. In essence, hydrogen fuel cells offer a holistic solution to many of the challenges faced by traditional energy systems. As technology matures and policy frameworks evolve, fuel cells will become an integral part of the global energy mix, contributing to a more resilient, efficient, and sustainable world. Through continued technological advancements and strong international cooperation, hydrogen fuel cells will help shape a cleaner, more decarbonized future for generations to come.

References

1 Singla MK, Nijhawan P, Oberoi AS. Hydrogen fuel and fuel cell technology for cleaner future: A review. *Environ Sci Pollut Res*. 2021 28:13 [Internet]. 2021 Feb 4 [cited 2025 Apr 1];28(13):15607–26. Available from: https://link.springer.com/article/10.1007/s11356-020-12231-8.

2 Haseli Y. Maximum conversion efficiency of hydrogen fuel cells. *Int J Hydrogen Energy*. 2018 May 3;43(18):9015–21.

3 Rajalakshmi N, Balaji R, Ramakrishnan S. Recent developments in hydrogen fuel cells: Strengths and weaknesses. In: *Sustainable Fuel Technologies Handbook*. 2021 Jan 1;431–56. Available from: https://doi.org/10.1016/B978-0-12-822989-7.00015-9.

4 Aminudin MA, Kamarudin SK, Lim BH, Majilan EH, Masdar MS, Shaari N. An overview: Current progress on hydrogen fuel cell vehicles. *Int J Hydrogen Energy*. 2023 Feb 5;48(11):4371–88.

5 Manoharan Y, Hosseini SE, Butler B, Alzhahrani H, Senior BTF, Ashuri T, et al. Hydrogen fuel cell vehicles: Current status and future prospect. *Appl Sci* [Internet]. 2019 Jun 4 [cited 2025 Apr 1];9(11):2296. Available from: https://www.mdpi.com/2076-3417/9/11/2296/htm.

6 Tanç B, Arat HT, Baltacıoğlu E, Aydın K. Overview of the next quarter century vision of hydrogen fuel cell electric vehicles. *Int J Hydrogen Energy*. 2019 Apr 19; 44(20):10120–8.

7 Manzo D, Thai R, Le HT, Venayagamoorthy GK. Fuel cell technology review: Types, economy, applications, and vehicle-to-grid scheme. *Sustain Energy Technol Assessments*. 2025 Mar 1;75:104229.

8 Abedin T, Pasupuleti J, Paw JKS, Tak YC, Mahmud M, Abdullah MP, et al. Proton exchange membrane fuel cells in electric vehicles: Innovations, challenges, and pathways to sustainability. *J Power Sources*. 2025 Jun 1;640:236769.

9 Xiong Y, Wang F, Zhao D, Sun X, Ren F, Zhao M, et al. Cradle-to-gate GHG emissions and decarbonization potentials of minivan-equipped hydrogen fuel cell system. *Resour Conserv Recycl*. 2025 Jan 1;212:107877.

10 Shan R, Kittner N. Sector-specific strategies to increase green hydrogen adoption. *Renew Sustain Energy Rev*. 2025 May 1;214:115491.

11 Li J, Wu T, Cheng C, Li J, Zhou K. A Review of the research progress and application of key components in the hydrogen fuel cell system. *Processes* [Internet]. 2024 Jan 24 [cited 2025 Apr 1];12(2):249. Available from: https://www.mdpi.com/2227-9717/12/2/249/htm.

12 Ferreira-Aparicio P, Conde JJ, Chaparro AM. Fundamentals and components of portable hydrogen fuel-cell systems. In: *Portable Hydrogen Energy Systems: Fuel Cells and Storage Fundamentals and Applications*. 2018 Jan 1;15–39. Available from: https://doi.org/10.1016/B978-0-12-813128-2.00002-4.

13 Petrovic S, Hossain E. Development of a novel technological readiness assessment tool for fuel cell technology. *IEEE Access*. 2020;8:132237–52.

14 Dincer I. Environmental and sustainability aspects of hydrogen and fuel cell systems. *Int J Energy Res* [Internet]. 2007 Jan 1 [cited 2025 Apr 1];31(1):29–55. Available from: https://onlinelibrary.wiley.com/doi/full/10.1002/er.1226.

15 Mo T, Li Y, Luo Y. Advantages and technological progress of hydrogen fuel cell vehicles. *World Electr Veh J* [Internet]. 2023 Jun 19 [cited 2025 Apr 1];14(6):162. Available from: https://www.mdpi.com/2032-6653/14/6/162/htm.

16 Lago Sari R, Fogue Robles A, Monsalve Serrano J, Cleary D. Techno-economic assessment of hydrogen as a fuel for internal combustion engines and proton exchange membrane fuel cells on long haul applications. *Energy Convers Manag*. 2024 Jul 1;311:118522.

17 Shaker LM, Mohammed JK, Basem A, Halbos RJ, Mahdi RR, Mohammed SA, et al. Comparative analysis of solar cells and hydrogen fuel: A mini-review. *Results Eng*. 2024 Sep 1;23:102507.

18 Laimon M, Yusaf T. Towards energy freedom: Exploring sustainable solutions for energy independence and self-sufficiency using integrated renewable energy-driven hydrogen system. *Renew Energy*. 2024 Feb 1;222:119948.

19 Soleimani A, Hosseini Dolatabadi SH, Heidari M, Pinnarelli A, Mehdizadeh Khorrami B, Luo Y, et al. Progress in hydrogen fuel cell vehicles and up-and-coming technologies for eco-friendly transportation: An international assessment. *Multiscale Multidiscip Model Exp Des* [Internet]. 2024 Sep 1 [cited 2025 Apr 1]; 7(4):3153–72. Available from: https://link.springer.com/article/10.1007/s41939-024-00482-8.

20 Menesy AS, Sultan HM, Zayed ME, Habiballah IO, Dmitriev S, Safaraliev M, et al. A modified slime mold algorithm for parameter identification of hydrogen-powered proton exchange membrane fuel cells. *Int J Hydrogen Energy*. 2024 Oct 11; 86:853–74.

21 Zhang Y, Yu J, Zhao N, Xu Z, Yan Y, Wu D, et al. Particle swarm optimization for a hybrid freight train powered by hydrogen or ammonia solid oxide fuel cells. *Int J Hydrogen Energy*. 2024 Jun 27;72:626–41.

22 Zhang X, Zhao Q, Wang J, Zhang H. Synergizing alkaline fuel cell and thermocapacitive heat engine for hydrogen energy efficient utilization. *Process Saf Environ Prot*. 2024 Jun 1;186:706–14.

23 Biebl M, Roes J, Hoster H. Investigation of the technical potential of a hydrogen powered phosphoric acid fuel cell (PAFC) for CHP. *J Phys Conf Ser* [Internet]. 2024 Jan 1 [cited 2025 Apr 1];2689(1):012014. Available from: https://iopscience.iop.org/article/10.1088/1742-6596/2689/1/012014.

24 Lu Z, Zhang H, Duan L, Zhang R, Lu C, Desideri U. Experimental analysis of ammonia as input fuel for molten carbonate fuel cell. *Fuel*. 2024 Feb 1;357:130032.

25 Chi X, Chen F, Zhai S, Hu Z, Zhou S, Wei W. Real-time power optimization based on Q-learning algorithm for direct methanol fuel cell system. *Int J Hydrogen Energy*. 2024 Nov 4;89:1241–53.

26 Hydrogen and Fuel Cell Technologies Office | Department of Energy [Internet]. [cited 2025 Mar 20]. Available from: https://www.energy.gov/eere/fuelcells/hydrogen-and-fuel-cell-technologies-office.

27 Wee JH. Applications of proton exchange membrane fuel cell systems. *Renew Sustain Energy Rev*. 2007 Oct 1;11(8):1720–38.

28 Wu D, Peng C, Yin C, Tang H. Review of system integration and control of proton exchange membrane fuel cells. *Electrochem Energy Rev* [Internet]. 2020 Sep 1 [cited 2025 Apr 1];3(3):466–505. Available from: https://link.springer.com/article/10.1007/s41918-020-00068-1.

29 Alaswad A, Omran A, Sodre JR, Wilberforce T, Pignatelli G, Dassisti M, et al. Technical and commercial challenges of proton-exchange membrane (PEM) fuel cells. *Energies* [Internet]. 2020 Dec 29 [cited 2025 Apr 1];14(1):144. Available from: https://www.mdpi.com/1996-1073/14/1/144/htm.

30 Cheng S, Hu D, Hao D, Yang Q, Wang J, Feng L, et al. Investigation and analysis of proton exchange membrane fuel cell dynamic response characteristics on hydrogen consumption of fuel cell vehicle. *Int J Hydrogen Energy*. 2022 Apr 26; 47(35):15845–64.

31 Ralph TR. Proton exchange membrane fuel cells. *Platin Met Rev.* 1997 Jul 1; 41(3):102–13.

32 Sopian K, Wan Daud WR. Challenges and future developments in proton exchange membrane fuel cells. *Renew Energy.* 2006 Apr 1;31(5):719–27.

33 Tellez-Cruz MM, Escorihuela J, Solorza-Feria O, Compañ V. Proton exchange membrane fuel cells (PEMFCs): Advances and challenges. *Polymers* [Internet]. 2021 Sep 10 [cited 2025 Mar 21];13(18):3064. Available from: https://www.mdpi.com/2073-4360/13/18/3064/htm.

34 Singh M, Zappa D, Comini E. Solid oxide fuel cell: Decade of progress, future perspectives and challenges. *Int J Hydrogen Energy.* 2021 Aug 5;46(54):27643–74.

35 Kuterbekov KA, Nikonov AV, Bekmyrza KZ, Pavzderin NB, Kabyshev AM, Kubenova MM, et al. Classification of solid oxide fuel cells. *Nanomaterials* [Internet]. 2022 Mar 24 [cited 2025 Apr 1];12(7):1059. Available from: https://www.mdpi.com/2079-4991/12/7/1059/htm.

36 Wang K, Hissel D, Péra MC, Steiner N, Marra D, Sorrentino M, et al. A review on solid oxide fuel cell models. *Int J Hydrogen Energy.* 2011 Jun 1;36(12):7212–28.

37 Peng J, Huang J, Wu XL, Xu YW, Chen H, Li X. Solid oxide fuel cell (SOFC) performance evaluation, fault diagnosis and health control: A review. *J Power Sources.* 2021 Sep 1;505:230058.

38 Boldrin P, Brandon NP. Progress and outlook for solid oxide fuel cells for transportation applications. *Nat Catal* [Internet]. 2019 Jul 11 [cited 2025 Apr 1]; 2(7):571–7. Available from: https://www.nature.com/articles/s41929-019-0310-y.

39 Choudhury A, Chandra H, Arora A. Application of solid oxide fuel cell technology for power generation—A review. *Renew Sustain Energy Rev.* 2013 Apr 1;20:430–42.

40 Alaedini AH, Tourani HK, Saidi M. A review of waste-to-hydrogen conversion technologies for solid oxide fuel cell (SOFC) applications: Aspect of gasification process and catalyst development. *J Environ Manag.* 2023 Mar 1;329:117077.

41 Zhuo Y, Zhang W, Luo Y, Hao P, Shi Y. Progress of elevated-temperature alkaline electrolysis hydrogen production and alkaline fuel cells power generation. *J Fuel Chem Technol.* 2025 Feb 1;53(2):231–47.

42 Qasem NAA, Abdulrahman GAQ. A recent comprehensive review of fuel cells: History, types, and applications. *Int J Energy Res* [Internet]. 2024 Jan 1 [cited 2025 Apr 1];2024(1):7271748. Available from: https://onlinelibrary.wiley.com/doi/full/10.1155/2024/7271748.

43 Ramasamy P, Muruganantham B, Rajasekaran S, Babu BD, Ramkumar R, Marthanda AVA, et al. A comprehensive review on different types of fuel cell and its applications. *Bull Electr Eng Inform* [Internet]. 2024 Apr 1 [cited 2025 Apr 1]; 13(2):774–80. Available from: https://www.beei.org/index.php/EEI/article/view/6348.

44 Gupta P, Toksha B, Rahaman M. A critical review on hydrogen based fuel cell technology and applications. *Chem Rec* [Internet]. 2024 Jan 1 [cited 2025 Apr 1]; 24(1):e202300295. Available from: https://onlinelibrary.wiley.com/doi/full/10.1002/tcr.202300295.

45 Singh M, Singla MK, Safaraliev M, Singh K, Odinaev I, Abdel Menaem A. Advancements and challenges of fuel cell integration in electric vehicles: A comprehensive analysis. *Int J Hydrogen Energy.* 2024 Oct 28;88:1386–97.

46 Peng S. Overview of hydrogen fuel cells. *Hydrogen Fuel Cells* [Internet]. 2024 [cited 2025 Apr 1];1–15. Available from: https://link.springer.com/chapter/10.1007/978-981-96-0157-8_1.

47 Eapen DE, Suseendiran SR, Rengaswamy R. Phosphoric acid fuel cells. *Compend Hydrogen Energy Hydrogen Energy Conv*. 2016 Jan 1;3:57–70.

48 Saidi M, Atashkar H. Role of hydrogen in phosphoric acid fuel cells. In: *Hydrogen Utilization in Fuel Cells* [Internet]. CRC Press. 2024 Jan 1 [cited 2025 Apr 1];283–304. Available from: https://www.taylorfrancis.com/chapters/edit/10.1201/9781003382577-13/role-hydrogen-phosphoric-acid-fuel-cells-majid-saidi-hajar-atashkar.

49 Stenina IA, Yaroslavtsev AB. Prospects for the development of hydrogen energy. Polymer membranes for fuel cells and electrolyzers. *Membr Membr Technol* [Internet]. 2024 May 9 [cited 2025 Apr 1];6(1):15–26. Available from: https://link.springer.com/article/10.1134/S2517751624010050.

50 Khan MS, Khattak R, Khan A, Khan MS, Khattak R, Khan A. Fuel cell technology (FCT): An overview. In: *Chemical Kinetics and Catalysis - Perspectives, Developments and Applications* [Internet]. 2024 Apr 12 [cited 2025 Apr 1]; Available from: https://www.intechopen.com/chapters/1170847.

51 Ito H. Economic and environmental assessment of phosphoric acid fuel cell-based combined heat and power system for an apartment complex. *Int J Hydrogen Energy*. 2017 Jun 8;42(23):15449–63.

52 Baranak M, Atakül H. A basic model for analysis of molten carbonate fuel cell behavior. *J Power Sources*. 2007 Oct 25;172(2):831–9.

53 Martsinchyk A, Milewski J, Szczesniak A, Dybinski O, Shuhayeu P, Sienko A, et al. Operation of solid oxide fuel cell stack in reversible mode for hydrogen generation for molten carbonate fuel cell as power-to-gas process. *Chem Eng J*. 2025 Mar 1;507:160624.

54 Kulkarni A, Giddey S. Materials issues and recent developments in molten carbonate fuel cells. *J Solid State Electrochem*. 2012 16: 10 [Internet]. 2012 Jun 8 [cited 2025 Apr 1];16(10):3123–46. Available from: https://link.springer.com/article/10.1007/s10008-012-1771-y.

55 Contreras RR, Almarza J, Rincón L. Molten carbonate fuel cells: A technological perspective and review. *Energy Sources Part A Recovery Util Environ Effects* [Internet]. 2021 Dec 12 [cited 2025 Apr 1]; Available from: https://www.tandfonline.com/doi/abs/10.1080/15567036.2021.2013346.

56 Kumari A, Debnath S, Sumit, Borah A, Rajeshkhanna G. Metal–organic framework-derived zinc–cobalt oxide materials as high-performance anodes for direct methanol fuel cell application. *Langmuir* [Internet]. 2025 Mar 27 [cited 2025 Apr 1];41:8812–8821; Available from: https://pubs.acs.org/doi/abs/10.1021/acs.langmuir.5c00116.

57 Joghee P, Malik JN, Pylypenko S, O'Hayre R. A review on direct methanol fuel cells—In the perspective of energy and sustainability. *MRS Energy Sustain* [Internet]. 2015 Dec 1 [cited 2025 Apr 1];2(1):E3. Available from: https://www.cambridge.org/core/journals/mrs-energy-and-sustainability/article/abs/review-on-direct-methanol-fuel-cells-in-the-perspective-of-energy-and-sustainability/9EA45787BA50E9049A9061587C219D10.

58 Bai XT, Cao LH, Zhao F, Li SH. Arylsulfonate ionic hydrogen-bonded organic frameworks enable highly stable and superprotonic conductivity for enhancing direct methanol fuel cells. *ACS Mater Lett* [Internet]. 2024 Aug 5 [cited 2025 Apr 1];6(8):3351–7. Available from: https://pubs.acs.org/doi/abs/10.1021/acsmaterialslett.4c00953.

59 Gawel L, Parasinska D. Dynamic impedance measurements of the direct methanol fuel cell cathode at various operating temperatures. *Int J Hydrogen Energy*. 2024 May 20;67:83–90.

60 Cheng T, Goddard WA, An Q, Xiao H, Merinov B, Morozov S. Mechanism and kinetics of the electrocatalytic reaction responsible for the high cost of hydrogen fuel cells. *Phys Chem Chem Phys* [Internet]. 2017 Jan 25 [cited 2025 Apr 1]; 19(4):2666–73. Available from: https://pubs.rsc.org/en/content/articlehtml/2017/cp/c6cp08055c.

61 Eftekhari A, Fang B. Electrochemical hydrogen storage: Opportunities for fuel storage, batteries, fuel cells, and supercapacitors. *Int J Hydrogen Energy*. 2017 Oct 5;42(40):25143–65.

62 Ran A, Fan L, Tongsh C, Wang J, Qin Z, Du Q, et al. Molecular understanding of the role of catalyst particle arrangement in local mass transport resistance for fuel cells. *Adv Sci* [Internet]. 2025 Feb 1 [cited 2025 Apr 1];12(5):2409755. Available from: https://onlinelibrary.wiley.com/doi/full/10.1002/advs.202409755.

63 Rossetti I, Tripodi A. Catalytic production of renewable hydrogen for use in fuel cells: A review study. *Top Catal* [Internet]. 2022 Jan 27 [cited 2025 Apr 1]; 67(19):1286–305. Available from: https://link.springer.com/article/10.1007/s11244-022-01563-z.

64 Andrade TS, Thiringer T. Low platinum fuel cell as enabler for the hydrogen fuel cell vehicle. *J Power Sources*. 2024 Apr 1;598:234140.

65 Chen Y. Research progress on platinum catalysts for hydrogen-oxygen fuel cells. *E3S Web of Conf* [Internet]. 2025 Jan 21 [cited 2025 Apr 1];606:03012. Available from: https://www.e3s-conferences.org/articles/e3sconf/abs/2025/06/e3sconf_icnaoe2024_03012/e3sconf_icnaoe2024_03012.html.

66 Sanati S, Morsali A, García H. Metal-organic framework-based materials as key components in electrocatalytic oxidation and reduction reactions. *J Energy Chem*. 2023 Dec 1;87:540–67.

67 Xu L, Xu L, Shao Y, Zhang X, Hu Z, Li J, et al. Effects of hydrogen dilution on performance and in-plane uniformity of large-scale PEM fuel cell with low anode catalyst loading. *Appl Energy* 2025 Feb 1;379:124992.

68 Li R, Ma T, Guo H, Zhao J, Zhou J, Shi L, et al. Investigation on performance recovery and consistency of proton exchange membrane fuel cell stack under automotive accelerated stress test. *Appl Energy* 2025 Apr 1;383:125388.

69 Abu SM, Hannan MA, Rahman SA, Long CY, Ker PJ, Wong RT, et al. An effective optimization algorithm for hydrogen fuel cell-based hybrid energy system: A sustainable microgrid approach. *Int J Hydrogen Energy*. 2025 Jan 13;98:1341–55.

70 Hames Y, Kaya K, Baltacioglu E, Turksoy A. Analysis of the control strategies for fuel saving in the hydrogen fuel cell vehicles. *Int J Hydrogen Energy*. 2018 Jun 7;43(23):10810–21.

71 Hwang JJ. Effect of hydrogen delivery schemes on fuel cell efficiency. *J Power Sources*. 2013 Oct 1;239:54–63.

72 Okonkwo PC, Ben Belgacem I, Emori W, Uzoma PC. Nafion degradation mechanisms in proton exchange membrane fuel cell (PEMFC) system: A review. *Int J Hydrogen Energy*. 2021 Aug 10;46(55):27956–73.

73 Zarabi Golkhatmi S, Asghar MI, Lund PD. A review on solid oxide fuel cell durability: Latest progress, mechanisms, and study tools. *Renew Sustain Energy Rev*. 2022 Jun 1;161:112339.

74 Ferriday TB, Middleton PH. Alkaline fuel cell technology - A review. *Int J Hydrogen Energy* 2021 May 20;46(35):18489–510.

75 Sammes N, Bove R, Stahl K. Phosphoric acid fuel cells: Fundamentals and applications. *Curr Opin Solid State Mater Sci*. 2004 Oct 1;8(5):372–8.

76 Antolini E. The stability of molten carbonate fuel cell electrodes: A review of recent improvements. *Appl Energy*. 2011 Dec 1;88(12):4274–93.

77 Osman SH, Kamarudin SK, Shaari N, Hanapi IH, Elham OSJ, Aminudin MA. Review on direct methanol fuel cells: Bridging the gap between theory and application for sustainable energy solutions. *Energy Fuels* [Internet]. 2025 [cited 2025 Apr 1];39(12):5651-71. Available from: https://pubs.acs.org/doi/abs/10.1021/acs.energyfuels.4c05357.

78 Yan Q, Toghiani H, Causey H. Steady state and dynamic performance of proton exchange membrane fuel cells (PEMFCs) under various operating conditions and load changes. *J Power Sources*. 2006 Oct 20;161(1):492–502.

79 Chiara Massaro M, Pramotton S, Marocco P, Monteverde AHA, Santarelli M. Optimal design of a hydrogen-powered fuel cell system for aircraft applications. *Energy Convers Manag*. 2024 Apr 15;306:118266.

80 Pan Z, Xie F, Zhang Z, Zhao Z, Wu L, Li W, et al. Overall design of a gradient-ordered membrane electrode assembly for direct liquid fuel cells. *Adv Funct Mater* [Internet]. 2024 Nov 1 [cited 2025 Apr 1];34(45):2404710. Available from: https://onlinelibrary.wiley.com/doi/full/10.1002/adfm.202404710.

81 Zhou P, Wu CW, Ma GJ. Influence of clamping force on the performance of PEMFCs. *J Power Sources*. 2007 Jan 1;163(2):874–81.

82 Hou J, Yang M, Ke C, Wei G, Priest C, Qiao Z, et al. Platinum-group metal catalysts for proton exchange membrane fuel cells: From catalyst design to electrode structure optimization. *EnergyChem*. 2020 Jan 1;2(1):100023.

83 Wilberforce T, Olabi AG, Monopoli D, Dassisti M, Sayed ET, Abdelkareem MA. Design optimization of proton exchange membrane fuel cell bipolar plate. *Energy Convers Manag*. 2023 Feb 1;277:116586.

84 Dicks A, Rand DAJ. *Fuel Cell Systems Explained*. John Wiley & Sons. 2018.

85 Zhang F, Zu B, Wang B, Qin Z, Yao J, Wang Z, et al. Developing long-durability proton-exchange membrane fuel cells. *Joule* [Internet]. 2025 Mar 19 [cited 2025 Apr 1];9(3):101853. Available from: https://www.cell.com/action/showFullText?pii=S2542435125000340.

86 Dubau L, Castanheira L, Maillard F, Chatenet M, Lottin O, Maranzana G, et al. A review of PEM fuel cell durability: Materials degradation, local heterogeneities of aging and possible mitigation strategies. *Wiley Interdiscip Rev Energy Environ*. [Internet]. 2014 Nov 1 [cited 2025 Apr 1];3(6):540–60. Available from: https://onlinelibrary.wiley.com/doi/full/10.1002/wene.113.

87 Wang J. System integration, durability and reliability of fuel cells: Challenges and solutions. *Appl Energy*. 2017 Mar 1;189:460–79.

88 Abdelkareem MA, Elsaid K, Wilberforce T, Kamil M, Sayed ET, Olabi A. Environmental aspects of fuel cells: A review. *Sci Total Environ*. 2021 Jan 15; 752:141803.

89 Zhang H, Li X, Liu X, Yan J. Enhancing fuel cell durability for fuel cell plug-in hybrid electric vehicles through strategic power management. *Appl Energy*. 2019 May 1;241:483–90.

90 Wang J. Barriers of scaling-up fuel cells: Cost, durability and reliability. *Energy*. 2015 Feb 1;80:509–21.

91 Touil W, Li Z, Outbib R, Hissel D, Jemei S. A system-level modeling framework for predicting Pt catalyst degradation in proton exchange membrane fuel cells. *J Power Sources*. 2025 Jan 1;625:235628.

92 Meng X, Sun C, Mei J, Tang X, Hasanien HM, Jiang J, et al. Fuel cell life prediction considering the recovery phenomenon of reversible voltage loss. *J Power Sources* 2025 Jan 1;625:235634.

93 Schmittinger W, Vahidi A. A review of the main parameters influencing long-term performance and durability of PEM fuel cells. *J Power Sources*. 2008 May 15;180(1):1–14.

94 Jourdani M, Mounir H, El Marjani A. Compilation of factors affecting durability of proton exchange membrane fuel cell (PEMFC). *Proceedings of 2014 International Renewable and Sustainable Energy Conference*. IRSEC 2014. IEEE. 2014 Mar 12;542–7.

95 Chen Z, Shu C, Gan Z, Cao J, Qiu P, Sun X, et al. Research progress and perspectives on anti-poisoning hydrogen oxidation reaction electrocatalysts for hydrogen fuel cells. *Small* [Internet]. 2025 Feb 1 [cited 2025 Apr 1];21(7):2411049. Available from: https://onlinelibrary.wiley.com/doi/full/10.1002/smll.202411049.

96 Spazzafumo G, Sørensen B. *Hydrogen and Fuel Cells : Emerging Technologies and Applications* Ed. 3. Academic Press. 2018.

97 Fakhreddine O, Gharbia Y, Derakhshandeh JF, Amer AM. Challenges and solutions of hydrogen fuel cells in transportation systems: A review and prospects. *World Electr Veh J* [Internet]. 2023 Jun 13 [cited 2025 Apr 1];14(6):156. Available from: https://www.mdpi.com/2032-6653/14/6/156/htm.

98 Felseghi RA, Carcadea E, Raboaca MS, Trufin CN, Filote C. Hydrogen fuel cell technology for the sustainable future of stationary applications. *Energies* [Internet]. 2019 Dec 3 [cited 2025 Apr 2];12(23):4593. Available from: https://www.mdpi.com/1996-1073/12/23/4593/htm.

99 Ma Z, Eichman J, Kurtz J. Fuel cell backup power system for grid service and microgrid in telecommunication applications. *J Energy Resour Technol Trans ASME* [Internet]. 2019 Jun 1 [cited 2025 Apr 2];141(6). Available from: https://dx.doi.org/10.1115/1.4042402.

100 Rath R, Kumar P, Mohanty S, Nayak SK. Recent advances, unsolved deficiencies, and future perspectives of hydrogen fuel cells in transportation and portable sectors. *Int J Energy Res* [Internet]. 2019 Dec 1 [cited 2025 Apr 2];43(15):8931–55. Available from: https://onlinelibrary.wiley.com/doi/full/10.1002/er.4795.

101 Spazzafumo Giuseppe, Sørensen Bent. *Hydrogen and Fuel Cells : Emerging Technologies and Applications* Ed. 3. 2018.

102 Min J, Kim Y, Lee S, Ko K, Chougule SS, Chavan AA, et al. Design of anti-poisoning catalysts for hydrogen oxidation reaction in next-generation anion exchange membrane fuel cells. *Electrochem Commun*. 2025 Jan 1;170:107852.

103 Huang C, Wang F, Chen X, Li J, Shao M, Wei Z. Innovative strategies for designing and constructing efficient fuel cell electrocatalysts. *Chem Commun* [Internet]. 2025 Feb 6 [cited 2025 Apr 2];61(13):2658–83. Available from: https://pubs.rsc.org/en/content/articlehtml/2025/cc/d4cc05928j.

104 Ishaq M, Dincer I. Process development and simulation of a novel solar energy plant integrated with solid oxide fuel cell, hydrogen, heat recovery and carbon capture systems. *International Journal of Thermofluids*. 2025 Mar 1;26:101122.

105 Anser MK, Ali S, Umair M, Javid R, Tayab M. Optimizing hydrogen integration in vehicle fuel systems for sustainable development: A step towards economic decarbonization. *Int J Hydrogen Energy*. 2025 Jan 13;98:321–33.

106 Wei X, Sharma S, Waeber A, Wen D, Margni M, Maréchal F, et al. Environmental implications of solid oxide fuel cell system for hydrogen sustainability. *Resour Conserv Recycl*. 2025 Apr 1;215:108134.

107 Gatto A, Nuta AC, Marjerison RK. Sustainable, just and circular decarbonization from lithium and hydrogen production. *J Clean Prod* [Internet]. 2025 Feb 4 [cited 2025 Apr 2];507:144795. Available from: https://linkinghub.elsevier.com/retrieve/pii/S0959652625001453.

108 Dincer I. Environmental and sustainability aspects of hydrogen and fuel cell systems. *Int J Energy Res* [Internet]. 2007 Jan 1 [cited 2025 Apr 2];31(1):29–55. Available from: https://onlinelibrary.wiley.com/doi/full/10.1002/er.1226.

109 Dash SK, Chakraborty S, Roccotelli M, Sahu UK. Hydrogen fuel for future mobility: Challenges and future aspects. *Sustainability* [Internet]. 2022 Jul 6 [cited 2025 Apr 2];14(14):8285. Available from: https://www.mdpi.com/2071-1050/14/14/8285/htm.

110 İnci M. Future vision of hydrogen fuel cells: A statistical review and research on applications, socio economic impacts and forecasting prospects. *Sustain Energy Technol Assessments*. 2022 Oct 1;53:102739.

111 Chalk SG, Miller JF. Key challenges and recent progress in batteries, fuel cells, and hydrogen storage for clean energy systems. *J Power Sources*. 2006 Sep 13;159(1):73–80.

112 Ajanovic A, Haas R. Prospects and impediments for hydrogen and fuel cell vehicles in the transport sector. *Int J Hydrogen Energy*. Elsevier. 2021 Mar 3;46(16):10049–58. Available from: https://doi.org/10.1016/j.ijhydene.2020.03.122.

113 Kumar A, Sehgal M. Hydrogen fuel cell technology for a sustainable future: A review. *SAE Technical Papers [Internet]*. 2018 Apr 3 [cited 2025 Apr 2];2018-April. Available from: https://www.sae.org/publications/technical-papers/content/2018-01-1307/.

114 Jamal T, Shafiullah GM, Dawood F, Kaur A, Arif MT, Pugazhendhi R, et al. Fuelling the future: An in-depth review of recent trends, challenges and opportunities of hydrogen fuel cell for a sustainable hydrogen economy. *Energy Rep.* 2023 Nov 1;10:2103–27.

115 Waseem M, Amir M, Lakshmi GS, Harivardhagini S, Ahmad M. Fuel cell-based hybrid electric vehicles: An integrated review of current status, key challenges, recommended policies, and future prospects. *Green Energy and Intell Transp.* 2023 Dec 1;2(6):100121.

6

Hydrogen-powered Transportation and Industrial Applications

6.1 Introduction

The global transition toward cleaner energy sources has placed hydrogen at the forefront of efforts to decarbonize critical sectors, particularly transportation and heavy industry. As nations strive to achieve net-zero emissions by mid-century, hydrogen has emerged as a crucial alternative to fossil fuels due to its high energy density, versatility, and ability to be produced from a range of sources, including renewable energy. Unlike conventional fuels, hydrogen does not emit carbon dioxide when used in fuel cells or combustion engines, making it an attractive option for sectors that are difficult to electrify. Governments and industries worldwide are increasingly recognizing hydrogen's role in reducing dependence on fossil fuels, mitigating climate change, and ensuring energy security [1, 2].

The transportation sector is a major contributor to global greenhouse gas emissions, with road, air, and maritime transport accounting for nearly a quarter of total energy-related CO_2 emissions. While battery electric vehicles (BEVs) are well-suited for passenger cars and light-duty vehicles, hydrogen fuel cell electric vehicles (FCEVs) offer distinct advantages for applications requiring long-range, rapid refueling, and high payload capacity. Hydrogen-powered trucks, buses, trains, ships, and aircraft are already being developed and deployed, offering viable solutions for reducing emissions in sectors where battery technology faces limitations. One of the primary advantages of hydrogen fuel cell technology is its ability to provide extended range and fast refueling, making it ideal for heavy-duty transport and long-haul applications. Unlike BEVs, which require lengthy charging times and may suffer from range anxiety, hydrogen-powered vehicles can be refueled in minutes, similar to conventional gasoline or diesel vehicles. This advantage is particularly critical for freight transport, public transit, and logistics industries, where minimizing downtime is essential for operational efficiency [3]. In the maritime sector, hydrogen-based fuels, including ammonia and methanol, are being explored as alternatives to traditional marine fuels. The shipping industry, responsible for nearly 3% of global CO_2 emissions, faces increasing pressure to adopt cleaner technologies under international regulations such as the International Maritime Organization's (IMO's) emissions

Next-Generation Hydrogen Economy: Production, Storage, and Fuel Cell Technologies, First Edition.
Peeyush Phogat, Shreya Sharma and Satyam Rawat.

reduction targets [4]. Hydrogen-based propulsion systems are being integrated into ferries, cargo vessels, and even cruise ships, marking a significant step toward greener maritime transportation. Similarly, the aviation industry is actively exploring hydrogen as a long-term solution for sustainable air travel. Aircraft manufacturers and aerospace companies are developing hydrogen-powered propulsion systems that could revolutionize the industry, reducing its reliance on carbon-intensive jet fuel. Several experimental hydrogen aircraft prototypes have already been tested, and commercial hydrogen-powered flights are expected to be operational within the next two decades. With the rise of green hydrogen production and the development of liquid hydrogen storage technologies, aviation could experience a paradigm shift in decarbonization efforts [5].

Beyond transportation, hydrogen is playing a pivotal role in decarbonizing industrial sectors that rely on carbon-intensive processes. Heavy industries, including steel, cement, and chemicals, contribute a substantial portion of global CO_2 emissions, largely due to their dependence on fossil fuels for heat generation and chemical reactions. Hydrogen is emerging as a key alternative, offering a cleaner pathway for these industries to transition toward net-zero operations [6]. The steel industry, for example, has traditionally relied on coal-based blast furnaces to extract iron from ore, a process that generates significant CO_2 emissions. The introduction of hydrogen-based direct reduction of iron (DRI) provides a promising alternative, replacing coal with hydrogen to produce low-carbon or even carbon-free steel. Leading steel manufacturers in Europe, Japan, and the United States are investing in hydrogen-based steelmaking technologies, positioning the industry for a sustainable transformation [7].

Similarly, hydrogen is being integrated into refining and chemical industries, where it serves as both a feedstock and an energy source. Refineries use hydrogen for processes such as hydrocracking and desulfurization, which are essential for producing cleaner fuels. However, most of the hydrogen currently used in these industries is derived from natural gas through steam methane reforming (SMR), a process that generates CO_2 emissions. The shift toward green hydrogen, produced via electrolysis using renewable energy, is gaining momentum as industries seek to reduce their carbon footprint [8]. The cement industry, another major emitter of CO_2, is also exploring hydrogen-based solutions. Cement production requires high-temperature kilns, traditionally powered by coal or natural gas. By using hydrogen as a combustion fuel, manufacturers can significantly lower emissions while maintaining operational efficiency. Several pilot projects are underway to test hydrogen's feasibility in cement production, highlighting its potential as a key decarbonization tool [9].

The expansion of hydrogen applications in transportation and industry is closely linked to ongoing technological advancements in production, storage, and distribution. One of the most significant breakthroughs is the rapid development of electrolysis technology, which enables the production of green hydrogen using renewable electricity. Improvements in electrolyzer efficiency and cost reductions are making green hydrogen increasingly competitive with traditional hydrogen production methods, accelerating its adoption in key sectors [10].

Fuel cell technology has also seen remarkable progress, enhancing the performance and durability of hydrogen-powered vehicles. Modern fuel cells boast higher efficiency, longer lifespans, and lower costs, making FCEVs more attractive for commercial and industrial use [11]. Additionally, advancements in hydrogen storage technologies, such as high-pressure tanks, cryogenic storage, and hydrogen carriers like ammonia and liquid organic hydrogen carriers (LOHCs), are addressing logistical challenges related to transport and distribution [12]. Governments and private sector stakeholders are investing heavily in hydrogen infrastructure, including production hubs, refueling stations, and dedicated transport networks. Major economies, including Japan, Germany, South Korea, and the United States, have established hydrogen roadmaps that outline ambitious deployment targets and funding mechanisms. International collaboration is also playing a crucial role in standardizing regulations, ensuring supply chain security, and fostering a global hydrogen economy.

6.2 Hydrogen in Transportation Sector

The transportation sector is one of the largest contributors to global greenhouse gas (GHG) emissions, accounting for approximately 25% of total energy-related CO_2 emissions. As nations strive to achieve net-zero targets, hydrogen has emerged as a viable solution to decarbonize transportation, particularly in segments where direct electrification faces significant limitations. Unlike conventional fossil fuels, hydrogen offers a zero-emission alternative when used in fuel cells, producing only water and heat as byproducts. This makes it a compelling choice for reducing carbon footprints across various modes of transport, including road, rail, maritime, and aviation. Hydrogen's role in transportation is primarily driven by FCEV technology, which converts hydrogen into electricity through a chemical reaction, powering electric motors with high efficiency. While BEVs have gained traction in passenger car markets, hydrogen fuel cells offer distinct advantages in applications requiring extended range, rapid refueling, and high payload capacity. As a result, hydrogen is increasingly being explored for heavy-duty transportation, including trucks, buses, ships, and trains, where long operational hours and high energy demands make batteries less practical. Several leading economies, including Japan, Germany, South Korea, and the United States, have launched hydrogen mobility strategies aimed at integrating fuel cell technology into their transportation networks. These initiatives include large-scale investments in hydrogen production, refueling infrastructure, and vehicle deployment. Hydrogen-powered public transport systems, such as buses and trains, have already been introduced in multiple cities, showcasing the feasibility of hydrogen in mass transit [13]. As shown in Figure 6.1, hydrogen is finding applications across diverse modes of transportation.

Beyond road transport, hydrogen is gaining momentum in aviation and maritime industries, where the transition to low-carbon alternatives is more complex. Hydrogen-based propulsion systems, including hydrogen combustion and fuel

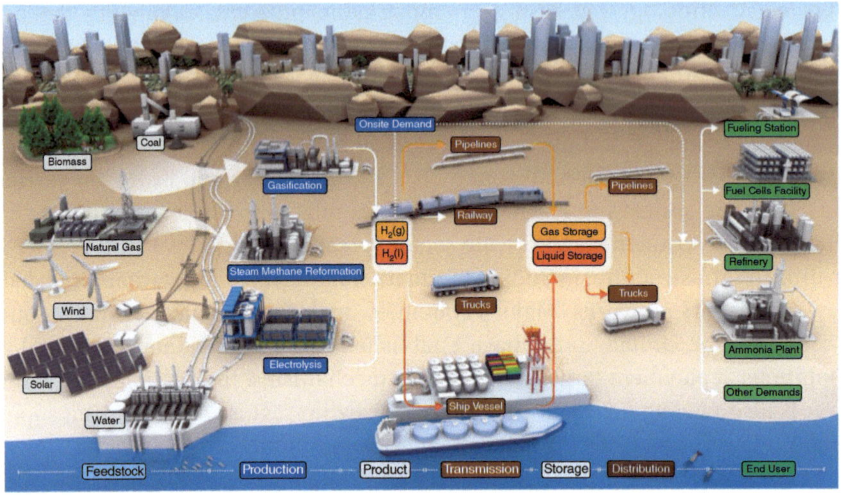

Figure 6.1 Hydrogen applications across transportation modes. Reproduced with permission from [16]/Elsevier/CC BY-NC-ND 4.0.

cells, are being developed for aircraft, with leading aerospace companies exploring liquid hydrogen storage and hydrogen-electric hybrid designs. Similarly, the shipping industry is examining hydrogen and hydrogen-derived fuels like ammonia and methanol to replace conventional marine fuels, in line with the IMO's decarbonization goals [14]. Despite its potential, hydrogen adoption in transportation faces several challenges, including infrastructure development, cost competitiveness, and energy efficiency concerns. The expansion of hydrogen refueling stations (HRS), the reduction of fuel cell costs, and advancements in hydrogen storage technologies will be critical in overcoming these barriers [15]. Policy support, including subsidies, carbon pricing mechanisms, and research funding, will play a key role in accelerating hydrogen deployment in transport. As hydrogen technology continues to evolve, its integration into transportation is expected to expand, contributing significantly to emission reductions and energy diversification. The following sections explore the two major applications of hydrogen in transportation: FCEVs and hydrogen for heavy-duty transport, highlighting their technological progress, benefits, and challenges.

6.2.1 Hydrogen FCEVs

Hydrogen FCEVs represent a transformative approach to decarbonizing road transportation by providing a zero-emission alternative to conventional internal combustion engine vehicles. Unlike BEVs, which rely on lithium-ion batteries for energy storage, FCEVs utilize hydrogen fuel cells to generate electricity on demand, offering advantages in refueling time, driving range, and operational efficiency. As governments and automakers accelerate the transition toward sustainable mobility, FCEVs are emerging as a key technology in achieving carbon-neutral transportation. The core of an FCEV is its proton exchange

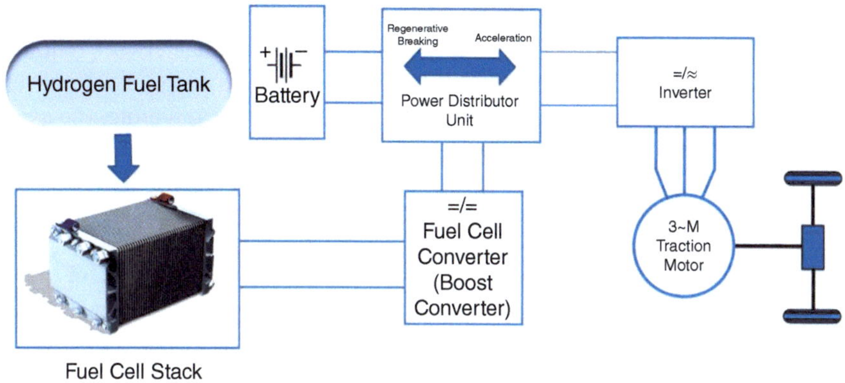

Figure 6.2 Working principle of hydrogen FCEV. Reproduced with permission from [24]/ Elsevier.

membrane fuel cell, which facilitates an electrochemical reaction between hydrogen and oxygen, generating electricity, heat, and water as byproducts. The electricity produced powers an electric motor, eliminating the need for combustion and significantly reducing GHG emissions. Hydrogen is stored in high-pressure tanks, typically at 350–700 bar, ensuring sufficient fuel capacity for extended driving ranges. Unlike BEVs, which require hours for recharging, FCEVs can be refueled in under five minutes, making them a practical alternative for drivers requiring quick turnaround times [17]. Figure 6.2 illustrates the working principle of a typical hydrogen fuel cell vehicle.

One of the key benefits of FCEVs is their long driving range. Current hydrogen-powered models, such as the Toyota Mirai and Hyundai Nexo, offer ranges exceeding 600 kilometers on a single tank, making them well-suited for long-distance travel. Additionally, FCEVs perform well in extreme weather conditions, where battery performance may degrade, providing an edge in regions with harsh climates. Another significant advantage is their scalability across vehicle segments. While BEVs dominate the passenger car market, FCEVs are increasingly being explored for buses, trucks, and other commercial vehicles that require high energy output and rapid refueling. Hydrogen-powered buses have already been deployed in cities such as London, Tokyo, and Los Angeles, reducing emissions in urban transit systems. Furthermore, FCEVs contribute to energy diversification, reducing dependency on fossil fuels. When powered by green hydrogen produced from renewable sources, FCEVs offer a truly sustainable transportation solution, aligning with global climate goals. They also have the potential to integrate with emerging smart grid technologies, where excess renewable energy can be stored as hydrogen and later used for mobility applications. Despite their advantages, the widespread adoption of FCEVs faces several challenges, primarily related to cost and infrastructure. The production of hydrogen remains expensive, particularly for green hydrogen derived from electrolysis. Additionally, fuel cell stacks rely on costly materials such as platinum, increasing vehicle prices compared to their battery electric counterparts. Infrastructure limitations also hinder market penetration. HRS are scarce, with limited availability outside select regions like California, Japan, and parts of Europe. Expanding the refueling network requires significant investment, which remains a bottleneck for mass adoption. Moreover, energy efficiency concerns persist. While FCEVs offer quick refueling, the overall energy conversion efficiency from hydrogen production to vehicle propulsion is lower than that of direct battery electric systems. Addressing these efficiency gaps through technological advancements and process optimization is crucial for enhancing the competitiveness of hydrogen fuel cell technology [18–21].

Several countries are actively promoting hydrogen mobility through policy support and investment in hydrogen infrastructure. Japan, a pioneer in hydrogen technology, has ambitious targets to expand its FCEV fleet and refueling networks by 2030. The European Union's Hydrogen Strategy also emphasizes FCEV deployment as part of its decarbonization roadmap. Automakers are advancing fuel cell technology, improving efficiency, and reducing costs through innovation in materials and system design. Companies like Toyota, Hyundai, and Honda continue to invest in next-generation fuel cells, while collaborations between governments and private industry aim to accelerate the development of hydrogen supply chains. In the long term, the success of FCEVs will depend on achieving cost parity with conventional and BEVs, expanding hydrogen refueling networks, and scaling up green hydrogen production. As advancements continue, hydrogen fuel cell vehicles have the potential to play a crucial role in achieving sustainable and emissions-free transportation [22, 23].

6.2.2 Hydrogen for Heavy-duty Transport

The transition to hydrogen as a fuel source for heavy-duty transport, such as trucks, buses, trains, and even ships, represents a significant opportunity in the decarbonization of the transportation sector. Unlike passenger vehicles, which primarily rely on battery electric technologies, heavy-duty transport faces greater challenges due to the higher energy demands and longer operating ranges required. Hydrogen, with its higher energy density compared to batteries, offers a promising alternative to reduce emissions from industries such as freight, logistics, and public transportation. Hydrogen-powered FCEVs are well-suited for heavy-duty transport because they can deliver the energy required to power large vehicles while maintaining a longer range and shorter refueling time compared to BEVs. This is crucial for sectors like trucking, where operational efficiency and minimal downtime are essential. Hydrogen FCEVs have the ability to refuel in a matter of minutes, unlike battery electric trucks, which can take hours to recharge, particularly for long-haul routes. Additionally, hydrogen's high energy density means that fuel cell trucks are capable of carrying heavier loads over longer distances, making them viable for long-haul freight and shipping applications. One of the primary advantages of hydrogen in heavy-duty transport is its potential to address the challenges of energy consumption and range anxiety. The weight and size of batteries needed to power large trucks for long distances significantly reduce the available cargo capacity, which could affect operational viability. Hydrogen fuel cells, on the other hand, can be installed without compromising payload capacity, thus making them more suitable for freight and cargo industries. As such, hydrogen fuel cell trucks can effectively compete with diesel-powered vehicles, especially for long-distance and freight-heavy transportation. Various global initiatives have started to explore the use of hydrogen in heavy-duty transport [25]. For example, in Europe, projects like the European Clean Hydrogen Alliance and the Hydrogen for Clean Transport initiative are paving the way for the adoption of hydrogen-powered buses and trucks. In the United States, companies such as Nikola Motor and Cummins are working on developing fuel cell trucks capable of covering long-haul distances. These efforts are complemented by investments in hydrogen refueling infrastructure, which is crucial to ensure the widespread adoption of hydrogen-powered heavy-duty vehicles [26].

However, challenges remain in the widespread adoption of hydrogen for heavy-duty transport. The high upfront costs of fuel cell vehicles, combined with the cost of hydrogen production, storage, and refueling infrastructure, are significant barriers. The scaling of hydrogen production to meet the needs of a growing fleet of hydrogen-powered trucks is also a critical issue. Moreover, hydrogen's dependence on renewable energy for its "green" status adds another layer of complexity to the economic feasibility of hydrogen fuel cells. Additionally, the need for robust refueling stations and the logistics of delivering hydrogen to remote or less-developed areas could slow down the adoption in certain regions. In conclusion, hydrogen has the potential to be a game-changer for the heavy-duty transport sector, providing a clean and efficient alternative to fossil fuels. As technologies improve and infrastructure expands, the role of hydrogen in decarbonizing freight and logistics will continue to grow. Investments in research and development, along with policy incentives, will be key to overcoming current challenges and unlocking the full potential of hydrogen as a sustainable fuel source for heavy-duty transportation [27]. Table 6.1 compares FCVs with BEVs for heavy-duty transport applications. Hydrogen FCVs offer higher energy density, enabling significantly longer ranges (400–800 km) and faster refueling times (5–15 minutes), which are crucial for long-haul operations. In contrast, BEVs exhibit lower energy density, requiring larger batteries that increase vehicle weight and limit range (250–400 km), alongside longer recharging times. FCVs are structurally lighter, making them more suitable for heavy payloads, while BEVs often compromise cargo capacity due to battery mass. Infrastructure development for FCVs is still limited, requiring specialized hydrogen refueling stations, whereas BEVs benefit from more widely available electric charging networks. Operationally, FCVs excel in long-distance, heavy-duty roles, whereas BEVs are more efficient in short-range and urban contexts. From a cost perspective, FCVs demand higher upfront investment but potentially offer lower long-term operational costs, while BEVs, though cheaper initially, may incur higher maintenance due to battery degradation. Both technologies boast zero tailpipe emissions when powered by green sources, aligning with global decarbonization goals. However, their environmental benefit hinges on the source—hydrogen must be renewably produced, and electricity must come from clean grids to ensure sustainable operation across both vehicle platforms.

Table 6.1 Comparison of hydrogen FCVs vs. BEVs for heavy-duty transport.

Aspect	Hydrogen FCVs	BEVs
Energy density	Higher energy density, allowing for longer ranges [28]	Lower energy density, requiring larger and heavier batteries for long ranges
Refueling/ Recharging time	Fast refueling time (5–15 minutes) [29]	Long charging time (1–4 hours depending on battery size and charging infrastructure) [30]
Range	Typically 400–800 km per full tank [31]	Limited to 250–400 km, depending on battery capacity and vehicle type [32]
Weight and payload	Lighter and better suited for carrying heavy loads [33]	Heavier due to large batteries, reducing cargo capacity [34]
Infrastructure development	Requires HRS [35]	Requires electric charging stations [36]
Operational efficiency	More suitable for long-haul and heavy-duty transport [37]	Best suited for short-range or urban applications [38]
Cost	High initial cost, but lower operating costs over time [39]	Moderate initial cost, with higher operating costs due to charging time and battery degradation [40]
Emissions	Zero tailpipe emissions if hydrogen is green [41]	Zero tailpipe emissions if powered by renewable energy [42]
Energy source	Requires hydrogen production from renewable sources [43]	Powered by electricity from renewable or grid sources [44]

6.3 Hydrogen Refueling Infrastructure: Current Progress and Challenges

The widespread adoption of hydrogen-powered transportation, particularly in heavy-duty sectors such as freight, public transit, and logistics, is intrinsically linked to the development of a robust and accessible hydrogen refueling infrastructure. Just as electric vehicles depend on a reliable network of charging stations, FCEVs require strategically distributed HRS to support their operations and ensure user confidence. Without a well-developed refueling network, even the most advanced hydrogen vehicles will struggle to gain traction in the mainstream market. Consequently, hydrogen refueling infrastructure has become a critical bottleneck in the transition to a hydrogen-based mobility system, influencing both consumer adoption and industry investment decisions [45].

Currently, the deployment of HRS is largely concentrated in a few leading economies, with Japan, Germany, South Korea, China, and California (United States) at the forefront. Japan has taken a pioneering role, integrating hydrogen into its long-term energy strategy and committing to the construction of 1,000 HRS by 2030. As of now, the country has established over 160 stations, supporting a growing fleet of hydrogen-powered vehicles including the Toyota Mirai and various public transport buses. Germany, backed by the European Union's Hydrogen Strategy, has developed the second-largest hydrogen station network in Europe, with nearly 100 public stations in operation and several more planned under the H_2 MOBILITY initiative. South Korea, too, has been aggressive in rolling out hydrogen infrastructure, aiming for more than 310 HRS by 2030 to support its domestic FCEV goals. California remains the North American leader in hydrogen mobility, hosting over 60 operational refueling stations concentrated primarily around urban hubs like Los Angeles and San Francisco. However, even in these leading regions, coverage remains sparse outside metropolitan areas, limiting long-range travel and heavy-duty logistics routes [46, 47].

In other parts of the world, the progress has been slower. Many countries are still in the early stages of hydrogen infrastructure development, constrained by high costs, uncertain demand, and regulatory hurdles. In emerging economies, the lack of foundational energy and transport infrastructure makes hydrogen deployment even more challenging. Moreover, the economics of HRS installation can be prohibitive. A single hydrogen refueling station can cost between $1 million and $2 million depending on its capacity, technology, and location. This cost includes not only the refueling equipment but also hydrogen storage systems, compression units, safety systems, and integration with existing infrastructure. High capital investment requirements, combined with low vehicle utilization rates in the early phases of adoption, deter private investment and slow infrastructure scaling [48]. As shown in Figure 6.3, the distribution of HRS remains uneven across regions. While a few countries have made significant progress, vast areas of the world lack access to even basic hydrogen refueling services. This disparity limits cross-border mobility for hydrogen vehicles and hampers efforts to build global supply chains for hydrogen-powered freight and logistics. Moreover, the limited density of stations leads to what is known as the "chicken-and-egg" problem: consumers are hesitant to buy hydrogen vehicles without available refueling options, and investors are reluctant to build stations without sufficient vehicle demand.

Another major challenge lies in the hydrogen supply chain itself. Delivering hydrogen to refueling stations involves complex logistics, including high-pressure gas or cryogenic liquid transport, or on-site production using electrolysis or SMR. Each method has trade-offs in terms of cost, scalability, and carbon footprint. On-site electrolysis, while potentially greener when powered by renewable energy, remains expensive and limited by electricity availability. In contrast, hydrogen delivery via tube trailers or pipelines poses challenges in terms of energy efficiency, cost, and safety regulations. The development of regional hydrogen hubs, where production, storage, and distribution are co-located, has been proposed as a solution to mitigate some of these issues, but such hubs are still in nascent stages. Standardization is another critical factor. Hydrogen refueling systems must adhere to strict safety, pressure, and flow rate standards to ensure compatibility with diverse vehicle types

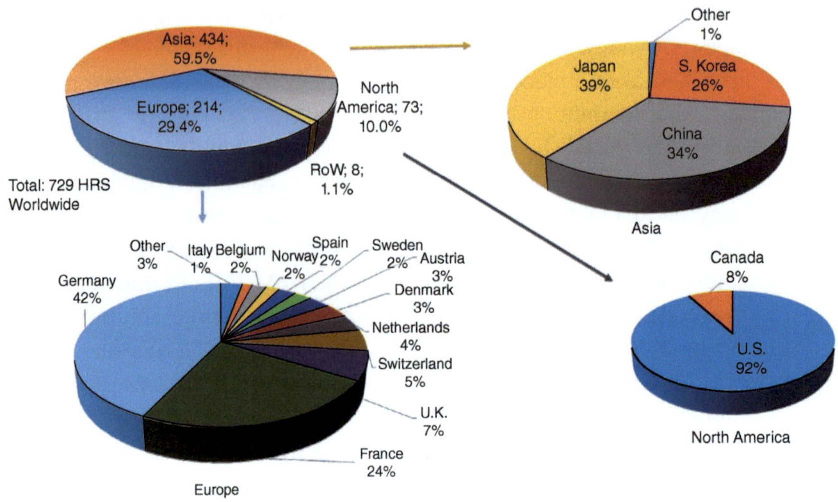

Figure 6.3 Global development status of HRS, highlighting major deployment regions and infrastructure gaps. [51]/MDPI/CC BY 4.0.

and international regulations. The global hydrogen industry is working toward common protocols, such as ISO 19880-1, which governs station safety and performance. However, differences in regulatory frameworks, especially between North America, Europe, and Asia, create additional barriers to international harmonization. Addressing these regulatory and technical discrepancies is essential for enabling cross-border hydrogen vehicle operations and integrated supply chains [49].

Looking ahead, scaling hydrogen refueling infrastructure will require coordinated public and private investment. Governments can play a pivotal role by offering financial incentives such as subsidies, tax credits, and low-interest loans to encourage station development. Public-private partnerships can also help de-risk early investments and build confidence in hydrogen markets. Policies that set clear deployment targets, streamline permitting processes, and mandate hydrogen station integration in new transport corridors will be critical in accelerating the rollout of infrastructure. Furthermore, integrating hydrogen infrastructure with existing natural gas or electricity networks – such as repurposing pipelines or colocating HRS with electric vehicle (EV) charging stations – can reduce costs and leverage existing assets. In addition to road transport, the hydrogen refueling network must also expand to support emerging applications in rail, shipping, and aviation. For instance, hydrogen-powered trains require larger-scale hydrogen storage and high-flow dispensing systems at railway depots. Similarly, ports and airports will need specialized hydrogen delivery infrastructure to fuel hydrogen-electric ships and aircraft. These use cases, though still in early development, underscore the need for flexible and scalable refueling infrastructure tailored to multiple transport modalities. Ultimately, the development of a widespread and reliable hydrogen refueling infrastructure is central to the success of hydrogen-powered mobility. Without it, the deployment of hydrogen vehicles will remain confined to niche applications and limited regions. However, with the right mix of policy support, innovation, and investment, hydrogen infrastructure can evolve from a bottleneck into a catalyst—unlocking the full potential of hydrogen as a clean, scalable, and sustainable fuel for the future of transportation [50].

6.4 Hydrogen in Industrial Applications

Hydrogen is emerging as a critical component in the transition to sustainable industrial practices, particularly in hard-to-abate sectors where direct electrification is challenging. Industries such as steel, chemicals, cement, and refining are among the largest contributors to global carbon emissions due to their reliance on fossil fuels for energy-intensive processes. The integration of hydrogen, particularly green hydrogen produced through electrolysis using renewable energy, offers a promising pathway to decarbonizing these sectors while maintaining efficiency and output. The industrial sector accounts for nearly 30% of global energy consumption and over 20% of CO_2 emissions, making decarbonization an urgent priority. Hydrogen provides a versatile solution by serving as both a clean energy carrier and a feedstock in various industrial processes. Unlike battery storage or direct renewable energy applications, hydrogen can be used for high-temperature operations, making it suitable for industries where alternative low-carbon technologies remain inadequate. Several countries and corporations are investing in hydrogen-powered industrial transformation. The European Union's Hydrogen Strategy, Germany's National Hydrogen Plan, and Japan's roadmap for hydrogen-based industries highlight the growing recognition of hydrogen's potential. Meanwhile, industrial giants such as ArcelorMittal, Thyssenkrupp, Shell, and Linde are actively developing pilot projects to integrate hydrogen into large-scale manufacturing [52].

The role of hydrogen in industrial applications can be broadly categorized into:

a. **Fuel Substitution**: Replacing coal, natural gas, or other fossil fuels with hydrogen for high-temperature heating processes.
b. **Feedstock Replacement**: Using hydrogen as a cleaner alternative to fossil-derived hydrogen in chemical and refining industries.
c. **Process Innovation**: Developing new hydrogen-based production techniques that minimize emissions and enhance efficiency.

Although hydrogen presents a viable decarbonization pathway, its large-scale adoption faces key challenges, including high production costs, infrastructure limitations, and the need for technological advancements. The economic feasibility of hydrogen in industry is closely tied to the cost of green hydrogen production, which is expected to decrease with improved electrolyzer efficiency, large-scale deployment, and declining renewable energy costs. Government policies, carbon pricing mechanisms, and financial incentives will play a crucial role in accelerating the adoption of hydrogen across industrial applications. As international regulatory frameworks increasingly favor low-carbon alternatives, industries that proactively integrate hydrogen solutions will be better positioned for a sustainable future. The following subsections will explore the specific applications of hydrogen in the steel and iron industry, the refining and chemical industries, and the cement sector, highlighting current advancements, challenges, and future prospects. Table 6.2 provides a concise overview of hydrogen's multifaceted

Table 6.2 Key applications of hydrogen in the industrial sector.

Industry	Hydrogen application	Potential benefits
Steel and iron	Hydrogen used as a reducing agent in DRI ore	Eliminates CO_2 emissions from blast furnaces, creating green steel
Chemical industry	Hydrogen used in ammonia production, refining processes, and as a feedstock for petrochemicals	Reduces reliance on fossil fuels and enhances efficiency in chemical production
Refining	Hydrogen used in hydrocracking, desulfurization, and hydroprocessing	Improves fuel quality and reduces emissions in petroleum refining
Cement industry	Hydrogen combustion used for cement kilns	Potential to decarbonize one of the most carbon-intensive industries
Energy production	Hydrogen in turbines for electricity generation and grid balancing	Decarbonizes power generation and stabilizes renewable grid supply

role across key industrial sectors. In the steel and iron industry, hydrogen replaces carbon as a reducing agent in ore processing, significantly cutting CO_2 emissions to produce "green steel." The chemical industry utilizes hydrogen in ammonia synthesis and refining operations, minimizing fossil fuel dependence and enhancing production efficiency. In petroleum refining, hydrogen plays a vital role in hydrocracking, desulfurization, and hydroprocessing, improving fuel quality while reducing pollutant emissions. The cement industry, one of the highest carbon emitters, can decarbonize operations through hydrogen combustion in kilns. Additionally, hydrogen-powered turbines support energy production by decarbonizing electricity generation and enhancing grid stability, particularly with intermittent renewables. These applications underline hydrogen's potential to transform high-emission industries into cleaner, more sustainable systems.

6.4.1 Hydrogen in Steel and Iron Industry

The steel and iron industry is one of the most carbon-intensive sectors, contributing nearly 8% of global CO_2 emissions due to its heavy reliance on coal and coke in conventional blast furnace processes. As the demand for low-carbon steel grows, hydrogen is emerging as a transformative solution to decarbonize the sector. By replacing carbon-based reducing agents with hydrogen, steel manufacturers can significantly lower emissions while maintaining production efficiency. Traditionally, steel production relies on blast furnace-basic oxygen furnace (BF-BOF) technology, where coal-derived coke acts as a reducing agent to extract iron from its ore. This process emits large amounts of CO_2. An alternative method, known as DRI technology, typically uses natural gas to produce sponge iron, which is later processed into steel using an electric arc furnace (EAF). The integration of green hydrogen in DRI-EAF technology replaces natural gas, enabling carbon-free steel production. One of the most promising approaches is hydrogen-based direct reduction (H_2-DRI), where hydrogen reacts with iron ore to produce water vapor instead of CO_2, making it a zero-emission process when powered by renewable energy. Another emerging method is plasma reduction, where hydrogen plasma reduces iron ore without intermediate steps, increasing efficiency and lowering energy consumption. A transitional approach known as hydrogen-enriched blast furnaces involves co-injecting hydrogen with pulverized coal, which helps reduce overall carbon emissions in traditional steelmaking [53].

Leading steel manufacturers and governments are investing in hydrogen-based technologies to decarbonize the industry. Sweden's HYBRIT project, a collaboration between SSAB, LKAB, and Vattenfall, is pioneering fossil-free steel production by replacing coke with hydrogen in direct reduction. The project successfully delivered its first batch of green steel in 2021. Another Swedish initiative, H_2 Green Steel, is developing a large-scale hydrogen-powered steel plant that is expected to reduce emissions by 95% compared to conventional steelmaking. Germany is also making significant progress, with Salzgitter AG's SALCOS (Salzgitter Low CO_2 Steelmaking) project transitioning to hydrogen-based steel production using DRI and electrolysis. Similarly, ArcelorMittal is investing in hydrogen injection at its European blast furnaces and constructing green hydrogen DRI facilities in Germany and Spain. Meanwhile, POSCO in South Korea is exploring hydrogen-based iron ore reduction as part of its long-term decarbonization strategy. These initiatives demonstrate a growing commitment to hydrogen steel production as a means to meet global climate targets [54].

Despite its potential, hydrogen-based steelmaking faces several challenges. One of the most significant barriers is the high cost of hydrogen production. Currently, green hydrogen remains expensive compared to natural gas and coal, making large-scale deployment economically challenging. Additionally, infrastructure limitations pose a hurdle, as hydrogen storage, transportation, and distribution networks need significant expansion to support industrial demand. Another key challenge is the high energy requirements for producing green hydrogen. Electrolysis, the primary method of generating green hydrogen, requires vast amounts of renewable electricity, necessitating large-scale expansion of solar, wind, or nuclear energy capacity [55]. Furthermore, technology readiness remains a concern, as full-scale hydrogen-based steelmaking is still in its early stages and requires further optimization. Governments must also play a crucial role by implementing policy and regulatory support, including financial incentives, carbon pricing, and green steel certification, to drive demand and competitiveness in the market. The transition to hydrogen-based steelmaking is expected to accelerate in the coming decades, driven by falling renewable energy costs, technological advancements, and stringent emissions regulations. According to industry forecasts, hydrogen-based steel production could become cost-competitive by 2030 if green hydrogen prices drop below \$2/kg. Governments and industries are introducing green procurement policies and carbon border adjustment mechanisms to incentivize low-emission steel production. Additionally, corporate demand for sustainable steel is rising, with industries like automotive and construction seeking low-carbon

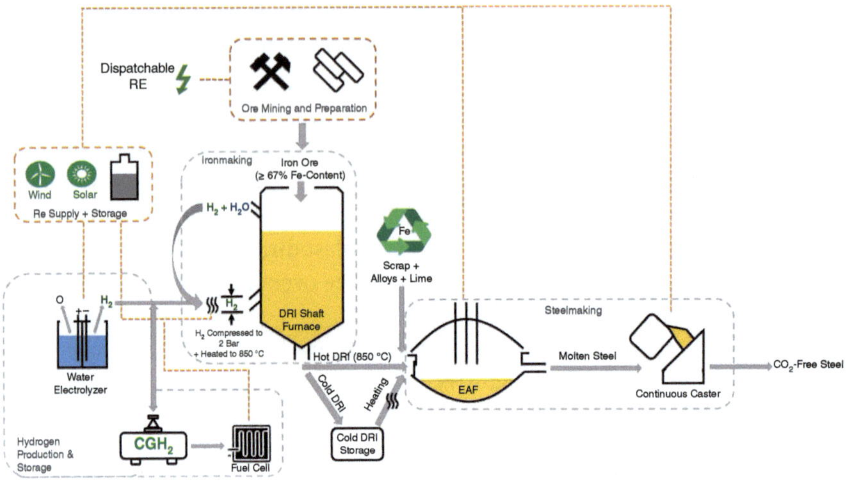

Figure 6.4 Schematic of H_2-DRI for green steel production. [57]/Springer Nature/ CC BY 4.0.

materials to meet sustainability goals. While challenges remain, hydrogen represents a game-changing solution for the decarbonization of steel and iron production. With continued investments, policy support, and innovation, hydrogen-based steel could become a mainstream reality by mid-century, reshaping one of the world's most essential industries [56]. Figure 6.4 depicts the green steel production process using hydrogen-based direct reduction.

6.4.2 Hydrogen in Refining and Chemical Industries

The refining and chemical industries are among the largest consumers of hydrogen, utilizing it for various processes, including hydrocracking, desulfurization, and ammonia production. However, the majority of hydrogen used in these industries today is grey hydrogen, produced from fossil fuels like natural gas, leading to significant carbon dioxide (CO_2) emissions. Transitioning to green hydrogen – produced via electrolysis using renewable energy – offers a path to deep decarbonization in these critical sectors. Refineries use hydrogen primarily for hydroprocessing operations, which improve fuel quality and meet stringent environmental regulations. Two key processes where hydrogen is essential include:

a. **Hydrodesulfurization (HDS):** This process removes sulfur from crude oil fractions, reducing sulfur dioxide (SO_2) emissions when fuels are burned. With tightening global sulfur regulations (e.g., IMO 2020 for marine fuels), demand for hydrogen in refineries is expected to rise.
b. **Hydrocracking:** Hydrogen is used to break down heavy crude fractions into lighter, more valuable products like diesel, jet fuel, and gasoline, improving fuel yields and quality.

Traditionally, refineries produce their hydrogen through SMR, a carbon-intensive process. However, as governments implement carbon pricing and stricter emission standards, refineries are exploring green and blue hydrogen (hydrogen produced via SMR but with carbon capture and storage) to reduce their carbon footprint. Replacing gray hydrogen with green hydrogen could reduce refinery CO_2 emissions by up to 90% [58].

The chemical industry is the largest consumer of hydrogen, using it as a feedstock for producing ammonia, methanol, and other essential chemicals. Transitioning to low-carbon hydrogen in this sector is crucial for achieving global emission reduction targets. Ammonia (NH_3) is a key component of fertilizers, with over 70% of global ammonia production used in agriculture. Traditionally, ammonia is produced using the Haber-Bosch process, which relies on grey hydrogen derived from natural gas, contributing to high CO_2 emissions. The shift to green ammonia, produced using green hydrogen, is gaining momentum as companies and governments seek sustainable agricultural solutions. Major ammonia producers, such as Yara, CF Industries, and Fertiberia, are investing in green hydrogen-based ammonia plants to supply fertilizers with minimal carbon impact. Additionally, green ammonia is being explored as a zero-carbon fuel for shipping, further increasing its strategic importance in global decarbonization efforts. Methanol is another major hydrogen-based chemical used in plastics, paints, adhesives, and synthetic fuels. Conventional methanol production emits significant CO_2, but green hydrogen enables the production of low-carbon or e-methanol. Green methanol is also being considered for marine fuel applications, providing a potential alternative to fossil-based shipping fuels. Companies like OCI, Maersk, and Methanex are developing green methanol projects to support cleaner industrial operations and transportation. In addition, carbon capture and utilization (CCU) technologies are being integrated into methanol production, using CO_2 as a feedstock to produce synthetic fuels and chemicals [59].

Several strategies are being explored to transition refining and chemical industries toward low-carbon hydrogen solutions:

a. **Electrolysis-based Hydrogen Production:** Using renewable electricity to split water into hydrogen and oxygen eliminates CO_2 emissions. Electrolysis-based hydrogen is increasingly favored as the cost of renewable energy declines.
b. **Carbon Capture and Storage (CCS):** For industries still using SMR hydrogen, CCS helps capture and store CO_2 emissions, producing "blue hydrogen" as a transitional step toward full decarbonization.
c. **Sector Coupling and Circular Economy:** Integrating hydrogen production with other sectors, such as utilizing captured CO_2 from refineries to produce synthetic fuels and chemicals, enhances efficiency and sustainability.

Governments are supporting this transition through policy incentives, carbon pricing mechanisms, and funding for hydrogen infrastructure. For example, the European Union's "Fit for 55" policy and the US Inflation Reduction Act are accelerating investments in green hydrogen and sustainable chemical production. While the transition to low-carbon hydrogen in refining and chemicals presents significant opportunities, several challenges remain. Green hydrogen remains more expensive than grey hydrogen due to the high costs of electrolysis and renewable energy. Continued cost reductions in electrolyzers and scaling up production will be key to widespread adoption. Scaling hydrogen production requires massive infrastructure investments in transportation, storage, and hydrogen pipelines, particularly for refineries and chemical plants operating on a global scale. While green ammonia and e-methanol technologies are advancing, large-scale deployment requires further innovation and demonstration projects. Despite these challenges, hydrogen is set to play a transformative role in decarbonizing the refining and chemical sectors. As renewable energy costs decline and global policies favor low-carbon industrial solutions, hydrogen adoption is expected to accelerate, paving the way for a cleaner, more sustainable future [60].

6.4.3 Hydrogen in Cement Industry

The cement industry is one of the largest industrial sources of carbon dioxide (CO_2) emissions globally, accounting for approximately 8% of total anthropogenic CO_2 emissions. These emissions primarily result from the calcination process, where limestone ($CaCO_3$) is heated to produce lime (CaO), releasing CO_2. The high energy consumption of cement production and the reliance on fossil fuels, such as coal, for heating, have made it a significant contributor to climate change. However, hydrogen offers a promising pathway to decarbonize this sector by replacing fossil fuels in certain processes and enabling more sustainable practices in cement manufacturing. One of the most direct applications of hydrogen in the cement industry is as a substitute for traditional fuels used in the high-temperature cement kiln process. In traditional cement production, fuels like coal, petroleum coke, and natural gas are burned to heat the kilns to temperatures exceeding 1,450 °C, necessary for clinker production. The combustion of these fuels results in significant CO_2 emissions. Hydrogen can be utilized as a clean alternative in this high-temperature process. When burned, hydrogen produces only water vapor as a byproduct, eliminating CO_2 emissions from the combustion process. This substitution could substantially reduce the overall carbon footprint of cement production. A pilot project at LafargeHolcim and other major cement producers has shown promising results in replacing a portion of the conventional fuels with hydrogen, demonstrating that it is technically feasible to integrate hydrogen as a fuel for cement kilns. However, the high costs of green hydrogen production and the need for specialized infrastructure for hydrogen storage, distribution, and handling present challenges to scaling this technology [61].

Beyond serving as a direct fuel replacement, hydrogen can also contribute to CCU strategies within the cement industry. Hydrogen can be used in reforming reactions to produce methanol or synthetic fuels from CO_2 emissions captured at cement plants. This approach is aligned with the circular economy model, where CO_2 emissions are not only captured but transformed into valuable products. The integration of hydrogen in CCU systems helps mitigate the CO_2 emissions associated with cement production while generating useful byproducts that can be used in other industrial applications or even as alternative fuels. For example, a cement plant could capture its emissions of CO_2, use hydrogen to convert this CO_2 into synthetic methane or methanol, and then use this fuel to power parts of the cement production process. Such systems could significantly lower the carbon footprint of the industry while creating a closed-loop system for both energy and emissions management. Several companies are investigating the viability of these systems, and they are expected to become more commercially feasible as both carbon capture technologies and hydrogen production scale up [62].

Another potential application of hydrogen in the cement industry involves the decarbonization of the raw materials processing stage, particularly during the calcination process. The calcination of limestone releases large amounts of CO_2, a process that is inherently difficult to decarbonize due to the chemical nature of the reaction. However, recent research has shown that it may be possible to use hydrogen in combination with other technologies, such as electrification and carbon capture, to minimize or even eliminate the emissions from this stage of production. One promising development is the concept of hydrogen-based calcination, where hydrogen is used as a reductant to convert limestone ($CaCO_3$) into lime (CaO) without releasing CO_2. This would effectively decarbonize one of the most carbon-intensive steps in cement manufacturing. Although this technology is still in the early stages of research and development, it holds great potential for reducing emissions in the cement industry. If successfully commercialized, it could serve as a crucial part of the sector's transition to net-zero emissions [63]. Despite the promising potential of hydrogen in the cement industry, several challenges need to be addressed for widespread adoption. First, hydrogen production – particularly green hydrogen produced from renewable energy sources – is still cost-prohibitive compared to traditional fossil fuel-based options. Although costs are expected to decline with technological advances and scaling, the economics of hydrogen in cement production remain a significant barrier to its immediate implementation. Second, the integration of hydrogen infrastructure into existing cement plants requires significant investment in storage, distribution, and safety measures. Cement plants would need to adapt their infrastructure to accommodate hydrogen as a fuel, which involves retrofitting kilns and other equipment. In addition, logistical challenges related to hydrogen transportation and storage are considerable, especially when large-scale hydrogen infrastructure is not yet available in many regions. Finally, the cement industry faces a broader challenge of reducing its overall energy consumption and increasing energy efficiency. While hydrogen offers a potential clean fuel alternative, it will be most effective when integrated into broader decarbonization strategies that include energy efficiency measures, renewable energy adoption, and carbon capture technologies [64]. Hydrogen's integration in cement manufacturing is illustrated in Figure 6.5.

Government policies and regulations will play a crucial role in determining how quickly hydrogen can be adopted in the cement industry. In the European Union, for example, the European Green Deal and the EU Hydrogen Strategy set ambitious targets for decarbonization, including the use of hydrogen in industrial applications. National governments, such as in Germany and the UK, are offering financial incentives and subsidies for companies to explore and adopt low-carbon technologies, including hydrogen. Looking ahead, the cement industry is likely to see a gradual adoption of hydrogen-based technologies, starting with hydrogen-based fuel replacement in kilns and moving toward more integrated hydrogen solutions involving CCU. Although widespread implementation may take decades, the increasing

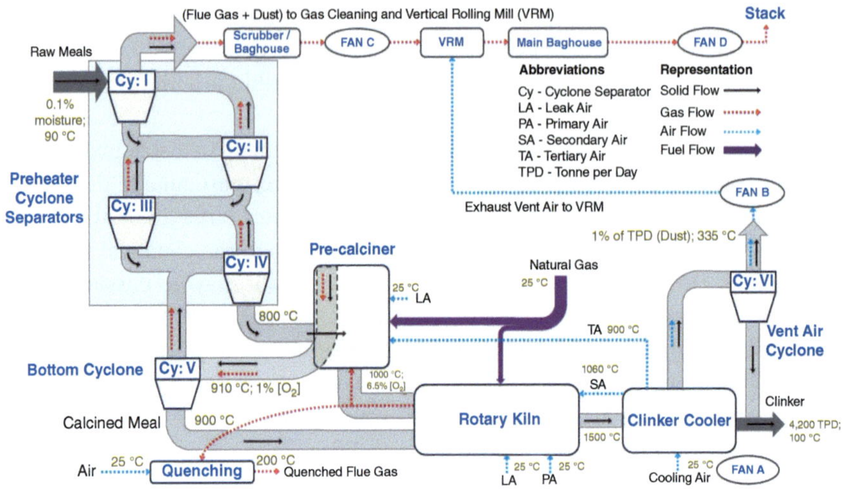

Figure 6.5 Cement production with hydrogen integration. Process flow diagram of cement production using hydrogen as a clean fuel in kilns. [66]/with permission of Elsevier/ CC BY-NC-ND 4.0.

focus on sustainable industrial practices and the rapid development of the hydrogen economy suggest that hydrogen will play a crucial role in decarbonizing the cement industry in the coming years [65].

6.5 Key Technological and Economic Drivers for Hydrogen in Transportation and Industry

The development and adoption of hydrogen technologies in both the transportation and industrial sectors are being shaped by a variety of technological and economic drivers. These factors, which span advancements in hydrogen production, storage, and distribution technologies, as well as shifts in policy, market demand, and cost structures, are pivotal to the realization of hydrogen's potential as a clean energy source. As governments, industries, and consumers increasingly recognize the need to transition to low-carbon energy systems, hydrogen is emerging as a key enabler in the push towards a sustainable energy future. In this section, we examine the primary technological and economic factors that are driving the adoption of hydrogen across transportation and industry [67].

One of the most significant technological drivers of hydrogen adoption is the advancement in hydrogen production technologies, particularly the development of green hydrogen. Green hydrogen, produced through the electrolysis of water using renewable energy sources such as wind, solar, or hydropower, has gained attention for its ability to produce hydrogen without emitting carbon dioxide. The cost of renewable electricity has significantly declined over the past decade, making green hydrogen a more cost-competitive option. Advances in electrolyzer efficiency and scalability are further driving down the costs associated with green hydrogen production, positioning it as a critical element in the decarbonization of sectors like transportation and heavy industries. In addition to production, improvements in hydrogen storage and distribution technologies are helping address one of the primary logistical challenges facing the hydrogen sector. Hydrogen has a low energy density by volume, which makes storage and transportation more complex compared to traditional fuels. Innovations in high-pressure tanks, cryogenic storage, and LOHCs are providing more efficient and scalable solutions for hydrogen storage [68]. For example, metal hydrides and advanced nanomaterials are being developed for solid-state storage, offering higher storage densities and safer handling. These technological advancements are not only facilitating the safe and efficient transport of hydrogen but are also enabling the global hydrogen supply chain to become more reliable and cost-effective. The integration of fuel cell technology into transportation vehicles is another key driver. Fuel cells convert hydrogen into electricity to power vehicles, offering an alternative to conventional combustion engines and reducing reliance on fossil fuels. The development of more affordable and efficient fuel cells has made hydrogen-powered vehicles more viable for commercial use, particularly in the heavy-duty transport sector, where hydrogen fuel cells can provide longer ranges and faster refueling times compared to BEVs. This has led to growing interest in hydrogen-powered buses, trucks, and trains, which are crucial for achieving zero-emission targets in urban transportation systems and freight logistics [69].

From an economic standpoint, hydrogen adoption is being driven by the need to address pressing issues such as energy security, resource scarcity, and environmental sustainability. In many regions, hydrogen offers an opportunity to diversify energy sources and reduce dependence on imported fossil fuels, enhancing national energy security. Hydrogen can be produced domestically from renewable resources, which is particularly appealing for countries looking to reduce their exposure to fluctuating global energy prices. Additionally, hydrogen as a storage solution for excess renewable energy generated during periods of high generation (e.g., sunny or windy days) helps stabilize energy systems and make renewable power more reliable. The economic competitiveness of hydrogen is also improving due to declining costs in renewable energy and technological advancements in hydrogen production and fuel cells. As the cost of renewable electricity continues to fall, the economic feasibility of green hydrogen becomes more attractive. This creates a favorable environment for investment in hydrogen infrastructure, which will further lower costs over time. In industries such as steel, cement, and chemicals, where hydrogen can replace carbon-intensive processes, the ability to use green hydrogen as a feedstock or fuel can significantly reduce operational costs related to carbon emissions, while also allowing companies to meet stricter emissions regulations and carbon pricing mechanisms. In the transportation sector, hydrogen is also gaining traction due to the economic advantages it offers in heavy-duty applications, where battery electric technologies may face limitations. For example, hydrogen-powered vehicles can travel longer distances and be refueled more quickly than BEVs, making them more economically viable for industries such as freight, logistics, and public transportation. Moreover, hydrogen-powered vehicles have fewer maintenance requirements compared to internal combustion engines, which further reduces long-term costs [70].

Market demand for clean energy solutions is another important economic driver for hydrogen adoption. Governments around the world are increasingly implementing net-zero emissions goals, with many setting ambitious decarbonization targets for 2030, 2040, and beyond. This is creating a strong policy push for clean technologies, including hydrogen, as part of national energy transitions. Subsidies, grants, and carbon pricing mechanisms are encouraging private sector investment and driving innovation in hydrogen technologies. In Europe, countries such as Germany and the Netherlands have committed to investing heavily in hydrogen infrastructure, while the European Union has allocated substantial funding under its Hydrogen Strategy for a Climate-Neutral Europe. These policy measures provide the necessary financial incentives and regulatory frameworks to make hydrogen a more attractive investment opportunity. At the same time, government-industry partnerships are accelerating the deployment of hydrogen infrastructure. Public-private collaborations allow for risk sharing and can expedite the scaling of hydrogen projects, including the development of HRS, storage facilities, and distribution networks. The strategic role of hydrogen in green recovery post-pandemic and its ability to create green jobs is further motivating governments to promote hydrogen as part of broader economic growth strategies. Additionally, as more industries and transportation sectors transition to hydrogen, the establishment of a robust hydrogen economy will stimulate innovation, create new markets, and lower overall production costs, accelerating the uptake of hydrogen technologies across sectors [71].

6.6 Environmental Impacts and Sustainability of Hydrogen in Transportation and Industry

Hydrogen, as a clean energy carrier, has significant potential to decarbonize the transportation and industrial sectors, but its environmental impacts and sustainability depend on the methods used for production, storage, and application. This section explores the environmental benefits and challenges associated with hydrogen adoption in these sectors, focusing on its role in reducing GHG emissions, improving energy efficiency, and supporting the transition to a sustainable, low-carbon economy. The sustainability of hydrogen is not just measured by its direct emissions, but also by the broader environmental effects of its production, infrastructure, and usage [72].

One of the primary environmental benefits of hydrogen is its ability to reduce carbon emissions in sectors that are traditionally reliant on fossil fuels, such as transportation and industry. In transportation, hydrogen FCVs produce zero tailpipe emissions, emitting only water vapor as a byproduct. When hydrogen is produced through green methods, such as electrolysis powered by renewable energy sources (wind, solar, hydro), it offers a truly zero-emission solution for mobility. This makes hydrogen particularly attractive for applications where other low-emission alternatives, like BEVs, face limitations – particularly in heavy-duty transport such as trucks, buses, and trains, which require long-range capabilities and quick refueling times. In industry, hydrogen plays a crucial role in reducing emissions in high-temperature processes such as steel production, cement manufacturing, and chemical refining, where traditional fossil fuels like coal, coke, and natural gas have been the main energy sources. Hydrogen can replace these carbon-intensive fuels, leading to a significant reduction in CO_2 emissions. For example, in the steel industry, hydrogen-based direct reduction processes (using hydrogen to reduce iron ore instead of coke) have the potential to cut down CO_2 emissions by up to 90%. Similarly, hydrogen's use as a feedstock in the chemical industry, particularly for the production of ammonia (used in fertilizers), can lead to cleaner production methods with lower emissions compared to conventional processes that rely on natural gas. Additionally, hydrogen can support the integration of renewable energy into the grid by serving as a storage solution for excess renewable electricity, helping balance supply and demand. When renewable energy generation outpaces demand, excess energy can be used to produce hydrogen, which can be stored and utilized later when demand is higher or renewable generation is low. This use of hydrogen as a seasonal storage medium promotes a more resilient and flexible energy system, allowing for greater integration of intermittent renewable energy sources such as wind and solar [73].

Despite its many environmental benefits, hydrogen is not without its challenges. A major concern is the carbon intensity of hydrogen production. While hydrogen itself is a clean fuel, the sustainability of its usage heavily depends on how it is produced. The vast majority of hydrogen today is produced through SMR, a process that extracts hydrogen from natural gas but generates significant CO_2 emissions. This is referred to as grey hydrogen, and while it remains the most economical production method in many regions, its environmental impact is substantial, as it perpetuates fossil fuel reliance and contributes to climate change. A cleaner alternative is blue hydrogen, where hydrogen is produced through SMR, but the CO_2 emissions are captured and stored using CCS technologies. While blue hydrogen reduces the net emissions from hydrogen production, it still depends on natural gas, which raises concerns about long-term sustainability. Moreover, the scalability and effectiveness of CCS technologies are still under development and have not yet been widely deployed at the scale required for a global hydrogen economy. The environmental footprint of hydrogen production and transportation can also vary depending on the location of renewable energy sources used for green hydrogen production. In regions with abundant renewable energy resources, such as solar-rich deserts or areas with high wind capacity, green hydrogen production can be highly sustainable. However, in regions where renewable energy is scarce or where production facilities rely on grid electricity from fossil fuel-based sources, the environmental benefits of green hydrogen production may be reduced. Furthermore, the construction and operation of hydrogen infrastructure, such as electrolyzers, storage facilities, and distribution networks, may involve the use of materials and processes with environmental impacts, although these are relatively small compared to the overall benefits. In addition, the transportation and storage of hydrogen present environmental and logistical challenges. Hydrogen must be transported under high pressure or at very low temperatures, both of which require substantial energy input. While advancements in storage technologies, such as metal hydrides and LOHCs, promise to increase storage efficiency, they also come with their own environmental costs. For example, the production of cryogenic tanks and high-pressure storage systems requires energy and materials, which may have environmental impacts if the energy used is not from renewable sources [74].

As hydrogen adoption grows, the sustainability of its infrastructure becomes a key factor in its long-term viability. Building a global hydrogen economy requires extensive investment in HRS, pipelines, storage systems, and electrolyzers, all of which have environmental and economic implications. The embodied energy required to construct these systems – particularly for large-scale hydrogen production plants – must be considered when evaluating hydrogen's overall carbon footprint. One potential solution to enhance the sustainability of hydrogen infrastructure is to adopt a circular economy approach, where materials used in hydrogen technologies (such as metals, polymers, and components in fuel cells) are recycled and reused. This could help mitigate the environmental impacts associated with the manufacturing and disposal of hydrogen infrastructure, reducing waste and minimizing the need for virgin raw materials. Moreover, as hydrogen technology matures, improvements in energy efficiency and system integration will likely reduce the environmental impact of hydrogen production, storage, and use over time [32]. Table 6.3 highlights the environmental benefits of using hydrogen over fossil fuels across major industrial sectors. In steel and cement production, hydrogen significantly reduces CO_2 emissions compared to high-emitting coke, coal, and limestone decomposition. Chemicals manufacturing with hydrogen feedstock eliminates CO_2 and CO pollutants, unlike natural gas and oil-based processes that emit harmful sulfur and nitrogen oxides. Similarly, in petroleum refining, green hydrogen reduces emissions of CO_2, VOCs, and other particulates. Fossil fuels, by contrast, degrade air quality due to the release

Table 6.3 Environmental impact comparison of hydrogen vs. fossil fuels in industrial applications.

Industry sector	Hydrogen (CO_2 emissions)	Fossil Fuels (CO_2 emissions)	Energy source	Impact on air quality
Steel production	Zero emissions with green hydrogen	High emissions from coke and coal	Coal, natural gas	Significant particulate and gas pollutants from fossil fuels
Cement production	Low emissions with hydrogen usage	High emissions from limestone decomposition	Coal, natural gas	Major air pollution from particulate matter and CO_2
Chemicals manufacturing	Zero emissions with hydrogen feedstock	Significant CO_2 and CO pollutants	Natural gas, oil	Sulfur and nitrogen oxides from fossil fuel combustion
Refining	Low emissions when hydrogen is green	High CO_2 emissions in refining process	Oil, natural gas	VOCs, NOx, CO, and particulate emissions from fossil fuels

of greenhouse gases and toxic pollutants, emphasizing hydrogen's potential as a cleaner, more sustainable alternative for industrial decarbonization.

6.7 Case Studies: Global Adoption of Hydrogen in Transportation and Industry

The global push towards hydrogen adoption in transportation and industrial sectors is gaining momentum, as countries seek to decarbonize their economies and transition to cleaner, more sustainable energy sources. This section presents four key case studies from countries that are at the forefront of hydrogen innovation and implementation. These case studies demonstrate the potential of hydrogen technologies in reducing emissions, driving economic growth, and enabling energy transitions in the transportation and industrial sectors.

6.7.1 Japan's Hydrogen-powered Transport Systems

Japan has long been recognized as a leader in hydrogen technology, particularly in the transportation sector. The country's Hydrogen Society Vision, which envisions the widespread adoption of hydrogen FCVs across various sectors, has guided Japan's strategy for decarbonizing its economy. This commitment to hydrogen is backed by strong government support, private sector innovation, and investments in hydrogen infrastructure, making Japan one of the most ambitious players in the global hydrogen race. Japan's most notable achievements in hydrogen-powered transport are seen in its hydrogen FCVs. The Toyota Mirai, which was first introduced in 2014, is one of the most recognized hydrogen-powered cars in the world. The success of the Mirai has spurred the development of other hydrogen vehicle models, with major Japanese automakers such as Honda and Nissan also introducing FCVs to the market. FCVs offer a zero-emission alternative to conventional gasoline-powered vehicles, emitting only water vapor as exhaust. In addition to the Toyota Mirai, hydrogen-powered buses and trains are also becoming increasingly common in Japan, contributing to the country's goal of reducing urban transportation emissions. The Japanese government has been instrumental in creating a conducive environment for hydrogen adoption. Through initiatives such as the Basic Hydrogen Strategy and the Hydrogen Roadmap, Japan has committed to building the necessary infrastructure for hydrogen, including HRS. By 2030, Japan aims to have 1,000 HRS across the country, which will make hydrogen-powered transport more viable and convenient. Japan's investment in hydrogen infrastructure is complemented by its collaboration with international partners to develop a global hydrogen supply chain, particularly focusing on importing hydrogen from countries such as Australia. Moreover, Japan has been a pioneer in hydrogen-powered public transport, with hydrogen buses operating in several cities, including Tokyo, Fukuoka, and Osaka. These buses are part of the country's efforts to decarbonize urban transportation and reduce dependence on fossil fuels. The success of these projects in Japan has sparked interest in other countries looking to replicate Japan's model, providing a scalable solution for sustainable urban mobility [75].

6.7.2 Germany's Hydrogen in Industrial Applications

Germany has positioned itself as a global leader in hydrogen technology, not just in transportation, but in industrial applications as well. The country's National Hydrogen Strategy, launched in 2020, outlines a comprehensive plan for integrating hydrogen into key sectors of the economy, including industry, transport, and energy. Germany has recognized hydrogen as a crucial element in achieving its climate goals, particularly in heavy industries such as steel production, chemical manufacturing, and refining, where decarbonization remains a significant challenge. One of the most prominent examples of hydrogen use in industry is the H_2 Steel project led by Thyssenkrupp, one of the world's largest steel producers. Germany has been testing hydrogen-based direct reduction in steel production, a process that replaces the use of coal with hydrogen to reduce iron ore into iron. This process, known as H_2-based direct reduction, has the potential to reduce CO_2 emissions in steel production by as much as 95%. The German government has supported this innovation through substantial funding, recognizing the role that hydrogen can play in making the steel industry more sustainable. Germany aims to establish hydrogen-based steel production as a key pillar of its industrial decarbonization strategy, with several pilot plants already operational and plans to scale up production in the coming decades. In addition to the steel sector, hydrogen is also being explored as a feedstock in the chemical industry, particularly in the production of ammonia and methanol, two crucial chemicals in fertilizers and other products. Hydrogen-powered fuel cells are being used to generate electricity for industrial applications, reducing the reliance on natural gas and helping manufacturers reduce their emissions. The Hydrogen Alliance in Germany, a coalition of businesses, research institutions, and government entities, is working to develop large-scale hydrogen production and infrastructure for industrial uses, further boosting the country's hydrogen capabilities. Germany's commitment to green hydrogen, produced through renewable energy sources such as wind and solar, has also been central to the country's strategy. This emphasis on green hydrogen is seen as critical in meeting the European Union's carbon neutrality targets and advancing Germany's own ambitious climate goals [76].

6.7.3 United States: Advancements in Hydrogen for Heavy-duty Transport

The United States has increasingly turned its focus to hydrogen in the context of heavy-duty transport, including trucks, buses, and rail. While hydrogen adoption in passenger vehicles has been slower than in other countries like Japan, the United States is making significant strides in the hydrogen space, particularly in areas where battery electric solutions face challenges, such as long-range transportation and heavy loads. One of the key initiatives driving hydrogen adoption in the United States is the California Hydrogen Highway, a network of HRS that supports the state's hydrogen fuel cell vehicle fleet. California, as a leader in green energy, has been at the forefront of developing hydrogen infrastructure to support FCVs, with a particular focus on hydrogen trucks. The state has also implemented policies to promote the deployment of zero-emission heavy-duty trucks. In particular, California has partnered with companies such as Nikola Motor Company and Toyota to advance the development of hydrogen-powered trucks, which are seen as an essential part of decarbonizing the freight sector. The US federal government has also introduced several programs to support hydrogen innovation, including funding for research into hydrogen production, fuel cell development, and hydrogen infrastructure. This is complemented by private sector investments in hydrogen technology. Plug Power, a leading company in hydrogen fuel cells, has made significant progress in deploying hydrogen-powered fuel cells for forklifts and other industrial applications, helping to decarbonize material handling operations in warehouses and distribution centers. Looking forward, the United States is poised to become a key player in the global hydrogen economy. As the Biden administration works toward ambitious climate goals, hydrogen is expected to play a key role in decarbonizing hard-to-abate sectors such as heavy-duty transport and industry, contributing to the achievement of net-zero emissions by 2050 [77].

6.7.4 Australia's Hydrogen Industry and Export Potential

Australia is positioning itself as a major global exporter of hydrogen, particularly green hydrogen, produced using its abundant renewable energy resources. With vast sun-drenched deserts and strong winds along its coasts, Australia has the potential to become a hydrogen powerhouse. The Australian government's Hydrogen Strategy, released in 2019, outlines its vision to make the country a major hydrogen exporter, focusing on the production of green hydrogen for export to markets in Japan, South Korea, and beyond. In addition to its export ambitions, Australia is also making strides in domestic hydrogen use, particularly in heavy industries such as mining, transport, and energy production. Hydrogen is being explored as a potential fuel source for trains, buses, and trucks in remote regions, where the use of conventional fuels is expensive and infrastructure is lacking. Australia's hydrogen-powered trains are already in use in several areas, providing a glimpse into the potential for hydrogen to replace diesel in regional and freight transport. Australia's hydrogen export ambitions are supported by a range of initiatives, including the Asian Renewable Energy Hub in Western Australia, a large-scale renewable energy project that aims to produce green hydrogen for export to Asia. The country is also working to build the necessary infrastructure for hydrogen production, storage, and transportation, with significant investments in hydrogen export terminals and hydrogen-powered shipping. Australia's focus on green hydrogen production offers a low-carbon energy alternative to fossil fuels, helping to reduce global emissions while creating new economic opportunities. The country's hydrogen export potential is a critical component of its long-term climate strategy, as it seeks to diversify its economy and create jobs in the growing hydrogen sector [78].

6.8 The Future Outlook for Hydrogen in Transportation and Industry

The future of hydrogen in both the transportation and industrial sectors is poised for significant growth as countries, companies, and research institutions continue to invest in developing the necessary technologies and infrastructure to support its adoption. As a clean, versatile energy carrier, hydrogen is increasingly being recognized as a critical component of a sustainable energy future. The transition to hydrogen-powered systems across transportation and industry is driven by the need to decarbonize hard-to-abate sectors, reduce dependency on fossil fuels, and mitigate the impacts of climate change [79].

In the transportation sector, hydrogen is set to play a key role in decarbonizing modes of transport that are difficult to electrify using battery technologies. These include heavy-duty freight trucks, buses, trains, and ships. While hydrogen-powered passenger vehicles are already on the market, the real growth potential lies in long-distance transport and large-scale industrial fleets. Hydrogen FCVs offer longer ranges and shorter refueling times compared to EVs, making them more suitable for sectors where operational efficiency and rapid refueling are critical. The expansion of hydrogen infrastructure is one of the most important enablers of the hydrogen transition in transportation. Governments and private companies are investing heavily in the development of HRS, particularly in key regions such as Europe, Japan, and California, to support the deployment of hydrogen-powered vehicles. This infrastructure expansion is expected to gain momentum in the coming years, facilitating the widespread adoption of hydrogen-powered transport. Advancements in hydrogen production technologies – especially green hydrogen produced using renewable energy sources like wind and solar – will further boost the growth of hydrogen-powered transportation. As green hydrogen becomes more cost-competitive with traditional fossil fuels, hydrogen will become an increasingly attractive option for the transport sector, which is under pressure to meet stringent emissions reduction targets. In the next decade, hydrogen is likely to dominate sectors such as heavy-duty transport and public transit, with widespread adoption of hydrogen-powered buses, trucks, and rail systems. The integration of hydrogen into transport systems, particularly in urban mobility, will also contribute to cleaner cities with reduced air pollution, promoting sustainability and a higher quality of life [80].

In the industrial sector, hydrogen offers an essential solution for decarbonizing some of the most energy-intensive processes. Steel manufacturing, chemical production, cement production, and refining are sectors that rely heavily on fossil fuels and emit large quantities of CO_2. Hydrogen has the potential to decarbonize these industries by replacing carbon-intensive feedstocks, such as coke and natural gas, with low-carbon hydrogen. One of the most promising developments is the transition of steel production to hydrogen-based direct reduction, which eliminates the need for coke and significantly reduces carbon emissions. This shift will be crucial in helping countries meet their climate goals and reduce emissions from the steel industry, one of the largest industrial sources of CO_2 emissions globally. The H_2 Steel and HYBRIT projects in Europe, which are testing hydrogen as a replacement for coke, are expected to pave the way for widespread adoption of this technology in the coming decades. Similarly, hydrogen is poised to transform the chemical industry, particularly in the production of ammonia and methanol, which are vital for fertilizers, plastics, and other industrial products. Green hydrogen, produced through electrolysis powered by renewable energy, is seen as a critical enabler of decarbonizing the chemical sector. As hydrogen becomes more cost-effective, it will be increasingly integrated into chemical production processes, reducing emissions and enabling more sustainable industrial practices. The cement industry, which is responsible for significant carbon emissions globally, is also exploring the potential of hydrogen to decarbonize its production processes. Hydrogen can replace natural gas in cement kilns, significantly lowering emissions from cement manufacturing. The growing interest in hydrogen-fired cement kilns presents an opportunity for the cement sector to contribute to global decarbonization efforts, while also meeting the demand for green building materials [81].

As the hydrogen economy matures, ongoing advancements in hydrogen production technologies will drive cost reductions, making hydrogen a more competitive alternative to fossil fuels. The cost of green hydrogen, which has been a major barrier to its widespread adoption, is expected to fall significantly due to advancements in electrolyzer technology and economies of scale in renewable energy generation. The global focus on innovation in the hydrogen sector will make hydrogen more economically viable, especially for use in industries and transportation systems where high energy demands are prevalent. Hydrogen storage and distribution technologies will also improve over the coming years, with innovations in liquefaction and compression technologies making it easier and more cost-effective to store and transport hydrogen. These developments are crucial for establishing a global hydrogen supply chain, which is vital for supporting the widespread adoption of hydrogen across different sectors and regions [82].

The future of hydrogen in both transportation and industry will also be shaped by government policies and public-private partnerships. Governments around the world are increasingly recognizing the role of hydrogen in achieving net-zero emissions targets and are offering policy incentives and funding to support the development and deployment of hydrogen technologies. This includes subsidies for hydrogen infrastructure, tax incentives for hydrogen-powered vehicles, and funding for research and development. Large-scale investments from both governments and private sectors are expected to accelerate the transition to a hydrogen economy. Investment in hydrogen infrastructure such as refueling stations, pipelines, and production facilities is crucial for enabling the large-scale adoption of hydrogen. Furthermore, as more companies invest in hydrogen technologies and commercialize hydrogen solutions, economies of scale will further reduce costs and create a self-reinforcing cycle of growth. The future outlook for hydrogen in transportation and industry is undoubtedly promising. With continued advancements in technology, ongoing reductions in production costs, and robust government support, hydrogen is set to play a pivotal role in decarbonizing hard-to-abate sectors and enabling a clean energy transition. As hydrogen adoption increases across transportation, heavy industry, and other sectors, it will contribute to global efforts to combat climate change, reduce air pollution, and build sustainable, low-carbon economies.

While challenges remain, the future of hydrogen is one of innovation, collaboration, and transformative change in both the transportation and industrial landscapes.

6.9 Conclusion

Hydrogen is emerging as a critical element in the global effort to decarbonize key sectors, with its applications in transportation and industry offering promising solutions to the challenges of climate change and energy transition. As a clean energy carrier, hydrogen's versatility and potential to decarbonize hard-to-abate sectors, such as heavy-duty transport, steel manufacturing, cement production, and chemical industries, position it as a cornerstone of future sustainable energy systems. The transportation sector, particularly in heavy-duty freight and public transit, stands to benefit significantly from hydrogen fuel cell technologies. Hydrogen-powered vehicles offer distinct advantages in terms of range, refueling time, and operational efficiency compared to BEVs, making them well-suited for applications where electrification is less feasible. As infrastructure for HRS expands, and the cost of green hydrogen continues to fall, hydrogen-powered transportation systems are set to grow and contribute to cleaner urban environments. Similarly, the industrial sector is on the verge of a transformation driven by hydrogen. Industries that have long relied on fossil fuels – such as steel, cement, and chemicals – are turning to hydrogen as a cleaner alternative, offering a path to decarbonize some of the world's most energy-intensive processes. Notably, hydrogen's role in steel manufacturing and cement production could significantly reduce global industrial CO_2 emissions, enabling the realization of ambitious climate targets. Key technological advancements in hydrogen production, storage, and distribution will be essential for overcoming the current barriers to widespread adoption. As electrolyzer technology improves and the costs of green hydrogen decline, the economic feasibility of hydrogen as a mainstream energy solution becomes increasingly viable. At the same time, innovations in storage and transportation technologies will help establish a robust hydrogen supply chain capable of supporting global demand. Despite the clear promise of hydrogen, challenges remain. The development of hydrogen infrastructure, the scaling up of green hydrogen production, and policy and regulatory frameworks that foster investment and innovation will be critical to its success. Governments and industries must collaborate closely to create the necessary conditions for hydrogen to flourish, ensuring that it plays an integral role in achieving global decarbonization goals. Ultimately, hydrogen represents a transformative opportunity for the future of both transportation and industry. As the hydrogen economy continues to evolve, its potential to contribute to global sustainability, energy security, and economic growth cannot be overstated. The transition to a hydrogen-powered world will require ongoing innovation, investment, and collaboration, but the promise of a cleaner, more sustainable energy landscape makes this endeavor both urgent and achievable. With continued advancements and the right support, hydrogen will undoubtedly play a leading role in shaping the future of global energy systems.

References

1 Roque BAC, Cavalcanti MHC, Brasileiro PPF, Gama PHRP, dos Santos VA, Converti A, et al. Hydrogen-powered future: Catalyzing energy transition, industry decarbonization and sustainable economic development: A review. *Gondwana Res.* 2025 Apr 1; 140:159–80.

2 Veziroglu A. Hydrogen powered transportation. *Xlibris Corp*; 2017 Feb 28.

3 Turoń K. Hydrogen-powered vehicles in urban transport systems—Current state and development. *Transp Res Procedia.* 2020 Jan 1; 45:835–41.

4 Alavi-Borazjani SA, Adeel S, Chkoniya V. Hydrogen as a sustainable fuel: Transforming maritime logistics. *Energies.* [Internet]. 2025 Mar 3 [cited 2025 Apr 2];18(5):1231. Available from: https://www.mdpi.com/1996-1073/18/5/1231/htm.

5 Kolodziejski M. Review of hydrogen-based propulsion systems in the maritime sector. *Arch Thermodyn.* 2023;44(4): 335–80.

6 Qazi UY. Future of hydrogen as an alternative fuel for next-generation industrial applications; challenges and expected opportunities. *Energies.* [Internet]. 2022 Jun 28 [cited 2025 Apr 2];15(13):4741. Available from: https://www.mdpi.com/1996-1073/15/13/4741/htm.

7 Bararzadeh Ledari M, Khajehpour H, Akbarnavasi H, Edalati S. Greening steel industry by hydrogen: Lessons learned for the developing world. *Int J Hydrogen Energy.* 2023 Dec 5; 48(94):36623–49.

8 Pal M, Kumar S. Hydrogen in the chemical industry. Renewable hydrogen: Opportunities and challenges in commercial success. 2024 Jan 1; 221–49. Available from: https://doi.org/10.1016/B978-0-323-95379-5.00007-9.

9 Cavaliere P. Hydrogen applications. *Water Electrol Hydrogen Prod.* [Internet]. 2023 [cited 2025 Apr 2];653–727. Available from: https://link.springer.com/chapter/10.1007/978-3-031-37780-8_18.

10 Akdağ O. The operation and applicability to hydrogen fuel technology of green hydrogen production by water electrolysis using offshore wind power. *J Clean Prod.* 2023 Nov 1;425:138863.

11 Wu H, Gao J, Tan Q, Huang N, Gao F, Puranen P, et al. Hydrogen fuel cell technology: A sustainable solution for revolutionizing aviation and achieving significant carbon reduction in the industry. *Int J Hydrogen Energy.* 2025 Mar 20; 111:711–22.

12 Jianlin L, Guanghui L, Suliang M, Jie S, Jianlin L, Guanghui L, et al. An overview on hydrogen energy storage and transportation technology and its typical application in power system. *Modern Electr Power.* [Internet]. 2021 Sep 27 [cited 2025 Apr 2];38(5):535–45. Available from: http://xddl.ncepujournal.com/en/article/doi/10.19725/j.cnki.1007-2322.2021.0023.pdf.

13 Hwang J, Maharjan K, Cho HJ. A review of hydrogen utilization in power generation and transportation sectors: Achievements and future challenges. *Int J Hydrogen Energy.* 2023 Aug 29; 48(74):28629–48.

14 Jayabal R. Hydrogen energy storage in maritime operations: A pathway to decarbonization and sustainability. *Int J Hydrogen Energy.* 2025 Mar 14; 109:1133–44.

15 Halder P, Babaie M, Salek F, Haque N, Savage R, Stevanovic S, et al. Advancements in hydrogen production, storage, distribution and refuelling for a sustainable transport sector: Hydrogen fuel cell vehicles. *Int J Hydrogen Energy*. 2024 Jan 2; 52:973–1004.

16 Riera JA, Lima RM, Knio OM. A review of hydrogen production and supply chain modeling and optimization. *Int J Hydrogen Energy*. 2023 Apr 30; 48(37):13731–55.

17 Jiang Z. Is the promotion policy for hydrogen fuel cell vehicles effective? Evidence from Chinese cities. *Energy*. 2025 Apr 1;320:135469.

18 Tian MH, Hu YJ, Li C, Tao Y, Wang H. The development pathway for hydrogen fuel cell heavy duty trucks in China: An energy-environment-economy life cycle assessment approach. *Energy*. 2025 May 1;322:135797.

19 Alsahli AA. Exploring the potential of using hydrogen fuel cell technology for locomotives in Saudi Arabia. *Transp Res Procedia*. [Internet]. 2025 Jan 1 [cited 2025 Apr 2];84:464–71. Available from: https://linkinghub.elsevier.com/retrieve/pii/S2352146525001450.

20 Xu J, Geng C, Peng Z. Toward sustainable transport: Evaluating the economic feasibility of hydrogen fuel cell taxis. *Int J Hydrogen Energy*. 2025 Mar 4; 105:1133–42.

21 Cekerevac Z. Ecological and economic risks of using gasoline, electric, hybrid, and hydrogen-powered vehicles. *Mest J*. [cited 2025 Apr 2]; Available from: https://orcid.org/0000-0003-2972-2472.

22 Filina-Dawidowicz L, Sęk J, Trojanowski P, Wiktorowska-Jasik A. Conditions of decision-making related to implementation of hydrogen-powered vehicles in urban transport: Case study of Poland. *Energies*. [Internet]. 2024 Jul 13 [cited 2025 Apr 2];17(14):3450. Available from: https://www.mdpi.com/1996-1073/17/14/3450/htm.

23 Shet K HN, Moholkar VS. Comparative life cycle analysis of hydrogen fuel cell electric vehicles and battery electric vehicles: An Indian perspective. *Int J Hydrogen Energy*. 2025 Feb 17; 103:729–39.

24 Gurz M, Baltacioglu E, Hames Y, Kaya K. The meeting of hydrogen and automotive: A review. *Int J Hydrogen Energy*. 2017 Sep 7; 42(36):23334–46.

25 Berglas R, Kochhan R, Sinha A, Gjerek M, Gore-Brown N. February 2023 Final iMOVE hydrogen powered heavy vehicle demand and infrastructure assessment-report.

26 Wilson S. Hydrogen-Powered Heavy-Duty Trucks A review of the environmental and economic implications of hydrogen fuel for on-road freight Introduction 3. 2023 [cited 2025 Apr 2]; Available from: www.ucsusa.org/resources/hydrogen-powered-heavy-duty-truckshttps://doi.org/10.47923/2023.15274.

27 Gnann T, Link S, Speth D, Scherrer A. Powerfuels for heavy-duty road transportation. *Green Energy Technol*. [Internet]. 2025 [cited 2025 Apr 2];Part F3771:821–47. Available from: https://link.springer.com/chapter/10.1007/978-3-031-62411-7_28.

28 Zakaria Z, Shaari N. Role of hydrogen in the future development of transportation vehicles. In: *Accelerating the Transition to a Hydrogen Economy: Achieving Carbon Neutrality*. 2025 Jan 1; 257–78.

29 Melideo D, Desideri U, Andreas J, Kalisch T. Zero-dimensional modelling and optimization of hydrogen refuelling for type IV tanks: From validation to large-scale applications. 2025 [cited 2025 Apr 2]; Available from: https://papers.ssrn.com/abstract=5174543.

30 Silva Junior LH, da Silva AK. Global assessment of hydrogen production from the electrical grid aiming the Brazilian transportation sector. *Energy Sustain Dev.* 2025 Apr 1;85:101626.

31 Frenette G, Forthoffer D. Economic & commercial viability of hydrogen fuel cell vehicles from an automotive manufacturer perspective. *Int J Hydrogen Energy.* 2009 May 1; 34(9):3578–88.

32 Kurc B, Gross X, Szymlet N, Rymaniak Ł, Woźniak K, Pigłowska M. Hydrogen-powered vehicles: A paradigm shift in sustainable transportation. *Energies.* [Internet]. 2024 Sep 24 [cited 2025 Apr 2];17(19):4768. Available from: https://www.mdpi.com/1996-1073/17/19/4768/htm.

33 Bethoux O. Hydrogen fuel cell road vehicles: State of the art and perspectives. *Energies.* [Internet]. 2020 Nov 9 [cited 2025 Apr 2];13(21):5843. Available from: https://www.mdpi.com/1996-1073/13/21/5843/htm.

34 Cunanan C, Tran MK, Lee Y, Kwok S, Leung V, Fowler M. A review of heavy-duty vehicle powertrain technologies: Diesel engine vehicles, battery electric vehicles, and hydrogen fuel cell electric vehicles. *Clean Technol.* [Internet]. 2021 Jun 1 [cited 2025 Apr 2];3(2):474–89. Available from: https://www.mdpi.com/2571-8797/3/2/28/htm.

35 Bethoux O. Hydrogen fuel cell road vehicles and their infrastructure: An option towards an environmentally friendly energy transition. *Energies.* [Internet]. 2020 Nov 23 [cited 2025 Apr 2];13(22):6132. Available from: https://www.mdpi.com/1996-1073/13/22/6132/htm.

36 Ventura JA. Climate benefits advocated by the development of sustainable vehicles and charging infrastructures in the transport sector. *Energies.* [Internet]. 2023 Apr 28 [cited 2025 Apr 2];16(9):3772. Available from: https://www.mdpi.com/1996-1073/16/9/3772/htm.

37 Yu Y, Guo Y, Ma T. Prioritizing the hydrogen pathways for fuel cell vehicles: Analysis of the life-cycle environmental impact, economic cost, and environmental efficiency. *Energy.* 2023 Oct 15;281:128322.

38 Parikh A, Shah M, Prajapati M. Fuelling the sustainable future: A comparative analysis between battery electrical vehicles (BEV) and fuel cell electrical vehicles (FCEV). *Environ Sci Pollut Res.* [Internet]. 2023 Apr 1 [cited 2025 Apr 2]; 30(20):57236–52. Available from: https://link.springer.com/article/10.1007/s11356-023-26241-9.

39 Halder P, Babaie M, Salek F, Shah K, Stevanovic S, Bodisco TA, et al. Performance, emissions and economic analyses of hydrogen fuel cell vehicles. *Renew and Sustain Energy Rev.* 2024 Jul 1; 199:114543.

40 Albatayneh A, Juaidi A, Jaradat M, Manzano-Agugliaro F. Future of electric and hydrogen cars and trucks: An overview. *Energies.* [Internet]. 2023 Apr 3 [cited 2025 Apr 2];16(7):3230. Available from: https://www.mdpi.com/1996-1073/16/7/3230/htm.

41 Ugurlu A. An emission analysis study of hydrogen powered vehicles. *Int J Hydrogen Energy*. 2020 Oct 9; 45(50):26522–35.

42 Dulău LI. CO_2 Emissions of battery electric vehicles and hydrogen fuel cell vehicles. *Clean Technol.* [Internet]. 2023 Jun 1 [cited 2025 Apr 2];5(2):696–712. Available from: https://www.mdpi.com/2571-8797/5/2/35/htm.

43 Farooq A, Cora ÖN. Current status and future perspectives for mobility options toward sustainable transportation with a focus on fuel cells. *Wiley Interdiscip Rev Energy Environ.* [Internet]. 2024 Sep 1 [cited 2025 Apr 2];13(5):e539. Available from: https://onlinelibrary.wiley.com/doi/full/10.1002/wene.539.

44 Vengatesan S, Jayakumar A, Sadasivuni KK. FCEV vs. BEV—A short overview on identifying the key contributors to affordable & clean energy (SDG-7). *Energy Strategy Rev*. 2024 May 1;53:101380.

45 Chu Y, Wu Z, Yin Y. Development of a solar-assisted hydrogen-from-power refueling station: A financial incentive model under the Well-to-Wheel and life cycle cost analyses. *Int J Hydrogen Energy*. 2025 Feb 17; 103:288–99.

46 Chrysochoidis-Antsos N, Liu C, Van Wijk A. On-site wind powered hydrogen refuelling stations - From national level to a case study in Germany. *2018 International Conference on Smart Energy Systems and Technologies, SEST 2018—Proceedings*. 2018 Oct 17. Available from: https://doi.org/10.1109/SEST.2018.8495693.

47 Mukherjee U, Maroufmashat A, Ranisau J, Barbouti M, Trainor A, Juthani N, et al. Techno-economic, environmental, and safety assessment of hydrogen powered community microgrids; Case study in Canada. *Int J Hydrogen Energy*. 2017 May 18; 42(20):14333–49.

48 Gökçek M, Paltrinieri N, Liu Y, Badia E, Dokuz AŞ, Erdoğmuş A, et al. Optimum sizing of hybrid renewable power systems for on-site hydrogen refuelling stations: Case studies from Türkiye and Spain. *Int J Hydrogen Energy*. 2024 Mar 15; 59:715–29.

49 Li F, Liu D, Sun K, Yang S, Peng F, Zhang K, et al. Towards a future hydrogen supply chain: A review of technologies and challenges. *Sustainability*. [Internet]. 2024 Feb 25 [cited 2025 Apr 2];16(5):1890. Available from: https://www.mdpi.com/2071-1050/16/5/1890/htm.

50 Sakib AN, Mehjabin F, Schmidt JB, Haque M, Saha K, Monjur M, et al. Harnessing hydrogen: A comprehensive literature review on strategic launching initiatives in the global energy market. *Int J Energy Res*. [Internet]. 2024 Jan 1 [cited 2025 Apr 2];2024(1):3265065. Available from: https://onlinelibrary.wiley.com/doi/full/10.1155/2024/3265065.

51 Samsun RC, Rex M, Antoni L, Stolten D. Deployment of fuel cell vehicles and hydrogen refueling station infrastructure: A global overview and perspectives. *Energies (Basel)*. [Internet]. 2022 Jul 1 [cited 2025 Mar 24];15(14):4975. Available from: https://www.mdpi.com/1996-1073/15/14/4975/htm.

52 Jain IP. Hydrogen the fuel for 21st century. *Int J Hydrogen Energy*. 2009 Sep 1; 34(17):7368–78.

53 Okologume WC, Aimikhe VJ. Application of hydrogen in metallurgy. Hydrogen *Appl Technol*. [Internet]. 2024 Nov 6 [cited 2025 Apr 2];369–86. Available from:

https://www.taylorfrancis.com/chapters/edit/10.1201/9781003382560-18/application-hydrogen-metallurgy-wilfred-chinedu-okologume-victor-joseph-aimikhe.
54 Canat AN, Özkan C. Supplying hydrogen for green steel through renewable energy sources: A case study of Turkiye. *J Clean Prod*. 2024 Apr 15;450:141961.
55 Lagarinhos CAF, Azevedo LP. Challenges and opportunities of hydrogen economy in Industrial Revolution 4.0 era. Accelerating the transition to a hydrogen economy: Achieving carbon neutrality. 2025 Jan 1; 237–55.
56 Ahad MT, Bhuiyan MMH, Sakib AN, Becerril Corral A, Siddique Z. An overview of challenges for the future of hydrogen. *Materials*. 2023 Oct 13 [cited 2025 Apr 2]; 16(20):6680. Available from: https://www.mdpi.com/1996-1944/16/20/6680/htm.
57 Devlin A, Kossen J, Goldie-Jones H, Yang A. Global green hydrogen-based steel opportunities surrounding high quality renewable energy and iron ore deposits. *Nature Commun*. 2023 14: 1 [Internet]. 2023 May 4 [cited 2025 Mar 24];14(1):1–14. Available from: https://www.nature.com/articles/s41467-023-38123-2.
58 Lavate S, Yalamati HPS, Srivastava R. Application of hydrogen in various sectors. Solar-Driven Green Hydrogen Generation and Storage. 2023 Jan 1; 507–24. Available from: https://doi.org/10.1016/B978-0-323-99580-1.00002-9.
59 Behrendt F. The future of industrial hydrogen: Renewable sources and applications for the next 15 years. *Clean Energy*. [Internet]. 2025 Jan 22 [cited 2025 Apr 2];9(1):3–8. Available from: https://dx.doi.org/10.1093/ce/zkae103.
60 Chen G, Sun R, Wang B. Solar-powered hydrogen: Exploring production, storage, and energy integration strategies. *Clean Energy*. [Internet]. 2025 Jan 22 [cited 2025 Apr 2];9(1):123–46. Available from: https://dx.doi.org/10.1093/ce/zkaf005.
61 Alves C, Castro G, Coelho R, Hoccvar L, Alves C, Castro G, et al. Hydrogen technologies: Recent advances, new perspectives, and applications. [Internet]. 2024 Jun 5 [cited 2025 Apr 2]; Available from: https://www.intechopen.com/chapters/1183330.
62 Pedersen JL, Grue J, Dybdal Christensen J, Diab M, Georgiev IG, Krogh A, et al. Hydrogen, electrofuels, CCU and CCS in a Nordic context. Hydrogen, electrofuels, CCU and CCS in a Nordic context [Internet]. 2022 [cited 2025 Apr 2]; Available from: https://urn.kb.se/resolve?urn=urn:nbn:se:norden:org:diva-12316.
63 Yakubson KI. Prospects for using hydrogen in various branches of the world economy as one of the directions of its decarbonization. *Russ J Appl Chem* [Internet]. 2022 Mar 1 [cited 2025 Apr 2];95(3):309–40. Available from: https://link.springer.com/article/10.1134/S1070427222030016.
64 Razi F, Dincer I. Challenges, opportunities and future directions in hydrogen sector development in Canada. *Int J Hydrogen Energy*. 2022 Feb 19; 47(15):9083–102.
65 Abdin Z. Bridging the energy future: The role and potential of hydrogen co-firing with natural gas. *J Clean Prod*. 2024 Jan 10;436:140724.
66 Nhuchhen DR, Sit SP, Layzell DB. Decarbonization of cement production in a hydrogen economy. *Appl Energy*. 2022 Jul 1;317:119180.
67 Veziroglu A. *Hydrogen Powered Transportation*: 2017.
68 Hoelzen J, Silberhorn D, Zill T, Bensmann B, Hanke-Rauschenbach R. Hydrogen-powered aviation and its reliance on green hydrogen infrastructure—Review and research gaps. *Int J Hydrogen Energy*. Xlibris Corporation. 2022 Jan 15; 47(5):3108–30.

69 Marouani I, Guesmi T, Alshammari BM, Alqunun K, Alzamil A, Alturki M, et al. Integration of renewable-energy-based green hydrogen into the energy future. *Processes*. [Internet]. 2023 Sep 7 [cited 2025 Apr 2];11(9):2685. Available from: https://www.mdpi.com/2227-9717/11/9/2685/htm.

70 Aguilera RF, Inchauspe J. An overview of hydrogen prospects: Economic, technical and policy considerations. *Aust J Agric Resour Econ*. [Internet]. 2022 Jan 1 [cited 2025 Apr 2];66(1):164–86. Available from: https://onlinelibrary.wiley.com/doi/full/10.1111/1467-8489.12458.

71 Yalamati HPS, Vij RK, Srivastava R. Green hydrogen revolution: Mapping G20 nations' policies and geopolitical strategies in the net-zero emission era. *Wiley Interdiscip Rev Energy Environ* [Internet]. 2024 Sep 1 [cited 2025 Apr 2]; 13(5):e538. Available from: https://onlinelibrary.wiley.com/doi/full/10.1002/wene.538.

72 Acar C, Dincer I. The potential role of hydrogen as a sustainable transportation fuel to combat global warming. *Int J Hydrogen Energy*. 2020 Jan 29; 45(5):3396–406.

73 Adler EJ, Martins JRRA. Hydrogen-powered aircraft: Fundamental concepts, key technologies, and environmental impacts. *Prog Aerosp Sci*. 2023 Aug 1; 141:100922.

74 Ajiboye Y, Ojo OL, Kayode JF, Awolusi T, Jayesimi B, Afolalu SA, et al. The Future of green hydrogen in the global economy: Opportunities, challenges and implications. In: *International Conference on Science, Engineering and Business for Driving Sustainable Development Goals, SEB4SDG 2024*. 2024.

75 Nagashima M. *Japan's Hydrogen strategy and its economic and geopolitical implications études de l'Ifri monica nagasHima Centre for Energy*. France: Institut français des relations internationals; 2018.

76 Neuwirth M, Fleiter T, Manz P, Hofmann R. The future potential hydrogen demand in energy-intensive industries—A site-specific approach applied to Germany. *Energy Convers Manag*. 2022 Jan 15; 252:115052.

77 Camacho M de las N, Jurburg D, Tanco M. Hydrogen fuel cell heavy-duty trucks: Review of main research topics. *Int J Hydrogen Energy*. 2022 Aug 8; 47(68):29505–25.

78 Kar SK, Sinha ASK, Bansal R, Shabani B, Harichandan S. Overview of hydrogen economy in Australia. *Wiley Interdiscip Rev Energy Environ*. [Internet]. 2023 Jan 1 [cited 2025 Apr 2];12(1):e457. Available from: https://onlinelibrary.wiley.com/doi/full/10.1002/wene.457.

79 Balat M. Potential importance of hydrogen as a future solution to environmental and transportation problems. *Int J Hydrogen Energy*. 2008 Aug 1; 33(15):4013–29.

80 Sharma S, Ghoshal SK. Hydrogen the future transportation fuel: From production to applications. *Renew Sustain Energy Rev*. 2015 Mar 1; 43:1151–8.

81 Zou C, Li J, Zhang X, Jin X, Xiong B, Yu H, et al. Industrial status, technological progress, challenges, and prospects of hydrogen energy. *Natural Gas Industry B*. 2022 Oct 1; 9(5):427–47.

82 Singh S, Jain S, Ps V, Tiwari AK, Nouni MR, Pandey JK, et al. Hydrogen: A sustainable fuel for future of the transport sector. *Renew Sustain Energy Rev* 2015 Nov 1; 51:623–33.

7

Economics, Infrastructure, and Policy of Hydrogen Energy

7.1 Introduction

Hydrogen is increasingly recognized as a pivotal element in the global energy transition, offering a pathway to decarbonization across multiple sectors, from transportation and industry to power generation and heating. Unlike fossil fuels, hydrogen can be produced through clean energy sources, stored for long periods, and transported over vast distances, making it a versatile energy carrier. However, its widespread adoption depends not only on technological breakthroughs but also on the economic feasibility, infrastructure readiness, and policy support that shape its integration into mainstream energy systems. While hydrogen has long been used in industrial applications such as ammonia production and refining, its role as a clean energy vector is still evolving. The growing commitment of governments and industries worldwide to reducing carbon emissions has spurred unprecedented investment in hydrogen technologies. Nevertheless, key challenges remain – particularly in scaling up production, reducing costs, and establishing a seamless supply chain that connects production hubs with end users. Addressing these challenges requires a coordinated approach that bridges economic, infrastructural, and regulatory gaps [1].

Hydrogen's economic potential lies in its ability to drive a multi-trillion-dollar energy market, creating opportunities for innovation, job creation, and sustainable industrial development. According to various energy agencies, hydrogen could account for 12–20% of global energy demand by 2050, contributing significantly to net-zero targets. However, realizing this potential depends on addressing cost barriers associated with production, storage, and distribution. Currently, green hydrogen, produced via electrolysis powered by renewable energy, is still more expensive than fossil fuel-based alternatives like gray hydrogen (produced from natural gas without carbon capture) or blue hydrogen (produced with carbon capture and storage). The cost of green hydrogen is projected to decline as electrolyzer efficiencies improve, renewable energy costs decrease, and large-scale production benefits from economies of scale. According to IRENA (International Renewable Energy Agency),

Next-Generation Hydrogen Economy: Production, Storage, and Fuel Cell Technologies, First Edition.
Peeyush Phogat, Shreya Sharma and Satyam Rawat.

green hydrogen production costs could fall by 40–60% by 2030, making it competitive with conventional fuels. Beyond energy production, hydrogen has immense potential to revitalize industries. Countries with abundant renewable energy resources can establish themselves as global hydrogen exporters, while regions dependent on fossil fuel industries can transition their workforce and infrastructure toward hydrogen-based economies. The investment landscape is also shifting, with governments and private entities funneling billions into hydrogen innovation, production, and market expansion. The emerging hydrogen economy could generate millions of new jobs worldwide, spanning engineering, manufacturing, logistics, and research [2].

Hydrogen's economic viability is closely tied to the development of a robust infrastructure that supports its production, storage, transportation, and utilization. Unlike conventional fuels, hydrogen requires specialized infrastructure, including pipelines, high-pressure storage facilities, refueling stations, and conversion technologies to integrate seamlessly into existing energy systems. Currently, most hydrogen infrastructure is localized and limited to industrial clusters, restricting its widespread deployment. Expanding this network is essential for hydrogen to play a mainstream role in the energy transition. Governments and regulatory bodies play a crucial role in addressing these infrastructure challenges. Policy interventions, incentives, and strategic funding are needed to scale up hydrogen deployment and attract private investments. Countries leading in hydrogen innovation – such as Japan, Germany, South Korea, and Australia – have already implemented national hydrogen strategies, outlining subsidies, carbon pricing mechanisms, and research and development (R&D) funding to drive the sector forward. The European Union, for instance, has committed to installing 40 GW of electrolyzer capacity by 2030, reinforcing its ambition to become a global hydrogen leader. However, without clear regulatory frameworks and international standardization, hydrogen adoption could remain fragmented. Policies that establish carbon pricing, green hydrogen certification, and infrastructure development incentives will be instrumental in ensuring hydrogen's competitiveness against fossil fuels. Additionally, international cooperation is essential to create cross-border hydrogen supply chains, enabling efficient trade between energy-rich and energy-importing nations [3].

As hydrogen continues to evolve from a niche technology to a fundamental pillar of global energy systems, aligning economic priorities with infrastructure investments and regulatory support will be critical. The success of hydrogen will ultimately depend on how well policymakers, industries, and investors collaborate to create a sustainable and economically viable hydrogen ecosystem.

7.2 Economic Analysis of Hydrogen Production

Hydrogen production economics are at the core of its viability as a mainstream energy source. While hydrogen is abundant in nature, extracting it in a usable form requires energy-intensive processes that vary in cost, efficiency, and environmental impact. The economic feasibility of hydrogen is primarily influenced by production method, feedstock availability, energy source, infrastructure costs, and government policies. Understanding these factors is crucial for assessing hydrogen's potential to compete with fossil fuels and other renewables [4]. Currently, hydrogen production can be broadly categorized into gray, blue, and green hydrogen, each with distinct cost structures and market implications. Gray hydrogen, derived from natural gas without carbon capture, remains the cheapest and most widely used, accounting for nearly 95% of global hydrogen production. Blue hydrogen, which incorporates carbon capture and storage (CCS), offers a lower-carbon alternative but remains dependent on fossil fuel supply chains. Green hydrogen, produced via electrolysis powered by renewable energy, is the cleanest but also the most expensive due to high electricity and capital costs. The competitiveness of hydrogen is determined by its ability to achieve cost parity with fossil fuels in various applications. Market dynamics such as policy incentives, carbon pricing, technological advancements, and economies of scale play a key role in reducing production costs. Additionally, as global hydrogen demand increases, the development of large-scale production facilities, international trade agreements, and efficient transportation infrastructure will further shape its economic landscape [5].

For hydrogen to become a globally traded energy commodity, significant investment is needed in supply chain logistics, hydrogen liquefaction, pipeline expansion, and port infrastructure. Leading economies are already positioning themselves as future hydrogen exporters, leveraging their renewable energy resources to drive cost reductions. This shift will not only impact energy markets but also redefine global trade flows, potentially mirroring the oil and natural gas markets of today. As hydrogen scales up, its economic outlook will be shaped by a combination of production advancements, market policies, and industrial adaptation. Understanding these economic factors is essential for evaluating its role in the future energy economy, particularly in the context of energy security, industrial competitiveness, and sustainability goals [6].

7.2.1 Cost Breakdown Across Hydrogen Production Methods

The cost of hydrogen production varies significantly depending on the method used, the energy source, and technological efficiency. Several key factors influence the total cost, including feedstock costs, electricity prices, capital expenditures, operational expenditures, and carbon management expenses. Understanding these cost components is essential for assessing the economic viability of different hydrogen production pathways. Gray hydrogen, primarily produced through steam methane reforming (SMR) or coal gasification, is the most cost-effective but also the most carbon-intensive method. The major cost components include natural gas or coal feedstock, which accounts for 45–60% of the total cost, making gray hydrogen prices highly sensitive to fossil fuel market fluctuations. Capital costs contribute around 10–15%, while operational costs and energy losses add another 10–20%. Since gray hydrogen does not use carbon capture, there are no direct expenses for carbon management, but future carbon taxes could increase costs. Currently, the estimated cost of gray hydrogen ranges between \$1.00 and \$2.50 per kg H_2, depending on natural gas and coal prices [7].

Blue hydrogen follows the same production process as gray hydrogen but incorporates CCS to reduce emissions. This additional step increases costs but improves environmental sustainability. The cost structure of blue hydrogen is similar to gray hydrogen, but it includes added expenses for CCS integration. The natural gas feedstock still accounts for 40–50% of the total cost, while capital costs are higher, ranging from 15 to 25%, due to additional CCS infrastructure. Operational costs and energy penalties related to carbon capture further increase expenses. As a result, blue hydrogen costs range between $2.00 and $4.00 per kg H_2, depending on CCS technology and natural gas prices. Green hydrogen, produced via electrolysis using renewable electricity, has a cost structure highly dependent on electricity prices and electrolyzer efficiency. Electricity accounts for the largest share, making up 50–70% of the total cost. Capital costs for electrolyzer units contribute 20–30%, while water and operational costs are relatively low. Although green hydrogen is currently more expensive than fossil-fuel-based alternatives, ongoing advancements in electrolyzer efficiency and decreasing renewable energy costs are expected to make it more competitive. At present, green hydrogen costs range from $3.00 to $7.00 per kg H_2, but projections indicate that costs could fall below $2 per kg by 2030. Turquoise hydrogen, an emerging method based on methane pyrolysis, produces hydrogen while generating solid carbon instead of CO_2 emissions. This approach has the potential to be cost-effective while reducing environmental impact. Its cost structure includes natural gas feedstock expenses similar to gray and blue hydrogen, making up 40–50% of the total cost. Capital costs range between 20 and 30% for pyrolysis reactors, and operational costs add another 10–20%. However, the ability to sell the solid carbon byproduct, such as graphene or carbon black, could partially offset production costs. Currently, turquoise hydrogen costs between $2.00 and $3.50 per kg H_2, with potential reductions as carbon utilization technologies develop. Figure 7.1 illustrates the different hydrogen production methods (gray, blue, green, turquoise) along with their cost, efficiency, and carbon emissions [8].

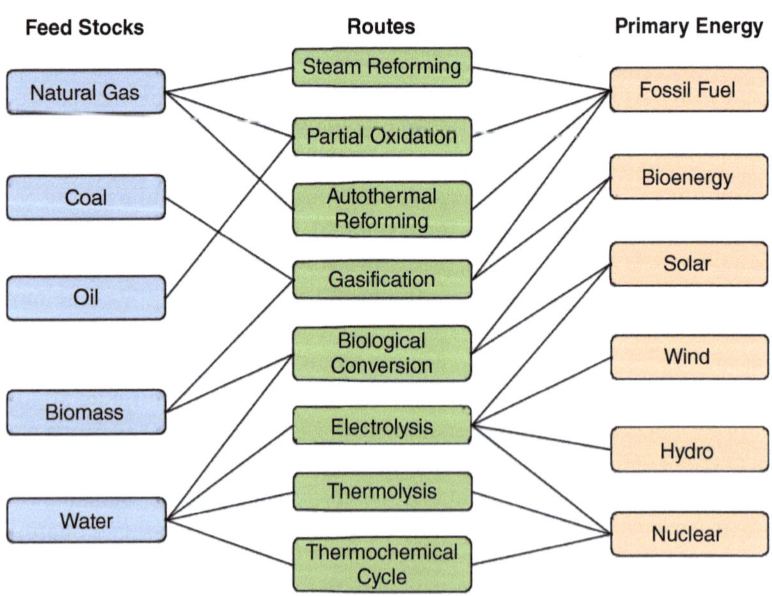

Figure 7.1 Hydrogen production methods comparison. [10]/with permission of Elsevier.

While gray hydrogen remains the cheapest option today, the declining costs of renewable energy, improvements in electrolyzer efficiency, and increasing carbon pricing policies are expected to make green hydrogen more competitive. Additionally, advancements in CCS for blue hydrogen and methane pyrolysis for turquoise hydrogen could influence the long-term economics of hydrogen production. Factors such as technology breakthroughs, declining renewable energy costs, stricter carbon regulations, and supportive government policies will shape the future cost trajectory of hydrogen production. The transition toward cost-competitive, sustainable hydrogen production is essential for its widespread adoption across energy, transportation, and industrial sectors [9]. Table 7.1 provides a comparative overview of hydrogen production methods.

7.2.2 Market Competitiveness and Price Parity with Fossil Fuels

The competitiveness of hydrogen as an energy source depends largely on its ability to achieve price parity with conventional fossil fuels. Currently, hydrogen remains more expensive than coal, natural gas, and oil-based fuels, limiting its large-scale commercial adoption. However, the trajectory of technological advancements, policy-driven incentives, and economies of scale are shaping a more competitive hydrogen market. One of the key challenges in achieving price parity is the cost disparity between different hydrogen production methods. Gray hydrogen, produced from natural gas via SMR without carbon capture, is currently the most affordable, typically ranging between $1 and $2 per kilogram. In contrast, blue hydrogen, which incorporates carbon capture, utilization, and storage (CCUS), costs between $2 and $3 per kilogram, depending on the efficiency of the capture technology and carbon pricing regulations. Green hydrogen, produced via electrolysis powered by renewable energy, remains the most expensive, often exceeding $4–$6 per kilogram, although recent innovations in electrolyzer efficiency and renewable energy expansion are driving costs downward. To compete with fossil fuels, hydrogen production costs must align with the levelized cost of energy (LCOE) of conventional energy sources. For instance, natural gas-based energy generation has an LCOE of around $40–$70 per megawatt-hour (MWh), while coal-fired power generation typically falls within $60–$120 per MWh. Hydrogen-based power generation, particularly from green hydrogen, currently exceeds $100 per MWh, highlighting the gap that needs to be bridged. However, projections indicate that with continued declines in renewable electricity prices and advancements in electrolyzer technology, green hydrogen could become cost-competitive with natural gas in several regions by 2030–2040 [16].

Market competitiveness is also influenced by regional energy economics and policy support. In areas where carbon pricing mechanisms and subsidies for clean hydrogen are in place, such as the European Union, Japan, and parts of the United States, green and blue hydrogen are increasingly competitive. Additionally, geopolitical shifts, such as fluctuations in natural gas prices and supply chain disruptions, can accelerate hydrogen's competitiveness by making fossil fuels more volatile in pricing. Beyond production costs, transportation, storage, and distribution costs impact hydrogen's price competitiveness. Unlike natural gas, which benefits from an extensive pipeline network, hydrogen infrastructure is still in development.

Table 7.1 Comparative overview of hydrogen production methods.

Production method	Energy source	Efficiency (%)	CO_2 emissions	Cost (per kg)	Scalability potential	Technological maturity
SMR [11]	Natural gas	65–75	High (if uncombined with CCS)	low (~$1.5–$3)	High (established industry)	High
Electrolysis (renewable) [12]	Wind, solar, hydro	60–80	None (if powered by renewables)	high (~$4–$6)	Moderate (growing demand)	Moderate
Coal Gasification [13]	Coal	55–70	High (if uncombined with CCS)	moderate (~$2.5–$4)	Moderate (not widely used)	Moderate
Biomass Gasification [14]	Biomass (organic materials)	45–60	Low to moderate	Moderate (~$3–$5)	Moderate (limited by feedstock)	Moderate
Algae-based Production [15]	Algae (biological feedstock)	25–30	Very low	High (~$5–$10)	High (emerging technology)	Low

Innovations in liquid organic hydrogen carriers (LOHCs), ammonia-based hydrogen transport, and dedicated hydrogen pipelines will be critical in lowering distribution costs. If infrastructure investments scale up efficiently, hydrogen's delivered cost could decline significantly, further narrowing the gap with fossil fuels. Ultimately, hydrogen's ability to reach price parity with fossil fuels depends on continued advancements in production efficiency, policy-driven incentives, and infrastructure development. While it is not yet directly cost-competitive in most applications, strategic investments and technological breakthroughs are accelerating the transition toward an economically viable hydrogen economy [17].

7.2.3 Economies of Scale: Cost Reductions Through Technological Advancements

Achieving cost competitiveness for hydrogen relies heavily on economies of scale and continuous technological advancements. As production capacity increases and technologies mature, the costs associated with hydrogen production, storage, and distribution are expected to decrease substantially. Economies of scale enable more efficient use of resources, lower capital expenditure per unit, and reduced operational costs, paving the way for hydrogen to become a viable alternative to fossil fuels. In hydrogen production, scaling up electrolyzer manufacturing is crucial for reducing costs. Mass production of electrolyzers – such as proton exchange membrane (PEM), alkaline, and solid oxide electrolyzers – has the potential to drive down capital costs by 40–60% over the next decade. Larger electrolyzer facilities benefit from standardized components, automated manufacturing processes, and reduced material waste, resulting in more affordable green hydrogen. Additionally, innovations in membrane materials, electrode catalysts, and cell design further enhance efficiency, contributing to lower overall production costs. Technological advancements in CCUS are also critical for cost-effective blue hydrogen production. Improved capture rates, enhanced storage methods, and innovative utilization pathways (such as converting captured CO_2 into valuable chemicals) reduce the costs and increase the feasibility of blue hydrogen. Large-scale CCUS projects, particularly in regions with existing fossil fuel infrastructure, benefit from shared facilities and transportation networks, further reducing the cost per kilogram of hydrogen produced [18].

Renewable energy integration is another area where economies of scale play a pivotal role. As the deployment of solar, wind, and hydropower expands, the cost of electricity – one of the primary expenses in green hydrogen production – continues to decline. Large-scale renewable energy projects, often exceeding 100 megawatts (MW), benefit from lower installation costs, improved grid integration, and optimized land use, translating into cheaper hydrogen production. For instance, offshore wind farms combined with on-site electrolysis are emerging as a promising model for scalable, cost-effective hydrogen generation. Innovations in hydrogen storage and transportation also contribute to cost reductions. Developing larger, more efficient storage solutions – such as high-pressure tanks, liquid hydrogen containers, and advanced metal hydrides – enhances storage density while reducing costs. In transportation, economies of scale are realized through the construction of dedicated hydrogen pipelines and the deployment of bulk carriers, such as those designed for LOHCs and ammonia. Scaling up these infrastructure elements reduces per-unit transportation costs, making hydrogen more accessible for widespread use. The integration of digital technologies, such as artificial intelligence (AI), machine learning, and blockchain, further enhances cost efficiency across the hydrogen value chain. Predictive analytics optimize production scheduling, maintenance, and energy consumption, while blockchain ensures transparent and secure hydrogen trading, reducing transaction costs. Additionally, digital twins – virtual models of physical hydrogen systems – enable real-time monitoring and optimization, minimizing operational expenses [19].

Government policies and public-private partnerships play a significant role in facilitating economies of scale. Initiatives such as the European Union's Hydrogen Strategy, Japan's Green Growth Strategy, and the United States' Hydrogen Earthshot Initiative support large-scale pilot projects and infrastructure development. These programs provide funding, tax incentives, and streamlined regulatory approvals, encouraging private sector investment and accelerating technology commercialization. Overall, the convergence of scaling effects and technological innovations is pivotal for reducing hydrogen costs and enhancing its market competitiveness. As production capacities expand and infrastructure networks mature, hydrogen is poised to become a cornerstone of the global energy landscape, contributing to a sustainable and resilient energy future [20].

7.2.4 Hydrogen as a Commodity: Global Trade and Export Potential

Hydrogen is rapidly emerging as a globally traded energy commodity, much like oil and natural gas. As nations strive to achieve decarbonization goals, hydrogen's role in international energy markets is expanding, creating new trade routes and economic opportunities. Countries with abundant renewable energy resources are positioning themselves as export hubs, while industrialized economies with high energy demands are investing in hydrogen import infrastructure. The development of a global hydrogen market depends on cost-effective production, efficient storage and transportation, and well-structured trade agreements. Several countries with surplus renewable energy potential – such as Australia, Chile, Saudi Arabia, and Namibia – are investing heavily in green hydrogen production to supply global markets. These nations benefit from low-cost solar and wind power, enabling the production of hydrogen at competitive prices. Australia, for instance, has launched multiple hydrogen export projects, with agreements to supply Japan, South Korea, and Europe. Similarly, the European Union's Hydrogen Strategy envisions partnerships with North African nations to import green hydrogen via pipelines and shipping routes. Blue hydrogen is also gaining traction in countries with extensive natural gas reserves, such as the United States, Russia, and the Middle East. With advancements in CCUS, these regions are leveraging existing fossil fuel infrastructure to produce and export low-carbon hydrogen. The Middle East, particularly Saudi Arabia and the UAE, is positioning itself as a leading hydrogen exporter by developing large-scale projects that combine natural gas reforming with CCUS and renewable-powered electrolysis [21–24].

One of the biggest challenges in hydrogen trade is cost-effective transportation. Since hydrogen has low volumetric energy density, it requires compression, liquefaction, or chemical conversion for efficient shipping. The three primary transportation methods being explored are:

a. **Liquefied Hydrogen (LH_2):** Hydrogen can be cooled to −253 °C to become a liquid, significantly increasing its energy density. However, liquefaction is energy-intensive and requires specialized cryogenic ships. Japan has pioneered this approach, with the world's first liquid hydrogen carrier, Suiso Frontier, demonstrating intercontinental hydrogen transport.
b. **Ammonia (NH_3) as a Carrier:** Converting hydrogen into ammonia provides a more efficient and cost-effective means of transportation. Ammonia is easier to store and transport than liquid hydrogen and can be reconverted back into hydrogen or used directly as a fuel. Major projects in Australia, Japan, and the Middle East are focusing on ammonia-based hydrogen supply chains.
c. **LOHCs:** LOHCs, such as methylcyclohexane (MCH), offer another promising approach for hydrogen storage and transport. These carriers can safely store hydrogen at ambient temperature and pressure, reducing infrastructure costs compared to liquefaction. Research into catalytic dehydrogenation is ongoing to improve efficiency and scalability.

For pipeline transport, hydrogen can be blended with natural gas or delivered through dedicated hydrogen pipelines. The European Hydrogen Backbone initiative is a prime example, aiming to repurpose existing natural gas pipelines to facilitate hydrogen transport across Europe [25].

To establish hydrogen as a traded commodity, transparent pricing mechanisms, standardized contracts, and market indices must be developed. Currently, hydrogen prices vary significantly based on production method, energy source, and location. The Hydrogen Council and various government initiatives are working toward creating benchmark prices to enable fair market competition. Hydrogen futures markets are beginning to take shape, with exchanges such as Japan's Hydrogen Value Chain Derivatives Market and the European Hydrogen Exchange (HyXchange) testing market-based hydrogen pricing mechanisms. A well-functioning commodity market for hydrogen would provide price transparency, liquidity, and long-term investment stability, making it a reliable energy source for industries worldwide.

Global trade in hydrogen requires harmonized regulations and policy frameworks to ensure safety, quality standards, and market efficiency. International bodies like the International Energy Agency (IEA), IRENA, and Hydrogen Council are working on establishing certification schemes, such as Guarantees of Origin, to differentiate between green, blue, and gray hydrogen.

Several international agreements are already fostering hydrogen trade partnerships:

a. **Japan and Australia:** The Hydrogen Energy Supply Chain (HESC) project has successfully transported liquid hydrogen from Australia to Japan.
b. **European Union and North Africa:** The European Union (EU) plans to import green hydrogen from Morocco and other North African countries via pipelines.
c. **South Korea and Saudi Arabia:** South Korea has signed multiple agreements to import hydrogen from Saudi Arabia's NEOM project.

Table 7.2 Global hydrogen market projections (2030–2050).

Year	Global hydrogen demand (million tons)	Investment in hydrogen technologies (billion USD)	Hydrogen share in global energy market (%)	Key regions leading in hydrogen production
2030	30	40	2.5	Europe, Japan, United States, China
2040	80	100	5	Europe, United States, Middle East, Australia
2050	200	250	12	Europe, United States, China, Australia, Middle East

These collaborations highlight the growing geopolitical importance of hydrogen, with nations competing to secure supply chains and establish market dominance. With the right policy support, infrastructure investments, and technological innovations, hydrogen is poised to become a mainstream global commodity, reshaping the energy landscape and contributing to the transition toward a low-carbon economy [26]. Table 7.2 shows the Global Hydrogen Market Projections (2030–2050).

7.3 Infrastructure Development and Logistics

The widespread adoption of hydrogen as a key energy carrier depends on a robust and efficient infrastructure that encompasses production, transportation, storage, and end-use delivery. Unlike conventional fossil fuels, hydrogen requires specialized logistics due to its low volumetric energy density, high diffusivity, and the need for advanced containment technologies. To ensure a reliable and cost-effective hydrogen supply chain, a comprehensive infrastructure network must be developed, balancing scalability, economic viability, and environmental sustainability. A well-integrated hydrogen ecosystem will require the establishment of centralized and decentralized production hubs, ensuring accessibility for industrial and consumer applications. Efficient transportation and distribution networks – whether via pipelines, liquid hydrogen shipping, or carriers like ammonia and LOHCs – are crucial for ensuring a steady supply. Additionally, hydrogen storage technologies will play a critical role in stabilizing supply-demand fluctuations, enabling long-term energy storage, and facilitating cross-border hydrogen trade. As the transportation sector transitions towards fuel cell electric vehicles (FCEVs), heavy-duty trucks, trains, and maritime applications, the development of hydrogen refueling stations (HRS) and their integration with existing fuel distribution networks will be essential. Likewise, aligning hydrogen infrastructure with smart grids can enhance energy security and optimize renewable energy utilization. The success of hydrogen infrastructure development depends on technological advancements, economies of scale, and policy support to ensure that hydrogen logistics become cost-efficient, scalable, and aligned with global decarbonization efforts. This section explores the critical components of hydrogen infrastructure, analyzing the challenges and opportunities that will shape the future of hydrogen-based energy systems [27].

7.3.1 Hydrogen Production Hubs: Centralized vs. Decentralized Approaches

Hydrogen production hubs are central to the effective deployment of hydrogen as a mainstream energy source. The choice between centralized and decentralized production models will have a significant impact on the cost, logistics, and scalability of hydrogen supply chains. Both approaches come with distinct advantages and challenges, influencing their potential in different regions and applications. In a centralized production model, hydrogen is produced in large-scale facilities located near sources of low-cost, clean energy such as renewable power plants or natural gas reforming plants. These hubs benefit from economies of scale, as large volumes of hydrogen can be produced more efficiently and at a lower cost per unit compared to smaller facilities. Centralized plants also benefit from the ability to produce hydrogen at locations with access to high renewable energy potential, such as coastal or desert regions with abundant solar and wind resources. However, transportation and distribution become critical in this model, as hydrogen needs to be delivered to end users across large distances, increasing logistics costs and infrastructure requirements [28].

In contrast, a decentralized production model involves producing hydrogen closer to where it is consumed, such as at local industries, transportation hubs, or even residential areas. This approach minimizes transportation costs and reduces the need for complex distribution networks. It is particularly suited for applications where hydrogen is required on-demand, such as HRS or small-scale industrial processes. Decentralized production can also be more adaptable and resilient, as local production facilities can adjust to changes in demand more flexibly. However, the key challenge lies in achieving cost-effective production at smaller scales, as decentralized facilities may lack the economies of scale that centralized plants enjoy. Furthermore, decentralization may not fully exploit renewable energy resources in areas that lack sufficient local generation capacity, potentially leading to reliance on more expensive or less sustainable energy sources. Both models can coexist in a hybrid system, where centralized hubs supply large quantities of hydrogen to key demand centers, while decentralized production meets local needs. The optimal approach will depend on factors such as regional energy resources, demand density, infrastructure readiness, and policy incentives. A well-designed hydrogen production network should balance the benefits of both centralized and decentralized strategies, enabling flexibility and cost efficiency while supporting the transition to a hydrogen-based economy [29].

7.3.2 Transportation and Distribution Networks

The transportation and distribution of hydrogen are key components in establishing a functional and scalable hydrogen economy. Given hydrogen's low volumetric energy density and unique properties, including its high diffusivity and flammability, transporting and distributing it presents distinct challenges compared to conventional fuels. Developing an effective hydrogen transportation network is essential for ensuring a steady and reliable supply of hydrogen from production hubs to end users across industries, transportation sectors, and energy systems. There are several methods for transporting hydrogen, each with its own set of considerations for cost, efficiency, and safety. Pipelines represent one of the most cost-effective and efficient means for transporting large volumes of hydrogen, particularly over land. Existing natural gas pipelines can be adapted for hydrogen transportation, a process known as hydrogen blending, where hydrogen is mixed with natural gas and transported through the same infrastructure. However, full-scale hydrogen pipelines dedicated solely to transporting pure hydrogen are necessary for larger hydrogen economies. These pipelines require advanced materials that can withstand the embrittlement effects hydrogen may have on metals, ensuring safety and longevity. Developing new pipeline networks or repurposing existing infrastructure involves substantial investment and regulatory coordination. For long-distance transportation, liquefied hydrogen can be a viable solution. Hydrogen is liquefied by cooling it to extremely low temperatures, allowing a higher volumetric density and making it easier to ship by sea or land. However, liquefaction is energy-intensive and requires specialized cryogenic tanks capable of maintaining extremely low temperatures. The transportation of liquid hydrogen by tanker ships has already been demonstrated in pilot projects, particularly for international trade, as it enables the export of hydrogen from regions with abundant renewable energy resources to areas with high demand [30].

Another method of transporting hydrogen is through hydrogen carriers, which are substances that can absorb and release hydrogen through chemical reactions. Ammonia and LOHCs are two key examples. Ammonia is already produced on a global scale for agricultural use, and its established infrastructure allows hydrogen to be transported safely and efficiently in ammonia form. Ammonia can then be converted back into hydrogen at the point of consumption. LOHCs, on the other hand, store hydrogen in a liquid form at ambient conditions, making transportation safer and more energy-efficient compared to cryogenic hydrogen. Once hydrogen is needed, it can be released through chemical reactions, often requiring a catalytic process. These carriers offer significant potential for hydrogen trade, particularly for countries without direct access to large-scale hydrogen production capabilities. Additionally, road and rail transport will play an important role in distributing hydrogen at smaller scales. Compressed hydrogen stored in high-pressure tanks can be transported by trucks to refueling stations and industrial sites. This form of transportation is most suited for shorter distances or where pipeline infrastructure is not feasible. It is cost-effective for local delivery but requires specialized vehicles and safety measures due to the high pressures involved [31].

The hydrogen distribution network also encompasses storage solutions at intermediate points, such as hydrogen hubs or distribution centers, where hydrogen can be stored temporarily before being transported further or used in different applications. Storage solutions, including high-pressure tanks, liquid storage, and chemical carriers, must be integrated into the distribution network to ensure a continuous, stable supply that matches demand fluctuations. Ultimately, creating a global hydrogen distribution system requires the integration of multiple transportation methods and infrastructure components, coupled with the development of standardized safety protocols, regulatory frameworks, and innovative technologies to ensure that hydrogen can be delivered safely, efficiently, and economically across borders and industries. The effectiveness of these transportation and distribution networks will play a critical role in enabling hydrogen to become a mainstream energy carrier, supporting the decarbonization of multiple sectors and contributing to energy security and sustainability worldwide [32].

7.3.3 Hydrogen Storage Technologies

Hydrogen storage remains one of the most significant challenges in realizing a fully functional hydrogen economy. Efficient storage technologies are necessary to ensure that hydrogen can be produced, transported, and utilized in a way that meets energy demands without compromising safety, efficiency, or cost. As hydrogen has a very low volumetric energy density, especially in its gaseous form, effective storage methods are essential to facilitate its widespread adoption across industries, transportation, and power generation. In this section, we explore the main storage technologies, categorized as physical, chemical, and solid-state storage, each of which offers unique advantages and challenges [30]. Figure 7.2 shows different hydrogen storage technologies.

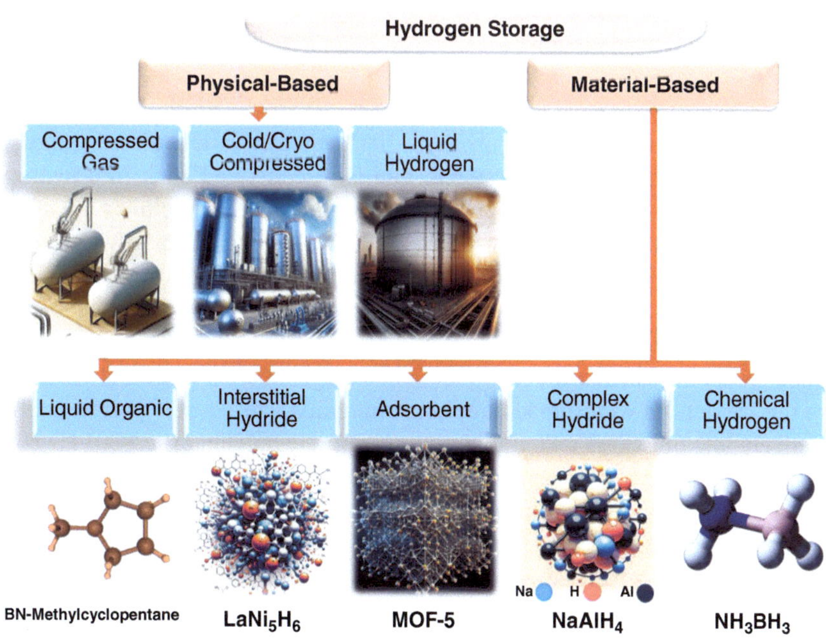

Figure 7.2 Illustrates different hydrogen storage methods (compressed gas, liquid hydrogen, ammonia, LOHCs, metal hydrides). [39]/MDPI/CC BY 4.0.

7.3.3.1 Physical Storage: Cryogenic and High-pressure Tanks

Physical storage techniques for hydrogen primarily involve compressing or liquefying hydrogen to increase its volumetric energy density. Cryogenic storage and high-pressure tanks are the two most commonly employed methods for storing hydrogen in its gaseous form, each with distinct characteristics and applications. Cryogenic storage involves liquefying hydrogen by cooling it to temperatures below its boiling point of −253 °C. At this temperature, hydrogen becomes a liquid, significantly increasing its density, making it more economical to store and transport in large quantities. This method is particularly useful for long-term storage and transportation over large distances, as liquefied hydrogen takes up far less space compared to its gaseous counterpart. However, the process of liquefaction requires a substantial amount of energy, and maintaining the low temperatures necessary for liquid hydrogen storage presents significant engineering challenges, such as ensuring minimal evaporation losses and ensuring the integrity of cryogenic tanks. On the other hand, high-pressure storage involves compressing hydrogen to several hundred bars (up to 700 bar) and storing it in tanks made from robust materials capable of withstanding the pressures. This method is simpler than cryogenic storage and has been widely used in FCEVs and stationary applications. High-pressure tanks are more energy-efficient than cryogenic storage in terms of the amount of energy required for compression. However, they still present safety concerns due to the high pressures involved, and their volumetric efficiency is lower than that of cryogenic storage. Both physical storage methods continue to evolve with advancements in materials science, such as the development of lightweight composite materials for tank construction and improved insulation systems to reduce energy losses during liquefaction and storage [33].

7.3.3.2 Chemical Storage: Ammonia and LOHCs

Chemical hydrogen storage involves storing hydrogen in chemical compounds that can be easily transported, stored, and later converted back into hydrogen when needed. Two of the most promising chemical storage methods are ammonia and LOHCs. Ammonia is a widely produced chemical, primarily used in the fertilizer industry. Ammonia contains a significant amount of hydrogen, approximately 17.6% by weight, making it an attractive candidate for hydrogen storage and transport. The key advantage of ammonia as a hydrogen carrier is its established global infrastructure for production, storage, and distribution, which lowers the barriers to widespread adoption. Ammonia can be transported safely at relatively low pressures and ambient temperatures, reducing the need for costly and energy-intensive cryogenic or high-pressure storage solutions. Hydrogen can be released from ammonia through a process called cracking, where ammonia is decomposed at high temperatures to release pure hydrogen. However, ammonia cracking requires advanced catalysts and high energy inputs, which can impact its overall efficiency. LOHCs are another innovative chemical storage solution. LOHCs are organic liquids that can absorb hydrogen through a reversible chemical reaction and release it when required. These carriers are typically stable at room temperature and can be transported and stored safely using existing infrastructure designed for liquid fuels. LOHCs are attractive for their safety, storage capacity, and compatibility with current fuel logistics. One of the main challenges with LOHCs is the energy required to hydrogenate and dehydrogenate the carrier, which can affect the overall efficiency of the system. Furthermore, the selection of appropriate LOHCs and development of efficient catalytic processes are key areas of ongoing research. Both ammonia and LOHCs present promising chemical storage methods but face challenges related to efficiency, cost, and energy requirements for hydrogen release. As research advances, these technologies have the potential to become viable solutions for large-scale hydrogen storage and transportation [34].

7.3.3.3 Solid-state Storage: Metal Hydrides, MOFs, and Advanced Nanomaterials

Solid-state hydrogen storage involves the absorption or adsorption of hydrogen into solid materials, offering high volumetric energy density and greater safety compared to gaseous or liquid storage. This method utilizes materials such as metal hydrides, metal-organic frameworks (MOFs), and advanced nanomaterials to store hydrogen. Metal hydrides are compounds formed by hydrogen bonding with metals, typically transition metals. These hydrides can absorb and release hydrogen reversibly, depending on the pressure and temperature conditions. The key advantage of metal hydrides is their high hydrogen density, making them one of the most promising solutions for compact, safe storage. However, metal hydrides typically require high temperatures or moderate pressure to release hydrogen, and their weight and volume can be limitations in certain applications like fuel cell vehicles. Recent advancements in alloy compositions and lightweight materials are improving the performance of metal hydrides, reducing these drawbacks. MOFs are another class of materials that show great potential for solid-state hydrogen storage. MOFs are highly porous materials made from metal ions and organic linkers, creating a vast network of pores capable of adsorbing hydrogen gas. These materials have extremely high surface areas and can store large amounts of hydrogen at relatively low pressures, making them ideal candidates for applications that require compact, energy-dense storage. However, the cost of producing MOFs and the need for more efficient hydrogen adsorption-desorption cycles are areas that require further development. The use of advanced nanomaterials is another exciting avenue in solid-state hydrogen storage. Nanostructured materials, such as carbon nanotubes, graphene, and nanocomposites, have shown the potential to store hydrogen through adsorption mechanisms at the molecular level. These materials offer extremely high surface areas and customizable properties, which can be fine-tuned to maximize hydrogen storage capacity. However, challenges related to scalability, cost, and hydrogen release efficiency need to be addressed before these materials can be deployed for commercial use. Solid-state storage technologies hold the promise of providing safe, efficient, and compact hydrogen storage solutions, particularly for applications requiring high energy density and enhanced safety. However, overcoming challenges related to cost, scalability, and energy efficiency remains a critical focus of ongoing R&D in the field of hydrogen storage [35].

7.3.4 Hydrogen Refueling Infrastructure

The development of a robust hydrogen refueling infrastructure is crucial for the widespread adoption of hydrogen as a clean energy carrier. Just as refueling stations for conventional fuels (gasoline, diesel) play a central role in ensuring the mobility of internal combustion engine vehicles, HRS are essential for enabling the mass deployment of FCEVs and other hydrogen-powered transportation systems. The deployment of HRS and their integration with existing energy and fuel networks will be central to scaling hydrogen usage in both transportation and industrial sectors. As hydrogen technology progresses, refueling infrastructure must evolve to support the specific needs of hydrogen vehicles and deliver the fuel efficiently, safely, and economically [36].

7.3.4.1 Deployment of HRS

The deployment of HRS is a critical element in establishing a hydrogen-powered transportation ecosystem. A well-distributed network of HRS will provide the necessary refueling infrastructure to support the growing number of hydrogen-powered vehicles. These stations must be strategically placed to minimize gaps in accessibility while ensuring optimal coverage for both urban and rural areas. HRS infrastructure typically involves the storage and dispensing of hydrogen to vehicles at high pressure, requiring careful design to meet safety standards and ensure efficiency in fuel delivery. The expansion of HRS is hindered by several challenges, including the high cost of station construction, hydrogen supply logistics, and regulatory requirements. HRS can either be independent, where hydrogen is produced on-site or delivered from a central production hub, or integrated, where hydrogen is delivered from a nearby centralized facility. On-site production technologies include electrolysis, which produces hydrogen from water using electricity, and SMR for a more cost-effective production method. However, for large-scale deployment, it is critical that the cost of building and maintaining HRS is reduced through technological advancements, economies of scale, and government incentives. The challenges of station placement are also influenced by hydrogen storage requirements and distribution logistics, which vary by location. Urban areas, where transportation demand is high, may see faster adoption of HRS, while rural regions may have slower infrastructure development due to lower vehicle density and higher costs of transportation infrastructure. The cooperation of public and private sectors is essential for establishing an extensive, accessible network of HRS that meets the needs of hydrogen-powered vehicles. Moreover, standardization is necessary for ensuring the compatibility of refueling infrastructure across different regions and countries. Protocols for refueling equipment, safety procedures, and vehicle connections must be aligned to ensure seamless refueling operations across diverse locations and hydrogen vehicles [37].

7.3.4.2 Integration with Existing Fuel Networks and Smart Grid Systems

Hydrogen refueling infrastructure will also need to be integrated with existing fuel and energy networks to enhance the efficiency and sustainability of the refueling process. The synergy between hydrogen and conventional fuel systems, as well as the integration with modern smart grid technologies, will offer flexibility and convenience to users while optimizing the supply chain. In many regions, HRS will rely on existing fuel networks for the transportation and distribution of hydrogen. This integration would involve adapting current pipeline, storage, and distribution infrastructures to accommodate hydrogen, ensuring safe and reliable delivery to refueling stations. The adaptation of existing natural gas pipelines to carry hydrogen, for instance, has already been explored in many parts of the world, with early pilot projects underway. Such an integration would help reduce costs by leveraging existing infrastructure and provide more seamless hydrogen distribution across wider geographic areas. Furthermore, hydrogen infrastructure can be enhanced by integrating with smart grid systems to optimize energy flow and storage. Smart grids, which use digital technology to monitor and manage energy production, distribution, and consumption, can provide a platform for the efficient integration of HRS. For instance, during times of low electricity demand, excess renewable energy from wind or solar can be directed to produce hydrogen through electrolysis, which can then be stored for later use at refueling stations. This integration reduces reliance on fossil fuels and increases grid resilience by balancing supply and demand more efficiently [31].

Smart grid technologies could also enable demand-side management for HRS, where the energy consumption of stations can be dynamically adjusted based on grid conditions and the availability of renewable energy. This would ensure that hydrogen production is optimized to reduce costs and minimize environmental impact, especially during periods of high renewable energy generation. In addition, smart grids can facilitate vehicle-to-grid (V2G) integration, where hydrogen-powered vehicles, as mobile storage units, can return excess electricity back to the grid when needed. This provides additional grid support, enabling better integration of intermittent renewable energy sources and ensuring greater energy security and resilience. The integration of hydrogen refueling infrastructure with existing fuel and energy networks is vital to reducing costs, optimizing energy use, and ensuring that hydrogen-powered vehicles can be seamlessly integrated into existing transportation ecosystems. Achieving this requires a collaborative effort between governments, energy providers, and the automotive industry to ensure that both the hydrogen supply chain and refueling infrastructure are sufficiently developed and interconnected [38].

7.4 Policy and Regulatory Frameworks

The successful development and scaling of hydrogen technologies hinge on a solid and supportive policy and regulatory framework. Governments around the world are beginning to recognize hydrogen as a key element in their energy transitions, driving efforts to align policies with sustainability goals and reduce dependence on fossil fuels. Effective policy frameworks must address key issues such as funding for R&D, infrastructure development, integration with existing energy systems, and the creation of market conditions that encourage innovation and investment. As hydrogen economies begin to take shape globally, the roles of government policies, international collaborations, and regulatory mechanisms become critical in determining the pace of adoption and the scale of technological breakthroughs. The creation of clear and coordinated policies is essential to providing market certainty, incentivizing private investment, and ensuring that hydrogen technologies are developed in a socially, environmentally, and economically responsible manner. Furthermore, international cooperation and the establishment of global hydrogen strategies will be pivotal in ensuring a sustainable, cost-effective, and interconnected hydrogen economy. The regulatory environment must promote cross-border collaboration, knowledge sharing, and the harmonization of standards, enabling the global hydrogen market to thrive. As nations seek to fulfill their climate commitments and develop clean energy systems, hydrogen will play an increasingly important role in shaping the future of energy policy [40].

7.4.1 Global Hydrogen Strategies: Leading Countries and Their Roadmaps

As the global push for clean energy transitions intensifies, several countries have begun implementing national hydrogen strategies, recognizing the critical role hydrogen can play in reducing carbon emissions and fostering sustainable economic growth. These strategies outline clear roadmaps for the development, production, infrastructure, and adoption of hydrogen technologies, with a focus on both green and blue hydrogen. Leading nations are working toward building hydrogen economies through innovative policies, strategic investments, and international partnerships, paving the way for the global deployment of hydrogen solutions.

The EU has established itself as a global leader in the push for hydrogen adoption. The EU Hydrogen Strategy, launched in 2020, aims to create a competitive hydrogen economy by 2050, emphasizing the use of hydrogen for decarbonizing key sectors such as transportation, industry, and power generation. The European Commission is committed to investing in hydrogen infrastructure, with a €430 billion budget allocated for the scaling of clean hydrogen production and its integration into the energy system. The EU's strategy includes a focus on green hydrogen, produced using renewable energy, while also supporting the development of blue hydrogen to complement green initiatives in the short term. Additionally, cross-border collaborations and hydrogen pipelines are planned to foster interconnection between hydrogen-rich regions [41].

Germany is one of the front-runners in the global hydrogen race, with its National Hydrogen Strategy introduced in 2020. The strategy focuses on accelerating the development of hydrogen as a cornerstone of Germany's energy transition. A key component of this strategy is its emphasis on scaling up green hydrogen, supported by renewable energy sources such as wind and solar power. The German government has committed €9 billion to hydrogen-related projects, aimed at expanding domestic production, developing hydrogen infrastructure, and forming strategic international partnerships. Germany's hydrogen ambitions include becoming a global hub for hydrogen technologies, particularly in heavy industries, transportation, and power generation, with a focus on decarbonizing sectors that are challenging to electrify [42].

Japan has long been a pioneer in hydrogen technologies and continues to lead the charge with its hydrogen roadmap. The country's Hydrogen Basic Strategy, introduced in 2017 and updated periodically, aims to make hydrogen an integral part of Japan's energy mix. Japan has set ambitious targets for hydrogen production and utilization, focusing on both domestic production and the importation of hydrogen from regions rich in renewable energy, such as Australia and the Middle East. Japan's approach includes fostering the development of hydrogen-powered transport, including FCEVs, as well as building a nationwide hydrogen refueling infrastructure. Moreover, Japan is also investing heavily in the development of hydrogen storage, transportation, and distribution technologies, ensuring the secure and cost-effective movement of hydrogen across its islands [43].

The United States has made significant strides toward integrating hydrogen into its energy strategy through both federal and state-level initiatives. In 2021, the Biden administration unveiled the American Jobs Plan, which includes a focus on developing a clean hydrogen economy as a key pillar of its climate strategy. The US Department of Energy (DOE) has been supporting the development of both green and blue hydrogen technologies through funding for research, demonstration projects, and the construction of hydrogen hubs across the country. The United States also recognizes the importance of hydrogen in decarbonizing heavy industries, including steel, cement, and chemicals, and is working to build a robust hydrogen infrastructure to enable large-scale adoption. The hydrogen roadmap is being shaped by a combination of public policy incentives, private investments, and regional initiatives, with the goal of achieving net-zero emissions by 2050 [44].

South Korea's hydrogen strategy, introduced in 2019, aims to position the country as a global leader in hydrogen production and use. The South Korean government has committed to a comprehensive strategy, investing over $30 billion in hydrogen-related technologies by 2040. South Korea focuses on hydrogen-powered transport, including the development of hydrogen buses, trucks, and passenger vehicles, and is planning to establish a large network of HRS across the country. Additionally, South Korea is leveraging its strong manufacturing base to develop and export hydrogen fuel cell technology. The country also emphasizes the development of green hydrogen, produced from renewable energy sources, and has been investing in the creation of hydrogen production hubs, fostering regional production while pursuing hydrogen imports from abroad [45].

Australia is positioning itself as a major player in the global hydrogen market, particularly in the production of green hydrogen. With abundant renewable energy resources, Australia's Hydrogen Strategy, launched in 2021, focuses on creating export opportunities for clean hydrogen. The strategy aims to reduce the cost of green hydrogen through technological innovation and scale, making hydrogen competitive in international markets. Australia is focusing on building hydrogen hubs in its rural and coastal regions, where renewable energy resources such as wind and solar are abundant. The country is also collaborating with key hydrogen importers, such as Japan and South Korea, to establish secure hydrogen supply chains and export networks [46].

China has rapidly developed its hydrogen strategy, focusing on using hydrogen for clean energy applications, particularly in the transport and industrial sectors.

The country's hydrogen roadmap, part of its broader energy transition plan, emphasizes hydrogen production, storage, and infrastructure development. China has set ambitious goals to increase the production of hydrogen from renewable energy sources and to establish a nationwide hydrogen refueling network to support the expansion of hydrogen-powered vehicles. Furthermore, China is focusing on technological advancements in hydrogen storage, transportation, and fuel cell systems. Given its large manufacturing base and expertise in renewable energy, China is poised to play a crucial role in the global hydrogen market, with ambitions to become a world leader in hydrogen technology [47].

These national hydrogen strategies illustrate the growing recognition of hydrogen as a key enabler of a sustainable, low-carbon future. While the specific approaches and timelines vary across countries, there is a shared vision for developing hydrogen as a critical energy source for decarbonizing industries, transportation, and power systems. The collaborative nature of these strategies, both within and between nations, is expected to accelerate the global transition to a hydrogen economy, with key players working toward scaling production, advancing technologies, and building the necessary infrastructure for hydrogen's widespread adoption [48]. Figure 7.3 shows leading countries in hydrogen policy.

7.4.2 Policy Incentives for Hydrogen Development

The successful integration of hydrogen into the global energy landscape requires robust policy incentives that encourage the development and scaling of hydrogen technologies. Governments across the world are recognizing the pivotal role of hydrogen in achieving net-zero emissions targets and are implementing a variety of policy measures to promote its production, infrastructure, and adoption.

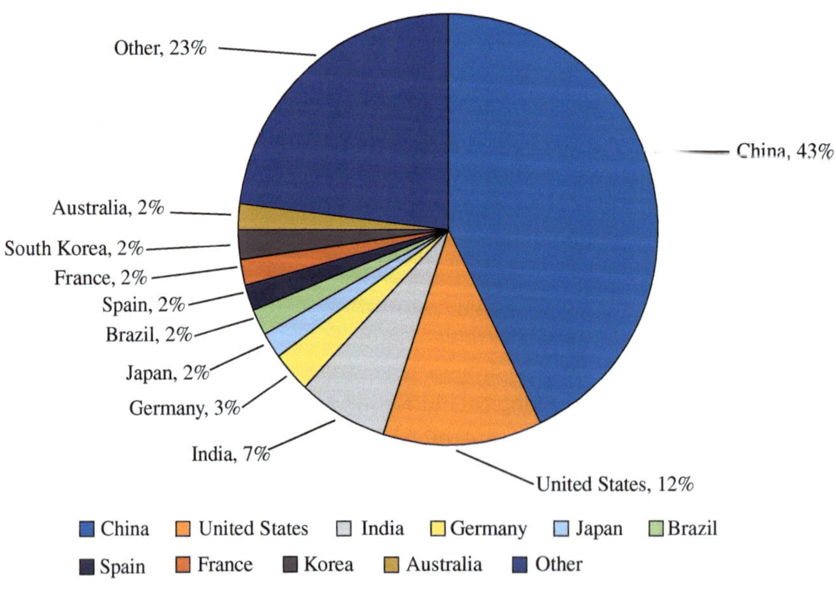

Figure 7.3 Infographic showcasing investment and policies from leading nations (EU, Japan, United States, Australia, South Korea). [49]/MDPI/CC BY 4.0.

These incentives include direct financial support, market-based mechanisms, and collaborative efforts between governments and industry stakeholders to accelerate innovation. The focus is on reducing costs, enhancing technological readiness, and creating an enabling environment for hydrogen to compete with traditional fossil fuels and other low-carbon energy alternatives.

Subsidies and grants for green hydrogen projects are among the most effective policy tools used to drive the development of hydrogen production technologies. As green hydrogen – produced from renewable energy sources like wind and solar – remains more expensive than conventional hydrogen production methods, these financial incentives help lower the cost of production and make green hydrogen more competitive in the market. Governments are providing direct funding to support research, development, and commercialization of green hydrogen projects, with particular emphasis on large-scale production facilities and technological innovations that can lower production costs over time. These subsidies and grants often target key areas, such as improving electrolyzer efficiency, expanding renewable energy capacity, and advancing hydrogen storage and distribution infrastructure. For instance, the EU's Clean Hydrogen Partnership, part of its Horizon Europe program, offers financial support for green hydrogen projects across the continent. In the United States, the DOE has launched several funding initiatives, including the Hydrogen Energy Earthshot initiative, which aims to reduce the cost of green hydrogen production by 80% over the next decade. These programs aim to catalyze the transition to clean hydrogen by making it more affordable and accessible.

Carbon pricing and emission reduction mechanisms play a critical role in creating a market-driven incentive for hydrogen adoption. By putting a price on carbon emissions, governments can internalize the environmental costs of fossil fuel use, making low-carbon alternatives like hydrogen more economically attractive. Carbon pricing mechanisms, such as carbon taxes or cap-and-trade systems, establish a financial incentive for industries to transition away from carbon-intensive technologies and adopt cleaner energy solutions like hydrogen. For example, the EU Emissions Trading System (EU ETS) has created a carbon market that charges emitters for each ton of CO_2 released into the atmosphere, thereby incentivizing the use of cleaner energy sources, including hydrogen. Similarly, countries like Canada and the UK have implemented carbon pricing policies to encourage the development of low-carbon technologies, including hydrogen production. These mechanisms help level the playing field between hydrogen and fossil fuels by incorporating the hidden environmental costs of the latter, which often go unaccounted for in conventional market models. Additionally, governments are integrating emission reduction targets with hydrogen adoption strategies, ensuring that hydrogen plays a key role in meeting long-term climate goals. As part of global commitments to the Paris Agreement, many countries are establishing emission reduction mechanisms that include hydrogen as a tool for decarbonizing industries such as steel, cement, and transportation.

Government-industry partnerships are essential to accelerate the commercialization of hydrogen technologies. Collaborative efforts between the public and private sectors help bridge the gap between research and large-scale deployment, allowing for the rapid scaling of hydrogen production, infrastructure, and end-use applications. These partnerships often involve joint funding arrangements, where governments provide financial support to de-risk early-stage investments, while industry stakeholders contribute expertise, technology, and commercial know-how. In many regions, innovation funding has been allocated to public-private partnerships aimed at advancing hydrogen technologies. Governments are establishing hydrogen R&D hubs, where industry players, research institutions, and startups can collaborate on the development of new technologies that drive down costs and improve performance. For example, the Hydrogen Council, a global initiative led by major energy and industrial companies, works in partnership with governments to foster innovation and secure investments in hydrogen infrastructure and production. The DOE has created several funding programs to support industry-led hydrogen projects, including the Hydrogen Hub initiative, which aims to establish regional hydrogen ecosystems to promote both production and demand for hydrogen. Similarly, the German government has partnered with private companies to develop hydrogen-powered transportation solutions and advance the deployment of HRS. These partnerships help leverage both public funding and private sector expertise to accelerate the commercialization of hydrogen technologies and ensure their wide-scale adoption [50]. Table 7.3 sums policy and regulatory support for hydrogen development in key regions.

7.4.3 Regulatory Standards and Safety Guidelines

As the hydrogen industry develops and scales globally, establishing rigorous regulatory standards and safety guidelines is paramount to ensuring the safe and sustainable production, transportation, storage, and use of hydrogen. Regulatory frameworks help manage risks, maintain public safety, and ensure that hydrogen technologies are implemented effectively and responsibly. These frameworks must address key concerns such as hydrogen purity, transport safety, environmental impacts, and the overall governance of hydrogen technologies. The integration of these standards is essential to gaining public trust, ensuring compliance with international regulations, and enabling the widespread adoption of hydrogen across various sectors. Regulatory standards for hydrogen span a broad range of issues, from technical specifications to safety protocols. Hydrogen purity and quality standards ensure that the hydrogen used in applications, such as fuel cells, meets specific requirements for efficiency and performance. Safety guidelines are crucial for handling hydrogen, given its flammability, volatility, and the need for specific handling protocols in production, storage, and transportation. Lastly, environmental impact assessments ensure that hydrogen production methods and infrastructure development align with sustainable practices, protecting ecosystems and minimizing environmental degradation. This section delves into the regulatory standards and safety guidelines that support the development of the hydrogen industry while prioritizing safety, efficiency, and environmental stewardship [51].

Table 7.3 Policy and regulatory support for hydrogen development in key regions.

Region/Country	Hydrogen strategy	Key policy incentives	Funding/Investment initiatives	2025 hydrogen production target	2025 hydrogen infrastructure goal
European Union	EU Hydrogen Strategy 2020	Green hydrogen funding, tax incentives	€100 billion investment in clean energy projects	10 million tons	1,000 HRS [44]
Japan	Hydrogen Roadmap 2020	Subsidies for hydrogen technologies, tax rebates	¥2.5 trillion investment over 10 years	3 million tons	160 refueling stations [43]
Germany	National Hydrogen Strategy 2020	Subsidies for production, R&D tax credits	€9 billion in public funding for green hydrogen	5 million tons	500 refueling stations [42]
United States	H2@Scale, Clean Hydrogen Program	Federal grants for hydrogen R&D, carbon tax credits	$8 billion in hydrogen infrastructure investment	6 million tons	250 refueling stations [41]
Australia	National Hydrogen Strategy 2021	Export-focused policy, investment incentives for low-cost production	A$300 million investment in green hydrogen infrastructure	1 million tons	100 refueling stations [46]

7.4.3.1 Hydrogen Purity and Quality Standards

Hydrogen purity and quality are essential factors in the effective application of hydrogen across various industries, particularly in sectors like fuel cells, industrial applications, and power generation. For example, hydrogen used in fuel cells for electric vehicles must meet strict purity levels to ensure maximum efficiency and prevent damage to fuel cell components. High-purity hydrogen ensures the optimal performance of the fuel cells and extends their lifespan, making it a crucial aspect of the hydrogen value chain. International standards for hydrogen purity, such as those set by the International Organization for Standardization (ISO) and the American Society for Testing and Materials, define the acceptable levels of impurities that hydrogen can contain for different applications. These standards specify acceptable levels of contaminants like oxygen, nitrogen, and moisture, which can significantly impact the performance of fuel cells or other hydrogen-based technologies. Additionally, the ISO 14687 standard outlines the purity requirements for hydrogen used in fuel cells for light-duty vehicles, establishing purity thresholds for different impurities. Hydrogen purity standards also play a vital role in industrial applications, where hydrogen is used in processes like refining, ammonia production, and metallurgy. The specific purity requirements for hydrogen in these industries are determined by the technical needs of the processes involved. For instance, in ammonia synthesis, hydrogen purity requirements are often higher, as even small impurities can compromise the efficiency of the process. Standardization of purity requirements across industries ensures that hydrogen meets the needs of diverse applications and enhances the overall reliability and performance of hydrogen technologies [52].

7.4.3.2 Safety Regulations for Production, Transport, and Storage

Given hydrogen's highly flammable nature and potential hazards in various stages of its lifecycle, safety regulations are critical in all areas of hydrogen production, transportation, and storage. Hydrogen is a highly reactive gas, and its properties require specific safety measures to mitigate risks such as leakage, explosions, and fires. Consequently, a comprehensive set of regulations must govern the handling of hydrogen across its value chain, from production through to end-use applications. **Production:** The production of hydrogen, especially through methods such as electrolysis and SMR, must adhere to stringent safety protocols to prevent accidents and ensure worker safety. In electrolysis, for example, safety protocols must prevent electrical hazards, while in reforming, safety measures are essential to address the risk of high-pressure systems and the handling of hazardous chemicals like natural gas. These safety regulations often focus on minimizing the risk of contamination, leakage, and system malfunctions. **Transport:** The transportation of hydrogen presents unique challenges due to the gas's low density and high flammability. Regulations for hydrogen transport address the safe compression and liquefaction of hydrogen, particularly when transported by pipeline, truck, or rail. For instance, safety protocols include requirements for reinforced tanks, pressure relief devices, and leak detection systems to prevent accidents during transportation. Regulations for hydrogen pipelines also involve strict guidelines to ensure the structural integrity of pipelines and to address risks of embrittlement, which could occur when hydrogen interacts with certain metals. **Storage:** Safe storage of hydrogen is another key aspect of hydrogen safety regulation. Various storage methods, such as high-pressure gas cylinders, liquid hydrogen tanks, and underground storage systems, require specific safety measures. Regulations are established for proper containment, venting systems, temperature control, and monitoring to ensure that hydrogen is stored safely and does not pose risks to surrounding areas. Additionally, the use of appropriate materials is crucial to prevent material degradation from hydrogen embrittlement, which can compromise the integrity of storage tanks and pipelines. International regulatory bodies such as the International Electrotechnical Commission and the National Fire Protection Association provide comprehensive safety standards for hydrogen production, storage, and transportation. The adoption of these safety guidelines is essential for minimizing the risk of accidents and ensuring that hydrogen systems are deployed with confidence [53].

7.4.3.3 Environmental Impact Assessments and Certification

Environmental impact assessments (EIAs) and certification processes are necessary to ensure that hydrogen production methods and related infrastructure development adhere to environmental protection standards. These assessments provide a comprehensive analysis of the potential environmental effects of hydrogen projects, evaluating issues such as land use, water consumption, emissions, and the potential impact on biodiversity. Hydrogen production, especially when derived from renewable energy sources, offers environmental benefits such as reduced greenhouse gas emissions compared to traditional fossil fuel-based methods. However, even green hydrogen projects must consider environmental concerns, such as the impact of large-scale renewable energy installations on local ecosystems or the potential for water scarcity in regions relying on electrolysis. Governments and regulatory bodies require developers to conduct EIAs for hydrogen production plants, refueling stations, and storage facilities to ensure that these projects do not negatively affect the environment. EIAs often include consultation with local communities, environmental organizations, and stakeholders to address concerns and mitigate adverse effects. Certification processes that align with international standards, such as those set by the Global Reporting Initiative or ISO 14001, ensure that hydrogen projects are conducted with sustainable practices that minimize their environmental footprint. In addition to EIAs, sustainability certifications, including carbon neutrality certifications, can enhance the credibility and marketability of hydrogen projects. These certifications verify that hydrogen production methods meet specific environmental performance criteria and align with broader climate and sustainability goals. The increasing emphasis on sustainable hydrogen production further reinforces the need for strong regulatory oversight and environmental responsibility [54].

7.5 Investment and Financing Strategies

As hydrogen technologies evolve from innovative concepts to commercial realities, substantial investment is required to drive their scaling and deployment across multiple sectors. Hydrogen represents a promising yet capital-intensive area of development, necessitating strategic financing to overcome challenges related to production, infrastructure, and market adoption. The role of investors, both public and private, is critical in accelerating the transition to a hydrogen economy by funding groundbreaking technologies, supporting infrastructure development, and mitigating financial risks associated with the early stages of hydrogen adoption. Investment in hydrogen technologies spans various stages, from R&D to large-scale infrastructure deployment. Financial strategies are multifaceted and require collaboration between private investors, governments, and international organizations. These investment frameworks are vital for ensuring that hydrogen can compete with established energy sources, become commercially viable, and support global energy transitions. The landscape for hydrogen investment is evolving rapidly, with various financing models emerging to support large-scale hydrogen projects, including public-private partnerships, green bonds, and venture capital for start-ups. However, while the hydrogen market presents vast opportunities, it also carries significant risks, particularly regarding the uncertain regulatory environment, fluctuating hydrogen prices, and technological challenges. Understanding the complexities of financing hydrogen technologies and managing investment risks is crucial for attracting sustained capital flow into the sector. This section delves into the current investment landscape for hydrogen technologies, explores financing models that have emerged to facilitate hydrogen deployment, and examines the associated risks and uncertainties that investors must navigate in this nascent but growing market [55].

7.5.1 Current Investment Landscape in Hydrogen Technologies

The investment landscape for hydrogen technologies has seen a significant transformation in recent years, with growing interest from both public and private sectors due to the increasing urgency of transitioning to clean energy solutions. Hydrogen is now recognized as a key player in decarbonizing industries, transport, and power generation, making it an attractive avenue for investment. However, the sector is still maturing, and substantial financial commitments are necessary to advance technologies, build infrastructure, and establish a viable hydrogen economy. Investment in hydrogen technologies is diverse, covering various aspects of the hydrogen value chain, including production, storage, transportation, and end-use applications like fuel cells. The current landscape involves a combination of government funding, venture capital, corporate investment, and institutional support. Many governments worldwide have introduced policies to incentivize private sector involvement in hydrogen development, often in the form of subsidies, tax credits, and green bonds aimed at funding clean energy projects. These public initiatives create a favorable investment climate, de-risking the financial commitment required for hydrogen technologies. In parallel, private investors and venture capital firms are increasingly recognizing the potential of hydrogen as a long-term investment opportunity. Companies specializing in electrolysis, fuel cells, and hydrogen-powered vehicles have attracted significant investments, including partnerships with energy giants and technology firms. Large corporations, particularly in the energy, automotive, and industrial sectors, are also ramping up their involvement by investing in hydrogen infrastructure and production facilities.

The role of multinational collaborations is also evident, as consortiums comprising governments, financial institutions, and private corporations come together to tackle the capital-intensive nature of hydrogen infrastructure. Such partnerships are essential to pool resources and expertise, ensuring the scalability of hydrogen projects. Major initiatives, like the EU's Green Deal and Japan's Basic Hydrogen Strategy, exemplify how governments can lead the charge by aligning funding efforts with national and international decarbonization goals. Moreover, the hydrogen investment landscape is expanding into emerging markets, where governments are keen to leverage hydrogen as a tool for energy security and economic growth. Countries with abundant renewable energy resources, such as Australia, the Middle East, and parts of Africa, are positioning themselves as key hydrogen exporters, making it a highly competitive market. These regions are attracting international investors who are interested in tapping into the export potential of green hydrogen, thus adding another layer of complexity to the global hydrogen market. While the financial backing for hydrogen technologies has increased, challenges remain. The cost of hydrogen production, particularly green hydrogen derived from renewable energy sources, remains high, which hampers its immediate competitiveness with traditional fossil fuels. Investors are also cautious about the regulatory uncertainties, such as unclear policy guidelines and inconsistent standards across regions, which could affect returns on investment. Despite these challenges, the momentum for hydrogen investment is growing, as countries, industries, and financial institutions recognize the long-term potential of hydrogen to play a central role in the global energy transition. This evolving landscape highlights that hydrogen is not just an energy resource; it is becoming a cornerstone of future energy systems, with significant investment flowing into the sector from multiple stakeholders. As hydrogen technologies continue to develop, the investment climate is expected to remain dynamic, with new players entering the market and new funding mechanisms emerging to meet the ever-growing demand for clean energy solutions [56].

7.5.2 Financing Models for Large-scale Hydrogen Deployment

Financing large-scale hydrogen deployment requires innovative and adaptable financing models to support the capital-intensive nature of infrastructure development, technological innovation, and commercial scaling. Hydrogen technologies, while crucial for achieving long-term decarbonization goals, currently face significant barriers to widespread implementation due to high upfront costs, uncertain returns, and the need for a coordinated global effort. To overcome these barriers, various financing models have emerged, drawing from a combination of public, private, and hybrid funding sources. Each model has its advantages and challenges, and their successful application depends on a country's regulatory framework, financial environment, and the stage of development of the hydrogen industry [57].

7.5.2.1 Public-Private Partnerships (PPP) in Hydrogen Infrastructure

PPP have become an essential tool for financing large-scale hydrogen infrastructure projects. These partnerships bring together government entities and private sector companies to pool financial resources, share risks, and capitalize on synergies between public policy goals and private sector efficiency and innovation. PPPs are particularly valuable in sectors like hydrogen, where the capital costs are high, and the risk-return profile may be too uncertain for private investors to undertake without some form of public support. Governments are typically instrumental in structuring these partnerships, providing the regulatory clarity, policy incentives, and, in some cases, initial funding necessary to kickstart large-scale hydrogen projects. In return, the private sector offers expertise, technology development, and operational efficiencies. This collaboration can be seen in many hydrogen projects, from green hydrogen production facilities to the development of hydrogen transport and refueling infrastructure.

One of the most significant benefits of PPPs in hydrogen is the sharing of risks associated with infrastructure development and the market uncertainty around hydrogen's long-term commercial viability. Governments can de-risk investments by offering financial support through subsidies, tax credits, or guaranteed purchase agreements, while private companies bring capital investment and technological innovation. For example, countries like Germany, the United Kingdom, and Japan have initiated several hydrogen infrastructure projects through PPP, in which private companies and energy providers work alongside governments to build hydrogen hubs, storage facilities, and refueling stations. Additionally, PPPs in hydrogen development can accelerate innovation in areas like electrolysis and fuel cell technologies, as the combination of government backing and private sector competition can spur faster advancements. Governments, with their long-term sustainability mandates, also have the capability to push for the standardization of hydrogen technologies and the creation of global hydrogen supply chains, further boosting investor confidence. However, the success of PPPs depends heavily on clear and supportive policy frameworks. Governments need to ensure stable and long-term policy commitments to avoid discouraging private investors who may be hesitant about fluctuating regulatory environments. Furthermore, transparency in financial arrangements and governance is critical for the success of these partnerships, as well as the willingness of private entities to share the operational risks with public stakeholders [58].

7.5.2.2 Green Bonds and Sustainable Finance Mechanisms

Green bonds have become an increasingly popular financial instrument to fund renewable energy and clean technology projects, including hydrogen. Green bonds are debt securities issued by corporations, governments, or financial institutions, where the raised capital is earmarked exclusively for projects with environmental benefits. Given hydrogen's potential as a key enabler of a low-carbon economy, green bonds are an ideal funding mechanism to finance large-scale hydrogen infrastructure and production. The issuance of green bonds allows governments and corporations to tap into a growing pool of environmentally conscious investors, who are specifically looking to fund projects that align with sustainability and climate action goals. These bonds are often structured with attractive interest rates and long-term investment horizons, making them appealing to both institutional and retail investors. As the hydrogen sector becomes more established, green bonds offer a way to finance the scaling up of hydrogen infrastructure, such as production facilities, storage, and transportation networks. The popularity of green bonds has also led to the development of more sophisticated sustainable finance mechanisms that incorporate hydrogen into broader environmental, social, and governance investment strategies. Financial institutions are increasingly aligning their portfolios with the Paris Agreement and other international climate targets, and hydrogen projects offer a promising opportunity to meet these goals. Investors are particularly interested in green bonds tied to hydrogen projects due to the growing demand for hydrogen as a clean fuel, the potential for hydrogen to be a large-scale decarbonization tool, and the global push for reducing greenhouse gas emissions.

The EU has been a pioneer in utilizing green bonds for hydrogen development, with initiatives such as the EU Green Bond Standard, which provides guidelines for the issuance of bonds linked to clean hydrogen projects. Similarly, countries like the United States and Japan are also encouraging the use of green bonds to fund hydrogen production facilities and other related infrastructure. However, challenges remain in the green bond market. Investors need reassurance that hydrogen projects will deliver the desired environmental outcomes and adhere to strict sustainability criteria. There is a risk that certain hydrogen projects – particularly those linked to gray hydrogen (hydrogen produced from fossil fuels) – may be classified under green bonds if clear sustainability metrics are not enforced. Regulatory bodies must ensure that green bonds are genuinely financing projects that contribute to carbon neutrality, thereby avoiding greenwashing and ensuring the credibility of the market [55].

7.5.2.3 Venture Capital and Private Equity in Hydrogen Startups

Venture capital (VC) and private equity (PE) investments play a critical role in the early-stage development of hydrogen technologies, including innovative solutions for hydrogen production, storage, and distribution. These forms of financing are often used to fund startup companies with high growth potential in the hydrogen sector. For early-stage hydrogen ventures, VC and PE investments can provide not only the financial backing necessary to commercialize groundbreaking technologies but also the strategic guidance and industry connections required to scale quickly. VC funding is typically focused on startups with promising new technologies or business models that have the potential to disrupt traditional energy markets. In the hydrogen sector, this might include innovations in electrolyzers for green hydrogen production, new methods for storing hydrogen at lower costs, or fuel cell technologies that can increase the efficiency of hydrogen usage in various applications. As hydrogen continues to gain recognition as a promising renewable energy source, venture capitalists are looking to invest in companies that can deliver cost-effective and scalable hydrogen solutions in a competitive market.

PE, on the other hand, generally comes into play when a company has moved past the initial startup phase and requires significant capital for expansion. For hydrogen, this might include the construction of hydrogen production plants or the development of large-scale refueling stations. PE firms often invest in these more mature companies with established business models, and their involvement brings not only funding but also operational expertise to accelerate growth. In recent years, both VC and PE investments in hydrogen startups have seen a significant increase. Leading firms have recognized the growing global hydrogen market, particularly in sectors like transport, industry, and power generation, which are in desperate need of decarbonization solutions. As governments worldwide intensify their commitments to hydrogen as part of their clean energy agendas, VC and PE investors are eager to fund innovative hydrogen startups poised to benefit from these policy shifts.

A notable example of successful VC-backed innovation in the hydrogen space is the development of hydrogen-powered vehicles and new electrolysis technologies. Companies like Plug Power, Ballard Power Systems, and Nel Hydrogen, which started with venture funding, have scaled significantly over the years and are now leaders in hydrogen fuel cell technologies and hydrogen production systems. Despite the increased interest from VC and PE investors, the hydrogen sector still carries risks. The technology is evolving rapidly, and there is uncertainty regarding which hydrogen production methods and business models will ultimately prove to be commercially viable. Additionally, investors may face challenges around regulatory hurdles, which can delay product development and scaling efforts. However, the potential of hydrogen to transform global energy systems has made it an attractive area for risk-tolerant investors, and the sector is poised to attract even more VC and PE investment as the hydrogen economy matures [59].

7.5.3 Risks and Uncertainties in Hydrogen Investment

Investing in hydrogen technologies, while promising significant environmental and economic benefits, comes with a set of risks and uncertainties that can complicate financial decision-making. These risks stem from a variety of factors, including technological development, market volatility, policy shifts, and the emerging nature of the hydrogen economy. As the global transition to hydrogen-based solutions gathers momentum, investors need to carefully evaluate both the opportunities and the challenges associated with hydrogen investments. This section highlights the key risks and uncertainties that investors must consider when financing hydrogen technologies.

One of the primary risks associated with hydrogen investment is technological uncertainty. While hydrogen has been a key focus of R&D for decades, several of the technologies required for large-scale hydrogen production, storage, transportation, and utilization are still in the early stages of commercialization. For instance, while electrolyzers for green hydrogen production are improving, they remain expensive, and large-scale deployment is not yet cost-competitive with traditional hydrogen production methods like SMR. Moreover, fuel cell technologies, which are essential for hydrogen's use in transportation and stationary power generation, face challenges in terms of efficiency, durability, and cost. Investors must recognize that hydrogen technologies are still evolving, and breakthroughs in efficiency, cost reduction, or new storage solutions could alter the investment landscape significantly. However, there is also the risk that certain technologies may fail to meet commercial expectations or that some hydrogen methods may ultimately prove to be less viable than anticipated. As such, investors must assess the technological maturity of each project and be prepared for the possibility of delays or technological underperformance.

The demand for hydrogen is largely tied to the broader energy transition, and market fluctuations in related sectors – such as fossil fuels, renewable energy, and electric vehicles – can directly impact the hydrogen market. For instance, the growth of hydrogen-powered vehicles and the industrial demand for hydrogen will depend on the pace of decarbonization in these sectors and the availability of alternative technologies, such as battery electric vehicles and electrified industrial processes. Any market shifts, including the wider adoption of competing energy solutions or unforeseen slowdowns in the hydrogen economy, could negatively affect the hydrogen market. The price of hydrogen is another key factor. Although the cost of green hydrogen is expected to decrease with technological advancements, price volatility in feedstock (e.g., electricity for electrolysis or natural gas for SMR) and fluctuating energy prices could pose challenges for investors. Price fluctuations in hydrogen could impact its competitiveness compared to other fuels, such as natural gas, and make it more difficult to project long-term returns on investment.

Hydrogen investment is highly dependent on supportive regulatory frameworks and policy incentives. Governments around the world have begun to set ambitious targets for hydrogen development, offering various incentives and subsidies to stimulate growth. However, these policies are not always stable or consistent. Sudden changes in policy, such as the rollback of subsidies, changes in carbon pricing, or the reduction of renewable energy incentives, can have a significant impact on the profitability of hydrogen projects. Investors must be prepared for the risk of policy uncertainty, especially in regions where political shifts could result in changes to national energy strategies. Furthermore, regulatory frameworks around hydrogen infrastructure, production methods, and safety standards are still developing, and countries vary widely in their approach. While some regions, such as the EU and Japan, have well-defined hydrogen roadmaps, others are still working to create the necessary regulatory frameworks. This uneven regulatory landscape can create challenges for cross-border investments in hydrogen infrastructure and complicate global market expansion.

Hydrogen infrastructure and projects require substantial capital investment, particularly in the early stages of development. The construction of large-scale hydrogen production plants, storage facilities, refueling stations, and distribution networks involves significant upfront costs. These capital-intensive projects often require long-term financing and may take many years to reach profitability, especially considering that hydrogen technologies are still maturing. For investors, this presents the challenge of securing sufficient capital while managing the risks associated with delayed or uncertain returns. The long payback periods of hydrogen projects also mean that investors must have the financial capacity to weather the initial phases of development. Smaller investors, in particular, may be deterred by the high initial costs and the risk of prolonged returns on investment. A crucial risk for hydrogen deployment is the development of the necessary infrastructure and supply chain. Hydrogen production, transportation, and storage require a comprehensive infrastructure network, which is currently lacking in many regions. Building this infrastructure presents both logistical and financial challenges. There are risks associated with scaling up hydrogen production in sufficient quantities, ensuring the secure transport of hydrogen across regions, and building storage solutions that can handle the fluctuating supply and demand of hydrogen. Moreover, the hydrogen supply chain is highly dependent on key materials, such as metals used in electrolysis and fuel cells, and any disruptions in the supply of these materials can delay projects and increase costs. These infrastructure and supply chain risks may not be immediately apparent but can have a significant impact on the successful deployment of hydrogen technologies.

Finally, the potential for competition from other emerging energy solutions presents an additional risk to hydrogen investment. While hydrogen offers unique advantages, it is not the only solution to decarbonizing sectors like transportation, industry, and power generation. Battery electric vehicles (BEVs), for example, are rapidly gaining market share in the automotive sector, potentially limiting hydrogen's role in this area. Similarly, advancements in other low-carbon technologies, such as direct electrification, CCS, or advanced biofuels, could reduce the demand for hydrogen in specific applications. As such, investors must consider the broader energy landscape and evaluate whether hydrogen technologies will be able to compete effectively with other solutions or if they will face challenges in carving out a significant market share. The success of hydrogen could depend on the degree to which it complements, rather than competes directly with, these alternative technologies [60].

7.6 Emerging Hydrogen Economies: Growth in India, Australia, and South Korea

As hydrogen emerges as a key energy carrier in the global transition towards decarbonization, India, Australia, and South Korea are positioning themselves as front-runners in the hydrogen economy. These three nations are leveraging their unique strengths – India's vast renewable energy potential, Australia's abundant natural resources and export-oriented hydrogen strategy, and South Korea's focus on hydrogen-powered transportation and fuel cell technology – to build sustainable hydrogen markets. Strategic policies, large-scale investments, and international collaborations are driving their progress toward establishing competitive hydrogen ecosystems.

7.6.1 India: National Hydrogen Mission and Domestic Market Expansion

India's hydrogen economy is gaining momentum under the government-led National Hydrogen Mission, which was launched in 2021 to promote green hydrogen production and position India as a global leader in hydrogen energy. The initiative aligns with India's broader commitment to achieving net-zero emissions by 2070 and reducing reliance on fossil fuel imports. India is uniquely positioned to scale green hydrogen production due to its abundant solar and wind energy resources. The country has one of the world's lowest renewable energy costs, making electrolysis-powered hydrogen production economically viable in the long term. The government has set a target of 5 million metric tons of green hydrogen production annually by 2030, with the aim of cutting production costs to below \$1 per kilogram through economies of scale and policy incentives.

To encourage private sector participation, India has introduced Production-Linked Incentives for electrolyzer manufacturing and green hydrogen production. Major Indian corporations, including Reliance Industries, Adani Group, and Indian Oil Corporation, are heavily investing in hydrogen infrastructure, aiming to develop large-scale hydrogen production hubs.

7.6.1.1 Industrial Applications and Hydrogen Adoption

Hydrogen is expected to play a crucial role in decarbonizing India's hard-to-abate industries, such as steel, cement, and fertilizer production. The Steel Authority of India (SAIL) and Tata Steel have begun pilot projects to replace coal with hydrogen in steelmaking, leveraging hydrogen's potential to reduce carbon emissions in heavy industries. Similarly, Indian fertilizer companies are exploring ammonia-based hydrogen for cleaner agricultural production.

7.6.1.2 Policy Framework and Government Initiatives

The Indian government has introduced multiple policy measures to accelerate hydrogen adoption:

- Hydrogen Purchase Obligations: Mandating industries to source a portion of their hydrogen consumption from renewable sources.
- Tax Incentives and Subsidies: Financial incentives for electrolyzer manufacturers and hydrogen production projects.
- Hydrogen Corridors: Establishing dedicated zones for hydrogen infrastructure and industrial usage.

India is also seeking international partnerships, with collaborations with Japan, the EU, and the United States for hydrogen technology transfer and infrastructure development [61, 62].

7.6.2 Australia: Export-oriented Hydrogen Strategy and Global Partnerships

Australia has emerged as a major player in the hydrogen economy, leveraging its vast renewable energy resources and existing expertise in energy exports. The country's National Hydrogen Strategy (2019) aims to position Australia as a leading hydrogen exporter, particularly to markets in Asia, such as Japan and South Korea.

Australia has a competitive advantage in green hydrogen due to its extensive solar and wind resources, particularly in Western and South Australia. The government has set ambitious goals, including producing up to 10 million tons of hydrogen annually by 2030 and driving down production costs through large-scale investment.

Key hydrogen projects in Australia include:

- The Asian Renewable Energy Hub in Western Australia, which plans to generate up to 26 GW of renewable energy for hydrogen production.
- The H2Perth project, a large-scale hydrogen and ammonia production facility targeting export markets.
- Sunrise Hydrogen Hub, which integrates solar and wind power with electrolysis technology.

7.6.2.1 Export Markets and Global Hydrogen Trade

Given its geographical proximity and strong trade relationships, Australia is prioritizing hydrogen exports to Japan, South Korea, and Europe. The country has already signed major hydrogen agreements with Japan's Kawasaki Heavy Industries and JERA, as well as South Korea's Hyundai and POSCO. The Australia-Japan Hydrogen Supply Chain (HESC) project is a landmark initiative demonstrating liquefied hydrogen transportation.

7.6.2.2 Government Support and Investment

The Australian government has committed over $500 million to hydrogen infrastructure projects through the Clean Energy Finance Corporation and the Australian Renewable Energy Agency. Policy measures include:

- Hydrogen Hubs: Dedicated industrial zones for hydrogen production, storage, and export.
- Tax Incentives: Subsidies for hydrogen projects and electrolyzer deployment.
- Research and Development: Investments in hydrogen innovations, including ammonia and LOHCs for efficient transportation [63, 64].

7.6.3 South Korea: Hydrogen Economy Roadmap and Fuel Cell Leadership

South Korea has taken an aggressive approach to hydrogen adoption, integrating hydrogen into its transportation, power generation, and industrial sectors. The government's Hydrogen Economy Roadmap (2019) aims to develop a fully integrated hydrogen ecosystem, with a focus on fuel cell technology and hydrogen-powered transportation.

South Korea is a leader in FCEV technology, with companies such as Hyundai Motor Company investing heavily in hydrogen-powered mobility solutions. The Hyundai NEXO, South Korea's flagship hydrogen fuel cell car, is among the world's best-selling hydrogen vehicles.

The government aims to have:

- 200,000 hydrogen vehicles on the road by 2025 and over 6 million by 2040.
- 1,200 HRS nationwide by 2040.
- Fuel cell-powered trains, ships, and aircraft integrated into transportation networks.

7.6.3.1 Hydrogen-based Power Generation

South Korea is investing in hydrogen fuel cells for electricity generation, aiming to replace coal-fired power plants with hydrogen-based alternatives. The country is home to some of the world's largest fuel cell power plants, including the Gyeonggi Green Energy fuel cell park.

7.6.3.2 Government Policies and Investments

To accelerate hydrogen adoption, South Korea has implemented:

- Subsidies for hydrogen vehicles and refueling infrastructure.
- Carbon pricing and tax incentives for hydrogen-based industries.

- International partnerships, including agreements with Saudi Arabia, Australia, and the EU for hydrogen imports.

South Korea is also exploring blue hydrogen production, utilizing CCS to produce hydrogen from natural gas while minimizing emissions.

India, Australia, and South Korea are rapidly emerging as global hydrogen leaders, each leveraging distinct strengths to build competitive hydrogen markets. While India focuses on domestic production and industrial decarbonization, Australia is driving hydrogen exports, and South Korea is pioneering hydrogen-powered transportation and fuel cell technology. Their efforts are reshaping the global hydrogen economy, demonstrating the transformative potential of hydrogen in energy security, industrial sustainability, and climate action. Through strategic investments, policy frameworks, and international collaborations, these nations are not only advancing their own hydrogen ecosystems but also contributing to the broader transition toward a cleaner, hydrogen-powered world [65, 66].

7.7 Challenges and Barriers to Hydrogen Commercialization

The commercialization of hydrogen as a sustainable energy solution presents significant opportunities, but it also comes with a series of challenges and barriers that need to be overcome. These challenges span technological, economic, infrastructural, and social dimensions, each influencing the speed and extent of hydrogen's adoption across industries. While hydrogen technologies have made remarkable advancements over the years, several technical hurdles remain that hinder the widespread adoption of hydrogen solutions. One of the key challenges is the cost and efficiency of hydrogen production methods. Electrolysis, which is used to produce green hydrogen, is still relatively expensive compared to conventional methods like SMR, especially when powered by renewable energy. Although the costs of electrolysis are expected to decrease with advancements in technology and economies of scale, it remains more expensive than the hydrogen produced from natural gas. Another technological challenge lies in hydrogen storage and transportation. Hydrogen, being the lightest and smallest molecule, is notoriously difficult to store and transport efficiently. Storing hydrogen at a sufficient density requires high-pressure tanks, cryogenic storage, or advanced materials-based solutions, each with its own set of challenges regarding cost, safety, and efficiency. Moreover, developing infrastructure capable of safely and efficiently distributing hydrogen is a significant technical hurdle. The development of pipelines, refueling stations, and distribution systems is still in its infancy and requires substantial investment and innovation to ensure reliability and safety. Furthermore, hydrogen fuel cells, a key application for hydrogen in transportation and stationary power generation, still face challenges related to efficiency, durability, and cost-effectiveness. The mass adoption of fuel cells is dependent on these technological hurdles being addressed to make hydrogen-powered vehicles and equipment competitive with battery electric alternatives [67].

One of the primary economic barriers to hydrogen commercialization is the high initial investment required for infrastructure development. Hydrogen production facilities, storage systems, and distribution networks involve substantial upfront capital costs. This financial burden is further compounded by the long payback periods associated with large-scale hydrogen projects, which can deter private investment without strong policy support or guaranteed returns. The cost of producing hydrogen also remains higher than fossil-based alternatives, particularly in the case of green hydrogen, which requires renewable energy sources for electrolysis. Additionally, the economic feasibility of hydrogen is highly dependent on the price of renewable energy. While hydrogen can be produced from a variety of feedstocks, the most sustainable form of hydrogen – green hydrogen – requires access to large amounts of low-cost renewable energy. The fluctuation in the price of renewable electricity can impact the competitiveness of hydrogen against other energy solutions, such as batteries or natural gas. At the same time, market competition from other low-carbon technologies, such as BEVs and CCS, presents economic pressure on hydrogen. As battery prices continue to decrease and BEVs become more widely adopted, the demand for hydrogen in certain sectors, such as transportation, could be undermined. Additionally, sectors like heavy industry may continue to rely on fossil fuels unless hydrogen becomes more economically viable, thereby delaying widespread adoption [26].

The infrastructure required to support a hydrogen economy is another significant barrier. A robust and expansive infrastructure for hydrogen production, storage, transport, and refueling is essential for large-scale commercialization. However, this infrastructure is still underdeveloped in many regions, especially compared to more established energy sectors like oil and natural gas. Developing this infrastructure requires massive investment in pipelines, storage facilities, and refueling stations, which presents a logistical and financial challenge for governments and private stakeholders. The lack of a unified and standardized global hydrogen infrastructure makes it even more difficult to scale up hydrogen technologies. Different regions may have different regulatory standards, safety protocols, and infrastructure requirements, which can slow the pace of international collaboration and trade. In addition, the lack of integration with existing fuel and energy networks in some countries may present operational challenges in scaling up hydrogen use in sectors like transportation or industry [68].

Government policies and regulations play a critical role in shaping the hydrogen market, yet there remains a lack of coordinated and harmonized regulations at the global level. Inconsistent policies across different countries, as well as uncertainty regarding future regulations, create challenges for both investors and developers of hydrogen projects. While some regions, such as the EU, Japan, and South Korea, have well-established hydrogen roadmaps, other countries are still in the process of formulating their strategies. Without clear, stable policies and long-term commitments, investors may be hesitant to allocate resources to hydrogen projects. In particular, hydrogen subsidies, carbon pricing mechanisms, and other policy incentives are crucial to reducing the cost of hydrogen production and making it competitive with fossil fuels. However, policy risks – such as the potential rollback of subsidies or the imposition of restrictive regulations – remain a concern. Additionally, aligning hydrogen policy with broader energy and climate objectives is critical to ensuring that hydrogen plays an effective role in decarbonizing various sectors [69].

Public acceptance and understanding of hydrogen technologies also represent a barrier to commercialization. The general public may have limited knowledge of hydrogen's benefits and its potential role in decarbonizing energy systems. Misconceptions or safety concerns related to hydrogen – such as its flammability and its use in fuel cells – could also slow public acceptance and hinder the adoption of hydrogen technologies. Moreover, the social license to operate for hydrogen production facilities, refueling stations, and infrastructure projects is a critical factor for successful commercialization. Local communities may express resistance to large-scale hydrogen projects, particularly if they perceive potential risks to safety or the environment. Gaining public trust and engaging in transparent dialogue with communities will be crucial for ensuring the smooth implementation of hydrogen projects [70].

7.8 Case Studies: Leading Hydrogen Initiatives

As hydrogen technologies continue to mature, various countries and regions have begun implementing strategies and initiatives to harness hydrogen's potential as a key pillar in their energy transition. These initiatives range from national roadmaps to localized projects aimed at decarbonizing specific sectors such as transportation, industry, and power generation. The following case studies highlight some of the leading hydrogen initiatives around the world, showcasing different approaches to integrating hydrogen into national and global energy systems.

7.8.1 Japan's Hydrogen Society: From Strategy to Implementation

Japan is one of the most proactive countries in embracing hydrogen as a cornerstone of its future energy landscape. The country has long recognized the strategic importance of hydrogen, not only as a tool for reducing carbon emissions but also as a means to enhance energy security in the face of limited domestic energy resources. Japan's hydrogen strategy is founded on the vision of creating a "Hydrogen Society," where hydrogen is a primary energy carrier in multiple sectors. The Japanese government laid the foundation for its hydrogen strategy with the release of its Basic Hydrogen Strategy in 2017, outlining the country's plans to create a hydrogen-based economy by 2050. The strategy emphasizes hydrogen's role in decarbonizing energy-intensive sectors such as transportation, industry, and power generation. Japan's strategy is based on four key pillars: promoting hydrogen production through diverse methods, expanding hydrogen infrastructure, scaling up fuel cell usage, and fostering international cooperation in hydrogen technology development. Japan's hydrogen production approach is diverse, incorporating both domestic and imported hydrogen. The country is investing in blue hydrogen produced from natural gas using CCS technology, as well as green hydrogen generated through renewable-powered electrolysis.

In addition to boosting domestic production, Japan is pursuing hydrogen imports through global supply chains, focusing on countries with abundant renewable energy resources, such as Australia and the Middle East [71].

A crucial aspect of Japan's hydrogen strategy is the development of hydrogen infrastructure. Japan is a global leader in the deployment of HRS, with the government providing financial incentives to encourage the establishment of refueling networks for hydrogen-powered vehicles. By 2020, Japan had over 100 HRS in operation, with plans for more stations to be installed in the coming years. Moreover, Japan is pushing for the widespread adoption of hydrogen fuel cells in transportation, particularly in buses, trucks, and passenger vehicles, with automakers like Toyota, Honda, and Nissan playing key roles in commercializing hydrogen fuel cell vehicles (FCVs). In addition to its domestic hydrogen initiatives, Japan is also investing in international partnerships to strengthen its hydrogen economy. One of the most notable collaborations is with Australia on the development of a large-scale hydrogen export supply chain. Japan's efforts to create a hydrogen society demonstrate a comprehensive, multifaceted approach, with an emphasis on technological development, infrastructure expansion, and international collaboration [72].

7.8.2 Germany's National Hydrogen Strategy: Lessons from Europe's Energy Transition

Germany, as a leader in Europe's energy transition, has developed a robust hydrogen strategy aimed at decarbonizing its economy while leveraging its industrial strengths. Germany's National Hydrogen Strategy, released in 2020, focuses on hydrogen as a key enabler of decarbonization across several sectors, including transportation, industrial production, and energy. This strategy aligns with Germany's broader climate goals of achieving carbon neutrality by 2050 and meeting the EU's Green Deal objectives. A distinguishing feature of Germany's strategy is its focus on the industrial use of hydrogen, particularly in sectors like steel, chemicals, and refining, where electrification is difficult. Hydrogen is seen as an essential part of Germany's efforts to transition its heavy industries away from fossil fuels. The strategy includes substantial investments in developing green hydrogen, primarily produced through renewable-powered electrolysis, and the scaling up of hydrogen infrastructure to support industrial and transport applications. Germany has committed €9 billion to its National Hydrogen Strategy, with significant funds dedicated to R&D of hydrogen technologies, including production, storage, and transportation. The strategy also supports PPPs, encouraging collaboration between government agencies, energy providers, and industrial stakeholders to build a sustainable hydrogen economy. Moreover, Germany is actively engaging in the European Hydrogen Backbone initiative, which aims to create a cross-border hydrogen infrastructure connecting European countries, promoting hydrogen trade, and ensuring security of supply [73].

Germany's approach also prioritizes international cooperation, particularly with countries like Norway, the Netherlands, and Spain. For example, Germany has developed a strong partnership with Norway for the production and transportation of hydrogen derived from renewable energy sources. These international initiatives help Germany diversify its hydrogen supply sources, contributing to a more resilient energy system. The key lesson from Germany's strategy is the integration of hydrogen into a broader energy transition that includes the decarbonization of multiple sectors. By focusing on the industrial application of hydrogen, Germany aims to address some of the most challenging aspects of its climate goals while strengthening its position as a global leader in hydrogen technology [74].

7.8.3 California's Hydrogen Roadmap: Decarbonizing Transportation and Industry

California, a global leader in clean energy and environmental policy, has made significant strides in the development and adoption of hydrogen technologies, particularly in the transportation sector. As part of its hydrogen roadmap, the state aims to accelerate the adoption of hydrogen fuel cells to decarbonize both light-duty and heavy-duty vehicles, as well as support other industrial applications. California's efforts reflect the state's commitment to achieving carbon neutrality by 2045 and reducing greenhouse gas emissions from its transportation sector. The California Hydrogen Roadmap, first published in 2016, identifies hydrogen as a critical solution for decarbonizing the state's transportation network, including zero-emission vehicles such as hydrogen FCEVs. The roadmap outlines a set of actions to address barriers to hydrogen adoption, such as infrastructure development, market creation, and cost reduction. It includes plans to expand California's network of HRS from the initial 40 locations to over 100 by 2030, providing greater access to FCEVs across the state. A key component of California's roadmap is the focus on heavy-duty vehicles, including buses, trucks, and rail, where BEVs face limitations in terms of range, refueling time, and payload capacity. California's commitment to hydrogen fuel cell trucks has led to numerous partnerships between the state, automakers like Toyota and Nikola, and infrastructure developers to deploy hydrogen-powered trucks on the state's highways [75].

In addition to transportation, California's hydrogen roadmap also emphasizes the industrial use of hydrogen in sectors like refining, cement, and steel production, where hydrogen can replace fossil fuels and reduce carbon emissions. The state has already seen pilot projects exploring the use of green hydrogen produced from renewable energy sources, with plans for further scaling. California is leveraging state and federal incentives to accelerate hydrogen adoption, including grants, tax credits, and subsidies for hydrogen infrastructure development and fuel cell vehicles. The state is also working closely with other US states and countries to build an integrated hydrogen supply chain. The success of California's hydrogen roadmap hinges on the continued development of its hydrogen infrastructure, the reduction of hydrogen costs, and the adoption of fuel cell vehicles across multiple sectors. California's experience offers valuable insights into how subnational governments can play a pivotal role in advancing hydrogen technology and promoting its widespread adoption [76].

7.8.4 Australia's Green Hydrogen Export Plan: A Model for Global Trade

Australia has positioned itself as a global leader in green hydrogen production, capitalizing on its vast renewable energy resources, including solar, wind, and hydroelectric power, to produce green hydrogen for both domestic use and export. The country's Green Hydrogen Export Plan envisions a future where Australia becomes a major exporter of hydrogen to global markets, particularly in Asia, where demand for clean energy is growing rapidly. Australia's green hydrogen strategy is built on the idea that the country's abundant renewable energy resources make it an ideal location for producing cost-competitive hydrogen. By using renewable-powered electrolysis, Australia plans to produce hydrogen at scale and export it to countries like Japan, South Korea, and China, where hydrogen demand is expected to surge in the coming decades. The export plan aims to establish a reliable, low-cost hydrogen supply chain, supported by partnerships between government agencies, energy providers, and international stakeholders. The Asia-Pacific region is a key target for Australia's hydrogen exports, as countries like Japan and South Korea have already established ambitious hydrogen roadmaps and are looking for reliable suppliers. Australia's ability to meet this demand is underpinned by the development of large-scale hydrogen production facilities, storage terminals, and specialized shipping infrastructure. These developments will allow Australia to produce green ammonia, a carrier for hydrogen, for export and transportation to global markets [77].

Australia has already secured partnerships with several countries, including Japan, where the Japan-Australia HESC project is underway. This project involves the transportation of hydrogen in the form of liquid ammonia from Australia to Japan, marking one of the first major efforts to develop an international hydrogen supply chain. The long-term goal of Australia's green hydrogen export plan is to establish hydrogen as a key export commodity that contributes to the country's economy, reduces global carbon emissions, and enhances regional energy security. With its vast renewable energy resources, Australia is well-positioned to lead the way in the development of an international hydrogen market, offering a model for other nations to follow [78].

These case studies of hydrogen initiatives from Japan, Germany, California, and Australia provide valuable insights into the diverse strategies and approaches countries are adopting to unlock hydrogen's potential. Each country's initiative reflects its unique energy needs, resources, and policy landscapes, but all share a common goal of harnessing hydrogen to accelerate the transition to a decarbonized global economy. By examining these successful efforts, policymakers, businesses, and researchers can draw inspiration for scaling up hydrogen technologies, refining regulatory frameworks, and fostering international collaboration in the development of hydrogen as a key energy resource for the future [79].

7.9 The Future of Hydrogen Economics and Policy

As the world seeks to transition to cleaner energy systems, hydrogen stands out as a promising solution to decarbonize multiple sectors. However, for hydrogen to fulfill its potential as a central component of global energy systems, significant advances in both economics and policy are needed. The future of hydrogen will depend on technological innovations, market developments, and strategic policy frameworks that enable efficient, large-scale production, distribution, and consumption. This section explores key trends and projections that will shape the future of hydrogen economics and policy, providing insights into the trajectory of the hydrogen economy from 2030 to 2050.

7.9.1 Global Hydrogen Market Projections (2030–2050)

The global hydrogen market is expected to experience significant growth over the next few decades, driven by the increasing demand for low-carbon energy sources and the expansion of hydrogen-based technologies. According to several market analyses, the hydrogen economy could be worth several trillion dollars by 2050. The IEA estimates that hydrogen could account for 18% of global energy demand by 2050, reducing CO_2 emissions by up to six gigatons annually. The global hydrogen market will likely see a shift in market share from fossil-based hydrogen (gray hydrogen) to cleaner hydrogen options, such as green hydrogen, produced via renewable-powered electrolysis, and blue hydrogen, produced from natural gas with CCS. By 2030, it is anticipated that the cost of green hydrogen will significantly decrease, making it competitive with traditional fossil fuels and setting the stage for mass adoption across industries.

Projections indicate that Europe, Asia-Pacific, and North America will be the primary hubs for hydrogen development. In Europe, hydrogen will play a crucial role in meeting the EU's ambitious Green Deal targets, with countries like Germany, France, and the Netherlands focusing on large-scale hydrogen infrastructure projects. In Asia, Japan, South Korea, and China are already implementing robust hydrogen strategies to decarbonize transportation and industry. Meanwhile, North America, particularly the United States and Canada, will focus on establishing cross-border hydrogen supply chains, particularly with Mexico. The demand for hydrogen will span multiple sectors, from transportation and industry to power generation and heating. Hydrogen-based FCVs, heavy-duty trucks, and buses will make up a large portion of hydrogen consumption, alongside industrial applications such as steel production, refining, and ammonia synthesis. In addition, hydrogen is expected to increasingly power renewable energy storage solutions, acting as an energy carrier for periods of high renewable generation.

7.9.2 Advancements in Low-cost Production and Storage Technologies

The future of hydrogen depends largely on advancements in production and storage technologies that can significantly reduce costs and improve efficiency. A key challenge to hydrogen's widespread adoption is the high production cost, especially for green hydrogen, which relies on renewable electricity for electrolysis. However, technological advancements and economies of scale are expected to dramatically reduce the cost of hydrogen production. One area of significant progress is in electrolysis technology, particularly PEM and alkaline electrolyzers. Innovations in electrolyzer design, improved efficiency, and economies of scale from larger production facilities are expected to make green hydrogen more competitive with fossil-fuel-based hydrogen. As renewable energy prices continue to fall, the cost of producing green hydrogen through renewable-powered electrolysis will continue to decline, enabling broader use of hydrogen across various sectors.

In addition to production, storage technologies will also see critical advancements. Hydrogen can be stored in high-pressure tanks, cryogenic containers, or chemical carriers, but each of these options presents challenges in terms of cost, efficiency, and energy density. Continued innovation in solid-state storage, such as the development of metal hydrides and MOFs, holds promise for more efficient, compact storage solutions. Furthermore, research into LOHCs and ammonia-based storage methods is expected to enable large-scale storage and transport of hydrogen in liquid form, which could facilitate global hydrogen trade. As these production and storage technologies continue to mature, the hydrogen sector will become increasingly cost-competitive with other energy sources, spurring greater adoption of hydrogen solutions across industries.

7.9.3 The Role of Digitalization, AI, and Blockchain in Hydrogen Trading

The integration of digital technologies, such as AI and blockchain, will play a transformative role in the future of hydrogen markets. As hydrogen production and distribution networks expand, digital tools will be essential for improving efficiency, reducing costs, and enhancing transparency in the hydrogen value chain. AI and machine learning can help optimize hydrogen production processes by improving the efficiency of electrolysis, forecasting energy demand, and managing supply chains. For example, AI can predict fluctuations in renewable energy generation and match them with hydrogen production schedules, improving the overall efficiency and reducing energy waste in the hydrogen supply chain. Another transformative technology is blockchain, which offers a transparent and secure method for recording transactions, particularly in the context of hydrogen trading. Blockchain can enable traceability of hydrogen production, ensuring that hydrogen is produced sustainably and that it meets the required quality standards. Blockchain-based platforms could also facilitate peer-to-peer hydrogen trading and support the creation of certification schemes for green hydrogen, allowing buyers to verify that the hydrogen they purchase is genuinely low-carbon. In the future, the convergence of digital technologies will create a more efficient and transparent hydrogen market, enabling seamless trade across borders and sectors. Smart grids and energy management systems, powered by digital solutions, will further enhance the integration of hydrogen into the broader energy system, supporting its role as a flexible energy carrier.

7.9.4 The Pathway to a Hydrogen-based Circular Economy

The future of hydrogen is deeply connected to the broader vision of a hydrogen-based circular economy. A circular economy aims to minimize waste, maximize resource efficiency, and promote sustainability by reusing, recycling, and regenerating products. Hydrogen's role in such an economy is multifaceted, with potential to close loops in several industries and create synergies across sectors. One key aspect of a hydrogen-based circular economy is the potential for hydrogen to serve as a bridge between various sectors. For example, hydrogen can be produced using excess renewable energy that would otherwise go to waste, such as during periods of high solar or wind generation. This hydrogen can then be used in transportation, heating, and industrial processes, effectively storing surplus renewable energy for later use. Additionally, waste materials such as agricultural and industrial waste can be used as feedstocks for hydrogen production, turning waste into a valuable resource. Hydrogen's role in industrial decarbonization is another critical aspect of the circular economy. By replacing fossil fuels with hydrogen in high-emission industries like steel production, cement manufacturing, and refining, hydrogen can help reduce emissions and close the loop on carbon-intensive processes. Furthermore, hydrogen can support circular materials management by serving as a clean fuel for the recycling of metals and other materials, creating more sustainable supply chains.

Green hydrogen can also contribute to a more circular economy by helping to decarbonize the chemical industry, where hydrogen is used in the production of ammonia, methanol, and other chemicals. Through the adoption of green hydrogen, the chemical industry could significantly reduce its reliance on fossil-derived hydrogen, creating a more sustainable production model. Ultimately, the pathway to a hydrogen-based circular economy will require coordinated action across various industries, government policies, and technological advancements. The adoption of hydrogen in combination with other sustainable practices such as renewable energy integration, CCS, and waste-to-energy solutions will form the foundation for a more resilient and sustainable economic model.

7.10 Conclusion

The economic and policy landscape surrounding hydrogen is rapidly evolving, with transformative potential for the global energy transition. As the world confronts the urgency of decarbonization, hydrogen stands out as a versatile and sustainable energy carrier, offering solutions to industries, transportation, and power generation. This chapter has explored the key economic factors, technological advancements, policy frameworks, and infrastructure requirements needed to drive the commercialization of hydrogen. From production methods and cost reductions to the creation of global hydrogen markets, each aspect of the hydrogen economy plays a crucial role in shaping its future. The continued development of hydrogen technologies, alongside supportive policy incentives and regulatory frameworks, will be essential to overcoming current barriers and scaling hydrogen solutions globally. Advances in production efficiency, storage solutions, and distribution infrastructure will pave the way for a competitive hydrogen market, making it an economically viable option across various sectors. Moreover, the integration of digital tools, the development of hydrogen trade networks, and the establishment of circular economy principles will further enhance hydrogen's role in a sustainable global energy system. While significant progress has been made, several challenges remain, including ensuring economic parity with fossil fuels, scaling infrastructure, and addressing financing gaps. The collaboration between governments, industries, and researchers will be pivotal in navigating these challenges and unlocking the full potential of hydrogen as a cornerstone of a decarbonized world. Looking ahead, hydrogen's future is intertwined with the broader goals of global sustainability, energy security, and climate resilience. The advancements in hydrogen technologies, coupled with robust policy support, will not only accelerate its deployment but also catalyze the global transition to a cleaner, more sustainable energy future. Hydrogen's promise as a key enabler of decarbonization is undeniable, and as we move toward 2030 and beyond, it will become increasingly central to achieving the ambitious climate targets set by nations worldwide.

References

1 Cader J, Koneczna R, Olczak P. The impact of economic, energy, and environmental factors on the development of the hydrogen economy. *Energies* (Basel) [Internet]. 2021 Aug 2 [cited 2025 Apr 3];14(16):4811. Available from: https://www.mdpi.com/1996-1073/14/16/4811/htm.

2 Dillman KJ, Heinonen J. A 'just' hydrogen economy: A normative energy justice assessment of the hydrogen economy. *Renew Sustain Energy Rev*. 2022 Oct 1;167: 112648.

3 Clark WW, Rifkin J, O'Connor T, Swisher J, Lipman T, Rambach G, et al. Hydrogen energy stations: Along the roadside to the hydrogen economy. *Util Policy*. 2005 Mar 1; 13(1):41–50.

4 Shibata Y. Economic analysis of hydrogen production from variable renewables. *IEEJ Energy Journal*. 2015;10(2): 26–46.

5 Muñoz Díaz MT, Chávez Oróstica H, Guajardo J. Economic analysis: Green hydrogen production systems. *Processes* [Internet]. 2023 May 4 [cited 2025 Apr 3]; 11(5):1390. Available from: https://www.mdpi.com/2227-9717/11/5/1390/htm.

6 Kaiwen L, Bin Y, Tao Z. Economic analysis of hydrogen production from steam reforming process: A literature review. Energy sources, part b: Economics, planning, and policy [Internet]. 2018 Feb 1 [cited 2025 Apr 3];13(2):109–15. Available from: https://www.tandfonline.com/doi/abs/10.1080/15567249.2017.1387619.

7 Yang B, Zhang R, Shao Z, Zhang C. The economic analysis for hydrogen production cost towards electrolyzer technologies: Current and future competitiveness. *Int J Hydrogen Energy*. 2023 Apr 30; 48(37):13767–79.

8 Varbanov PS, Wang Q, Zeng M, Seferlis P, Ma T, Klemeš JJ, et al. Low-carbon footprint hydrogen production from natural gas: A techno-economic analysis of carbon capture and storage from steam-methane reforming. [Internet]. 2020 [cited 2025 Apr 3];81:1015–20. Available from: https://sintef.brage.unit.no/sintef-xmlui/handle/11250/2673842.

9 Haghi E, Fwoler M, Raahemifar K. Economic analysis of hydrogen production in context of a microgrid. 2017 5th IEEE International Conference on Smart Energy Grid Engineering, SEGE 2017. 2017 Sep 27; 79–84.

10 Ji M, Wang J. Review and comparison of various hydrogen production methods based on costs and life cycle impact assessment indicators. *Int J Hydrogen Energy*. 2021 Nov 11; 46(78):38612–35.

11 Yan Y, Manovic V, Anthony EJ, Clough PT. Techno-economic analysis of low-carbon hydrogen production by sorption enhanced steam methane reforming (SE-SMR) processes. *Energy Convers Manag*. 2020 Dec 15;226:113530.

12 Nami H, Rizvandi OB, Chatzichristodoulou C, Hendriksen PV, Frandsen HL. Techno-economic analysis of current and emerging electrolysis technologies for green hydrogen production. *Energy Convers Manag*. 2022 Oct 1;269:116162.

13 Kaplan R, Kopacz M. Economic conditions for developing hydrogen production based on coal gasification with carbon capture and storage in Poland. *Energies* [Internet]. 2020 Sep 28 [cited 2025 Apr 3];13(19):5074. Available from: https://www.mdpi.com/1996-1073/13/19/5074/htm.

14 Salkuyeh YK, Saville BA, MacLean HL. Techno-economic analysis and life cycle assessment of hydrogen production from different biomass gasification processes. *Int J Hydrogen Energy*. 2018 May 17; 43(20):9514–28.

15 Chen Y. Global potential of algae-based photobiological hydrogen production. *Energy Environ Sci* [Internet]. 2022 Jul 13 [cited 2025 Apr 3];15(7):2843–57. Available from: https://pubs.rsc.org/en/content/articlehtml/2022/ee/d2ee00342b.

16 Lemus RG, Martínez Duart JM. Updated hydrogen production costs and parities for conventional and renewable technologies. *Int J Hydrogen Energy*. 2010 May 1; 35(9):3929–36.

17 Mansilla C, Avril S, Imbach J, Le Duigou A. CO_2-free hydrogen as a substitute to fossil fuels: What are the targets? Prospective assessment of the hydrogen market attractiveness. *Int J Hydrogen Energy*. 2012 Jun 1; 37(12):9451–8.

18 Balcı Y, Erbay C. Green hydrogen production from offshore wind: A techno-economic analysis for Türkiye. *Int J Hydrogen Energy*. 2025 Jan 6; 97:377–90.

19 Park J, Kang S, Kim S, Kim H, Cho HS, Lee JH. Comparative techno-economic evaluation of alkaline and proton exchange membrane electrolysis for hydrogen production amidst renewable energy source volatility. *Energy Convers Manag*. 2025 Feb 1;325:119423.

20 Lentschig H, Patonia A, Quitzow R. Multilateral governance in a global hydrogen economy: An overview of main actors and institutions, key challenges and future pathways. *Int J Hydrogen Energy*. 2025 Jan 6; 97:76–87.

21 Dyantyi-Gwanya N, Giwa SO, Ncanywa T, Taziwa RT. Exploring economic expansion of green hydrogen production in South Africa. *Sustainability* [Internet]. 2025 Jan 23 [cited 2025 Apr 3];17(3):901. Available from: https://www.mdpi.com/2071-1050/17/3/901/htm.

22 Guillot V, Assoumou E. Power and green hydrogen trade potential between North African and European countries: Conditions, challenges, and sustainability prospects. *Appl Energy*. 2025 Mar 15;382:125209.

23 Hassan MA, El-Amary NH. Economic and technical analysis of hydrogen production and transport: A case study of Egypt. *Sci Rep* [Internet]. 2025 Mar 15 [cited 2025 Apr 3];15(1):1–30. Available from: https://www.nature.com/articles/s41598-025-91589-6.

24 Kim H, Song G, Ha Y. Green hydrogen export potential in each Southeast Asian country based on exportable volumes and levelized cost of hydrogen. *Appl Energy*. 2025 Apr 1;383:125371.

25 Benalcazar P, Komorowska A. Techno-economic analysis and uncertainty assessment of green hydrogen production in future exporting countries. *Renew Sustain Energy Rev*. 2024 Jul 1; 199:114512.

26 Yatsenko O, Iatsenko O. Trends and prospects in international hydrogen trade in the face of new barriers and challenges to global cooperation. *Actual Problems of International Relations* [Internet]. 2024 [cited 2025 Apr 3];1(161):177–89. Available from: http://apir.iir.edu.ua/index.php/apmv/article/view/3946.

27 Hugo A, Rutter P, Pistikopoulos S, Amorelli A, Zoia G. Hydrogen infrastructure strategic planning using multi-objective optimization. *Int J Hydrogen Energy*. 2005 Dec 1; 30(15):1523–34.

28 Lundblad T, Taljegard M, Johnsson F. Centralized and decentralized electrolysis-based hydrogen supply systems for road transportation—A modeling study of current and future costs. *Int J Hydrogen Energy*. 2023 Feb 8; 48(12):4830–44.

29 Walker B, Klagge B, Dörry S, Schulz C. Infrastructure bottlenecks as opportunity for local development: The case of decentralized green-hydrogen projects. *Tijdschrift voor economische en sociale geografie* [Internet]. 2024 Dec 1 [cited 2025 Apr 3];115(5):643–59. Available from: https://onlinelibrary.wiley.com/doi/full/10.1111/tesg.12633.

30 Dagdougui H, Sacile R, Bersani C, Ouammi A. Hydrogen infrastructure for energy applications: Production, storage, distribution and safety. Academic Press. 2018 Jan 27; 1–156.

31 Gan W, Yan M, Yao W, Guo J, Fang J, Ai X, et al. Multi-network coordinated hydrogen supply infrastructure planning for the integration of hydrogen vehicles and renewable energy. *IEEE Trans Ind Appl*. 2022;58(2): 2875–86.

32 Parker N, Fan Y, Ogden J. From waste to hydrogen: An optimal design of energy production and distribution network. *Transp Res E Logist Transp Rev*. 2010 Jul 1; 46(4):534–45.

33 Domashenko A, Golovchenko A, Gorbatsky Y, Nelidov V, Skorodumov B. Production, storage and transportation of liquid hydrogen. Experience of infrastructure development and operation. *Int J Hydrogen Energy*. 2002 Jul 1; 27(7–8):753–5.

34 Preuster P, Papp C, Wasserscheid P. Liquid organic hydrogen carriers (LOHCs): Toward a hydrogen-free hydrogen economy. *Acc Chem Res* [Internet]. 2017 Jan 17 [cited 2025 Apr 3];50(1):74–85. Available from: https://pubs.acs.org/doi/full/10.1021/acs.accounts.6b00474.

35 Shaker LM, Al-Amiery AA, Al-Azzawi WK. Nanomaterials: Paving the way for the hydrogen energy frontier. *Discover Nano* [Internet]. 2024 Jan 3 [cited 2025 Apr 3];19(1):1–23. Available from: https://link.springer.com/article/10.1186/s11671-023-03949-8.

36 Li Z, Gao D, Chang L, Liu P, Pistikopoulos EN. Hydrogen infrastructure design and optimization: A case study of China. *Int J Hydrogen Energy*. 2008 Oct 1; 33(20):5275–86.

37 He C, Sun H, Xu Y, Lv S. Hydrogen refueling station siting of expressway based on the optimization of hydrogen life cycle cost. *Int J Hydrogen Energy*. 2017 Jun 29; 42(26):16313–24.

38 Vidas L, Castro R, Pires A. A review of the impact of hydrogen integration in natural gas distribution networks and electric smart grids. *Energies* [Internet]. 2022 Apr 26 [cited 2025 Apr 3];15(9):3160. Available from: https://www.mdpi.com/1996-1073/15/9/3160/htm.

39 Patel SKS, Gupta RK, Rohit MV, Lee JK. Recent developments in hydrogen production, storage, and transportation: Challenges, opportunities, and perspectives. *Fire* [Internet]. 2024 Jul 3 [cited 2025 Mar 25];7(7):233. Available from: https://www.mdpi.com/2571-6255/7/7/233/htm.

40 Gonzalez Hernandez S, Kirchofer A. Incentivizing hydrogen: A perspective review of lifecycle analysis methodology disparities affecting hydrogen incentives in policy frameworks. *Energy Climate Change*. 2025 Dec 1;6:100172.

41 Guillotin A, Bergaentzlé C, Dussartre V, Heggarty T, Massol O, Perez Y. Hydrogen subsidies under three pillar-frameworks: A Europe-United States multi-stakeholder comparison. *Renew Sustain Energy Rev*. 2025 Apr 1;211:115284.

42 Ringsgwandl LM, Schaffert J, Brücken N, Albus R, Görner K. Current legislative framework for green hydrogen production by electrolysis plants in Germany. *Energies* [Internet]. 2022 Feb 28 [cited 2025 Apr 3];15(5):1786. Available from: https://www.mdpi.com/1996-1073/15/5/1786/htm.

43 Lee Y, Pak S Bin. Advancing global cooperation toward the hydrogen economy: A case study of Japan–Korea cooperation. *Pacific Focus* [Internet]. 2025 [cited 2025 Apr 3]; Available from: https://onlinelibrary.wiley.com/doi/full/10.1111/pafo.12270.

44 Gatto A, Sadik-Zada ER, Lohoff T, Aldieri L, Vinci CP, Peitz KA. An exemplary subsidization path for the green hydrogen economy uptake: Rollout policies in the United States and the European Union. *J Clean Prod*. 2024 Feb 10;440:140757.

45 Cho A, Kim H, Park S. Resurgence of the hydrogen energy in South Korea's government strategies from 2005 to 2019. *Int J Hydrogen Energy*. 2024 May 2; 65:844–54.

46 Andeobu L, Wibowo S, Grandhi S. Renewable hydrogen for the energy transition in Australia - Current trends, challenges and future directions. *Int J Hydrogen Energy*. 2024 Oct 18; 87:1207–23.

47 Chen J, Yuan Z, Mei P, Adam NA, Chen X. Carbon dioxide emissions and mitigation in China's hydrogen energy supply chain: Insights for sustainability and policy development. *Int J Hydrogen Energy*. 2025 Mar 31; 114:378–91.

48 Khawaja MK, Al-Mohamad R, Salameh T, Alkhalidi A. Strategies to promote nuclear energy utilization in hydrogen production. *Int J Hydrogen Energy*. 2024 Sep 11; 82:36–46.

49 Tuluhong A, Chang Q, Xie L, Xu Z, Song T. Current status of green hydrogen production technology: A review. *Sustainability* [Internet]. 2024 Oct 19 [cited 2025 Mar 25];16(20):9070. Available from: https://www.mdpi.com/2071-1050/16/20/9070/htm.

50 Farrell N. Policy design for green hydrogen. *Renew Sustain Energy Rev*. 2023 May 1;178:113216.

51 Rey J, Segura F, Andújar JM. Green hydrogen: Resources consumption, technological maturity, and regulatory framework. *Energies* [Internet].

2023 Aug 27 [cited 2025 Apr 3];16(17):6222. Available from: https://www.mdpi.com/1996-1073/16/17/6222/htm.

52 Tholen L, Leipprand A, Kiyar D, Maier S, Küper M, Adisorn T, et al. The green hydrogen puzzle: Towards a German policy framework for industry. *Sustainability* [Internet]. 2021 Nov 15 [cited 2025 Apr 3];13(22):12626. Available from: https://www.mdpi.com/2071-1050/13/22/12626/htm.

53 Scheibe A, Poudineh, Rahmatallah. Regulating the future European hydrogen supply industry: A balancing act between liberalization, sustainability, and security of supply? 2023 [cited 2025 Apr 3]; Available from: https://www.econstor.eu/handle/10419/280131.

54 Osman AI, Mehta N, Elgarahy AM, Hefny M, Al-Hinai A, Al-Muhtaseb AH, et al. Hydrogen production, storage, utilisation and environmental impacts: A review. *Environ Chem Lett* [Internet]. 2021 Oct 6 [cited 2025 Apr 3];20(1):153–88. Available from: https://link.springer.com/article/10.1007/s10311-021-01322-8.

55 Harichandan S, Kar SK. Financing the hydrogen industry: Exploring demand and supply chain dynamics. *Environ Sci Pollut Res* [Internet]. 2023 Oct 9 [cited 2025 Apr 3];1:1–18. Available from: https://link.springer.com/article/10.1007/s11356-023-30262-9.

56 Khaleel M, Yusupov Z, Tahir Guneser M, Nassar Y, El-Khozondar H, Ahmed A, et al. Towards hydrogen sector investments for achieving sustainable electricity generation. *J Solar Energy Sustain Develop* [Internet]. 2024 Aug 22 [cited 2025 Apr 3];13(1):71–96. Available from: https://www.ajol.info/index.php/jsesd/article/view/276744.

57 Taghizadeh-Hesary F, Li Y, Rasoulinezhad E, Mortha A, Long Y, Lan Y, et al. Green finance and the economic feasibility of hydrogen projects. *Int J Hydrogen Energy*. 2022 Jul 8; 47(58):24511–22.

58 Gao J, Zhang T. Effects of public funding on the commercial diffusion of on-site hydrogen production technology: A system dynamics perspective. *Technol Forecast Soc Change*. 2022 Feb 1;175:121380.

59 Webb J, Longden T, Boulaire F, Gono M, Wilson C. The application of green finance to the production of blue and green hydrogen: A comparative study. *Renew Energy*. 2023 Dec 1;219:119236.

60 Palmer O, Radet H, Camal S, Girard R. Hedging hydrogen: Planning and contracting under uncertainty for a green hydrogen producer. *IEEE PES Innovative Smart Grid Technologies Conference Europe* [Internet]. 2024 Jul 31 [cited 2025 Apr 3]; Available from: https://arxiv.org/abs/2407.21574v2.

61 Sambasivam B, Sarma RN. India's National Green Hydrogen Mission: An analysis of the strategies, policies for net-zero emissions and sustainability. *Environ Res Energy* [Internet]. 2024 Dec 5 [cited 2025 Apr 3];1(4):045015. Available from: https://iopscience.iop.org/article/10.1088/2753-3751/ad8fb2.

62 Athia N, Pandey M, Saxena S. Evaluating the effectiveness of national green hydrogen mission in India. *Environ Dev Sustain* [Internet]. 2024 Dec 28 [cited 2025 Apr 3];1–23. Available from: https://link.springer.com/article/10.1007/s10668-024-05829-2.

63 Kolodziejczyk B. The role of hydrogen in Australia's decarbonization and export strategy. The clean hydrogen economy and Saudi Arabia: Domestic developments and international opportunities [Internet]. 2024 Apr 18 [cited 2025 Apr 3];308–28. Available from: https://www.taylorfrancis.com/chapters/oa-edit/10.4324/9781003294290-13/role-hydrogen-australia-decarbonization-export-strategy-bart-kolodziejczyk.

64 Curtis AJ, McLellan BC. Potential domestic energy system vulnerabilities from major exports of green hydrogen: A case study of Australia. *Energies* [Internet]. 2023 Aug 8 [cited 2025 Apr 3];16(16):5881. Available from: https://www.mdpi.com/1996-1073/16/16/5881/htm.

65 Stangarone T. South Korean efforts to transition to a hydrogen economy. *Clean Technol Environ Policy* [Internet]. 2021 Mar 1 [cited 2025 Apr 3];23(2):509–16. Available from: https://link.springer.com/article/10.1007/s10098-020-01936-6.

66 Khan H, Akhtar MS, Qyyum MA, Liu JJ. Post COVID-19 and Korea hydrogen economy roadmap 2040: Challenges and the way forward. *Int J Hydrogen Energy*. 2024 May 15. Available from: https://doi.org/10.1016/j.ijhydene.2024.05.147.

67 Krause T. An analysis of the barriers to deployment and supportive policies for the commercialization of hydrogen energy. 2021. Available from: https://jscholarship.library.jhu.edu/server/api/core/bitstreams/5c78f93d-6b9c-40c7-ad4c-c428b1e38644/content.

68 Dixon RK. Advancing towards a hydrogen energy economy: Status, opportunities and barriers. *Mitig Adapt Strateg Glob Chang* [Internet]. 2007 Mar 6 [cited 2025 Apr 3];12(3):305–21. Available from: https://link.springer.com/article/10.1007/s11027-006-2328-0.

69 Dou Y, Sun L, Ren J, Dong L. Opportunities and future challenges in hydrogen economy for sustainable development. Hydrogen economy: Processes, supply chain, life cycle analysis and energy transition for sustainability. 2023 Jan 1; 537–69.

70 Dixon RK, Li J, Wang MQ. Progress in hydrogen energy infrastructure development—Addressing technical and institutional barriers. Compendium of hydrogen energy: Hydrogen storage, distribution and infrastructure. 2016 Jan 1; 2: 323–43.

71 Hara D. Toward a hydrogen society—Introduction of representative projects in Japan. *ECS Trans* [Internet]. 2019 Jul 10 [cited 2025 Apr 3];91(1):3–7. Available from: https://iopscience.iop.org/article/10.1149/09101.0003ecst.

72 Nagashima M, Institut francais des relations internationales - Ifri 27 rue de la Procession, 75015 Paris (France). Japan's Hydrogen Society Ambition. 2020 Status and Perspectives. 2020 Sep 1 [cited 2025 Apr 3]; Available from: https://inis.iaea.org/records/jnj5d-68z22.

73 Pinto J. The key tenets of a hydrogen strategy: An analysis and comparison of the hydrogen strategies of the EU, Germany and Spain. *Global Energy Law Sustain.* https://doi.org/103366/gels20230094 [Internet]. 2023 Oct 27 [cited 2025 Apr 3];4(1–2):72–95. Available from: https://www.euppublishing.com/doi/10.3366/gels.2023.0094.

74 Ehret O, Bonhoff K. Hydrogen as a fuel and energy storage: Success factors for the German Energiewende. *Int J Hydrogen Energy*. 2015 Apr 27; 40(15):5526–33.

75 Vijayakumar V, Jenn A, Fulton L. Low carbon scenario analysis of a hydrogen-based energy transition for on-road transportation in California. *Energies*.

[Internet]. 2021 Nov 1 [cited 2025 Apr 3];14(21):7163. Available from: https://www.mdpi.com/1996-1073/14/21/7163/htm.

76 Vijayakumar V, Jenn A, Ogden J. Modeling future hydrogen supply chains in the western United States under uncertainties: An optimization-based approach focusing on California as a hydrogen hub. *Sustain Energy Fuels* [Internet]. 2023 Feb 28 [cited 2025 Apr 3];7(5):1223–44. Available from: https://pubs.rsc.org/en/content/articlehtml/2023/se/d3se00043e.

77 Kar SK, Sinha ASK, Bansal R, Shabani B, Harichandan S. Overview of hydrogen economy in Australia. *Wiley Interdiscip Rev Energy Environ* [Internet]. 2023 Jan 1 [cited 2025 Apr 3];12(1):e457. Available from: https://onlinelibrary.wiley.com/doi/full/10.1002/wene.457.

78 Rezaei M, Akimov A, Gray EMA. Techno-economics of renewable hydrogen export: A case study for Australia-Japan. *Appl Energy*. 2024 Nov 15;374:124015.

79 Wang F, Swinbourn R, Li C. Shipping Australian sunshine: Liquid renewable green fuel export. *Int J Hydrogen Energy*. 2023 May 5; 48(39):14763–84.

8

The Future of Hydrogen – Challenges, Innovations, and Sustainability

8.1 Introduction

As the global energy landscape shifts toward sustainability, hydrogen is emerging as a pivotal energy carrier with the potential to drive the clean energy transition. Hydrogen's versatility, ability to store and transport energy, and its role in reducing greenhouse gas emissions make it an attractive alternative to fossil fuels. However, despite its advantages, large-scale adoption of hydrogen technologies faces several challenges, including production efficiency, storage limitations, distribution infrastructure, and economic feasibility. Addressing these challenges requires an interdisciplinary approach, integrating advancements in materials science, artificial intelligence, quantum technologies, and policy frameworks. This chapter explores the future of hydrogen, highlighting emerging technologies, key barriers, and the innovations needed to unlock its full potential [1].

Hydrogen has long been recognized as a potential clean energy source, but recent advancements in production, storage, and fuel cell technologies have accelerated its adoption. Several next-generation hydrogen technologies are reshaping the energy sector, ensuring greater efficiency, lower costs, and reduced environmental impact. These include artificial intelligence (AI)-driven optimization in hydrogen research, quantum materials for hydrogen applications, bio-inspired catalysts, and hydrogen-powered smart cities. One of the most significant breakthroughs is the integration of AI and machine learning in hydrogen research. AI is revolutionizing catalyst discovery, material optimization, and process efficiency, enabling faster and more cost-effective advancements. AI-driven predictive modeling allows scientists to simulate thousands of material compositions, identifying the most efficient catalysts for hydrogen production through electrolysis and photocatalysis. Additionally, digital twin technology is being utilized to optimize hydrogen storage and distribution networks by creating virtual models of real-world hydrogen infrastructure [2].

Another transformative area is the development of quantum materials for hydrogen-related applications [3]. Quantum dots, high-efficiency catalysts, and advanced nanomaterials are being designed to enhance hydrogen production and storage. These materials offer higher surface areas and improved reaction kinetics, making them ideal for water splitting, hydrogen adsorption, and electrochemical

Next-Generation Hydrogen Economy: Production, Storage, and Fuel Cell Technologies, First Edition.
Peeyush Phogat, Shreya Sharma and Satyam Rawat.

applications [4]. Furthermore, quantum computing is expected to accelerate research in hydrogen energy by solving complex molecular simulations that were previously unattainable with classical computing [5]. Bio-inspired catalysts are also gaining attention as a means to improve hydrogen production efficiency. By mimicking enzymatic processes found in nature, researchers are developing synthetic catalysts that enable more sustainable hydrogen generation. Biological hydrogen production methods, such as microbial electrolysis and enzymatic catalysis, provide pathways to generate hydrogen from renewable and waste-derived sources, reducing reliance on fossil fuels [6].

Hydrogen's integration into smart cities represents another key development. Hydrogen-powered public transportation, microgrids, and fuel cell-based heating and cooling systems are being implemented in cities worldwide to reduce carbon footprints. Hydrogen refueling stations and infrastructure are expanding, making hydrogen-fueled vehicles more practical [7]. Additionally, off-grid applications of hydrogen, such as using fuel cells for remote communities and emergency power backup, are demonstrating its reliability as an alternative energy source [8]. By harnessing these emerging technologies, hydrogen has the potential to play a critical role in achieving global climate goals. However, to scale up hydrogen adoption, several significant challenges must be addressed.

8.1.1 Key Challenges Hindering Large-scale Adoption of Hydrogen

Despite the promise of hydrogen as a clean energy carrier, several barriers continue to slow its widespread adoption. These challenges span multiple domains, including technological, economic, infrastructural, and policy-related issues.

8.1.1.1 High Production Costs and Energy Inefficiency

Hydrogen is predominantly produced through steam methane reforming (SMR), a process that releases carbon dioxide. While cleaner alternatives such as electrolysis exist, they remain energy-intensive and costly, particularly when powered by renewable energy sources. The cost of green hydrogen, produced via water electrolysis using wind or solar power, is still significantly higher than fossil-fuel-based hydrogen. Advances in catalyst efficiency, materials discovery, and process automation are needed to make hydrogen production more economically viable [9].

8.1.1.2 Storage and Transportation Challenges

Hydrogen is the lightest element, making its storage and transportation complex. Compressed gas storage at high pressures (350–700 bar) requires expensive infrastructure, while cryogenic liquid hydrogen storage at extremely low temperatures (20K) demands advanced insulation technologies. Solid-state hydrogen storage, such as metal hydrides and MOFs (Metal-Organic Frameworks), is still in its early development stages, facing challenges in hydrogen uptake and release kinetics. Transporting hydrogen via pipelines or tankers is inefficient and expensive compared to conventional fuels, necessitating new solutions for hydrogen logistics [10].

8.1.1.3 Infrastructure Limitations

Hydrogen infrastructure is still underdeveloped, creating a major barrier to its commercialization. Hydrogen refueling stations are sparse, limiting the adoption of fuel cell vehicles. Similarly, industrial hydrogen applications require specialized pipelines, storage tanks, and distribution networks. Retrofitting existing natural gas pipelines for hydrogen transport remains a technological challenge due to hydrogen embrittlement, which weakens pipeline materials. Without significant investment in infrastructure, hydrogen adoption will remain constrained [11].

8.1.1.4 Scalability of Hydrogen Technologies

While hydrogen technologies have been demonstrated successfully in pilot projects, their large-scale implementation remains a challenge. Electrolysis, fuel cells, and hydrogen storage systems need mass production capabilities to achieve cost reductions. Additionally, hydrogen-powered industrial applications, such as steel and cement manufacturing, require substantial modifications to existing facilities. Scaling up these technologies requires coordinated efforts across research institutions, industries, and governments [12].

8.1.1.5 Policy and Market Uncertainties

Global hydrogen policies vary widely, creating uncertainty for investors and industries. Some regions, such as the European Union, Japan, and South Korea, have ambitious hydrogen roadmaps, while others lack clear regulatory frameworks. Carbon pricing, subsidies, and incentives for green hydrogen adoption play a crucial role in determining its economic competitiveness. Additionally, fluctuating energy prices, policy shifts, and competition from other renewable technologies impact market confidence in hydrogen investments [13, 14].

Addressing these challenges requires a collaborative and interdisciplinary approach that brings together researchers, policymakers, and industry stakeholders to drive innovation in hydrogen technologies.

To overcome the barriers to hydrogen adoption, a multidisciplinary approach is essential. Hydrogen research requires collaboration across chemistry, materials science, engineering, AI, economics, and policy-making to develop integrated solutions. Materials scientists are working on next-generation catalysts, membranes, and storage materials to improve efficiency and reduce costs. Advanced nanomaterials, high-performance electrolyzers, and new hydrogen carriers, such as ammonia and liquid organic hydrogen carriers (LOHCs), are paving the way for more practical hydrogen solutions [15]. Engineers and data scientists are leveraging AI and machine learning algorithms to optimize hydrogen production processes, predict system performance, and design smart hydrogen infrastructures. AI-driven digital twins are helping industries simulate real-world hydrogen operations before implementing them on a large scale [16]. Economists and policymakers are focusing on creating financial incentives, carbon pricing mechanisms, and regulatory frameworks to encourage hydrogen adoption. International collaboration is crucial to standardizing hydrogen production and distribution policies across borders. Industry partnerships are playing a crucial role in scaling up hydrogen technologies.

Automotive manufacturers, energy companies, and technology firms are investing in hydrogen fuel cells, refueling networks, and industrial hydrogen applications. Public-private partnerships are essential to bridging the gap between research and commercialization.

Hydrogen has the potential to revolutionize the global energy system, providing a clean, sustainable, and versatile energy source. However, to realize its full potential, significant technological, economic, and infrastructural challenges must be addressed. Emerging hydrogen technologies, including AI-driven research, quantum materials, bio-inspired catalysts, and hydrogen-powered smart cities, are paving the way for a hydrogen-based economy. By fostering interdisciplinary collaboration, investment in research, and strong policy frameworks, hydrogen can become a mainstream energy solution, contributing to global carbon neutrality and energy security. The following sections will delve deeper into the key innovations and future directions that will shape hydrogen's role in a sustainable future [17].

8.2 AI and Data-driven Optimization in Hydrogen Research

8.2.1 AI for Hydrogen Production and Storage

AI is revolutionizing hydrogen production and storage by enhancing efficiency, reducing costs, and accelerating material discovery. The application of AI, particularly machine learning (ML), is transforming key areas such as catalyst design, electrolysis process optimization, fuel cell advancements, and predictive modeling for hydrogen storage and transportation. These innovations are crucial for overcoming the current technological and economic challenges that hinder the large-scale adoption of hydrogen as a sustainable energy carrier [18].

One of the most significant areas where AI is making an impact is in the design and discovery of catalysts for hydrogen production. The efficiency of hydrogen production through electrolysis, photocatalysis, and thermochemical processes is highly dependent on the performance of catalysts. Traditional catalyst discovery is a time-consuming and costly process that involves extensive experimental trials. AI, particularly ML algorithms, is revolutionizing this field by predicting optimal catalyst compositions, analyzing material properties, and accelerating screening processes. ML models can analyze vast datasets containing information on material structures, electronic properties, and catalytic activity to identify the most promising materials for hydrogen production [19]. For example, AI-driven algorithms can predict how different metal alloys or nanostructured materials interact with reactants, helping researchers develop more efficient and cost-effective catalysts. This has led to the discovery of non-precious metal catalysts as alternatives to expensive platinum-group metals, significantly reducing the cost of electrolysis [20]. Additionally, AI-driven computational chemistry and quantum simulations enable scientists to model molecular interactions at an atomic scale, predicting reaction kinetics and optimizing catalytic efficiency [21]. These AI-powered approaches are not only accelerating material discovery but also helping to design highly stable and durable catalysts that can withstand the harsh operating conditions of hydrogen production systems [22].

AI is also being used to enhance the efficiency of electrolysis processes, which are crucial for producing green hydrogen. Electrolyzers – whether proton exchange membrane (PEM), alkaline, or solid oxide – require precise control over operating conditions such as voltage, temperature, and electrolyte concentration to maximize efficiency while minimizing energy consumption. AI-driven control systems can continuously monitor and optimize these parameters in real time, improving hydrogen yield and reducing operational costs. For example, deep learning algorithms can analyze data from electrolysis systems to predict performance trends and recommend adjustments that enhance efficiency. These AI systems can detect inefficiencies caused by fluctuations in renewable energy inputs (e.g., solar or wind), ensuring that hydrogen production remains stable even when power sources are intermittent. Reinforcement learning models are also being employed to develop self-optimizing electrolysis systems that adapt to changing conditions without human intervention [23].

Similarly, AI is playing a crucial role in fuel cell technology, where it is used to enhance membrane durability, optimize fuel cell architectures, and improve efficiency. Fuel cells operate by converting hydrogen into electricity through electrochemical reactions, and their performance is highly dependent on factors such as temperature, humidity, and reactant flow rates. AI-driven algorithms can predict degradation patterns in fuel cells, enabling predictive maintenance and extending the lifespan of fuel cell stacks. This has significant implications for applications such as hydrogen-powered vehicles, where reliability and longevity are critical [24].

Storage and transportation remain some of the biggest challenges in the hydrogen economy, and AI-driven predictive modeling is proving to be a valuable tool in addressing these issues. Hydrogen can be stored in several forms, including compressed gas, liquid hydrogen, metal hydrides, and chemical carriers such as ammonia or LOHCs. Each storage method presents unique challenges related to efficiency, safety, and cost-effectiveness. AI-powered models can analyze large datasets from past experiments and real-world applications to predict the best storage conditions, materials, and system configurations for different use cases. For example, AI algorithms can optimize pressure and temperature conditions for compressed hydrogen storage, minimizing energy losses during compression and expansion cycles. In solid-state storage, AI is being used to screen new materials (such as MOFs and complex hydrides) that exhibit high hydrogen uptake capacities while maintaining stability under varying environmental conditions [25].

In hydrogen transportation, AI-driven simulations are used to design efficient distribution networks, optimizing hydrogen flow through pipelines, tankers, and refueling stations. These predictive models consider factors such as distance, pressure drops, and demand fluctuations to determine the most cost-effective and energy-efficient transport strategies. AI is also being used in hydrogen leak detection and risk assessment by integrating real-time sensor data with predictive maintenance algorithms, enhancing safety across the supply chain [26].

The integration of AI in hydrogen production and storage is driving significant advancements that are making hydrogen more viable as a clean energy source. From accelerating catalyst discovery to optimizing electrolysis and fuel cell efficiency, AI is enabling breakthroughs that reduce costs and improve performance. Additionally, predictive modeling for hydrogen storage and transportation is addressing logistical challenges, making hydrogen distribution more efficient and reliable. As AI continues to evolve, its role in the hydrogen economy will become even more critical, paving the way for large-scale adoption and commercialization. By leveraging AI's capabilities, researchers and industries can unlock the full potential of hydrogen as a key component of the global energy transition [27].

8.2.2 Digital Twins and Computational Simulations

The integration of digital twins and computational simulations in the hydrogen economy is revolutionizing infrastructure planning, material interaction studies, and system maintenance. As hydrogen technologies continue to advance, the need for precise, data-driven models to predict performance, optimize operations, and reduce risks becomes increasingly important. Digital twins – virtual representations of physical hydrogen systems – along with AI-driven computational simulations, offer real-time monitoring, predictive analytics, and enhanced system optimization. These innovations are crucial in ensuring the safety, efficiency, and scalability of hydrogen-based solutions [28].

Digital twins are virtual models that replicate physical hydrogen infrastructure, including production facilities, storage tanks, pipelines, and refueling stations. These models are continuously updated using real-time data from sensors, allowing operators to simulate different scenarios and predict potential issues before they occur. The primary benefit of digital twins is their ability to enhance decision-making for infrastructure development, minimize risks, and optimize hydrogen distribution networks. For example, digital twins can be used to analyze hydrogen flow within pipelines, ensuring that pressure and temperature conditions are optimized for safe and efficient transportation. They can also model hydrogen refueling stations, predicting peak demand times and adjusting storage levels accordingly. Additionally, by integrating real-time weather and energy demand data, digital twins can help in optimizing hydrogen production from renewable sources, such as solar and wind, to match supply with demand dynamically. Moreover, digital twins enable regulatory compliance by simulating safety measures, testing the impact of potential failures, and helping industries develop proactive risk mitigation strategies. Governments and companies investing in hydrogen infrastructure can utilize digital twins to simulate large-scale hydrogen deployment scenarios, ensuring that expansion plans are both cost-effective and technologically viable [29].

Computational simulations at the molecular level play a critical role in understanding how hydrogen interacts with different materials, especially in storage systems and fuel cell components. Hydrogen storage technologies, including metal hydrides, MOFs, and carbon-based materials, rely on efficient adsorption and desorption processes, which must be optimized for practical applications [30]. Molecular dynamics (MD) simulations and density functional theory (DFT) calculations are widely used to predict hydrogen absorption capacities, diffusion rates, and reaction kinetics in various materials. For example, DFT calculations help in screening potential metal hydrides that can store hydrogen at lower pressures and moderate temperatures, improving energy efficiency. Similarly, MD simulations provide insights into how hydrogen molecules diffuse through MOFs, assisting in the design of materials with higher adsorption capacities [31]. Another important application of computational simulations is in fuel cell membrane development. Hydrogen fuel cells require high-performance PEMs that exhibit excellent proton conductivity and durability. AI-driven computational simulations are used to analyze polymer membrane structures, predict degradation patterns, and optimize ion transport mechanisms, leading to the development of more efficient and longer-lasting fuel cells. Molecular simulations are also beneficial in hydrogen embrittlement studies, where hydrogen atoms infiltrate metals, weakening their structural integrity. By simulating how hydrogen interacts with steel pipelines or fuel cell components, researchers can develop protective coatings and alternative materials that minimize embrittlement risks. These simulations significantly reduce the time and cost associated with trial-and-error experimental testing, accelerating the commercialization of advanced hydrogen technologies [32].

The deployment of AI-powered predictive maintenance in hydrogen infrastructure is a game-changer for ensuring system reliability and operational efficiency. Hydrogen production, storage, and transportation systems involve complex machinery and high-pressure environments, making real-time monitoring and failure prediction essential for preventing accidents and downtime. AI-driven predictive maintenance uses sensor data, historical performance records, and real-time analytics to detect early signs of equipment degradation or system inefficiencies. By applying ML algorithms, hydrogen plant operators can forecast when components, such as electrolyzers, compressors, or fuel cell stacks, require maintenance or replacement. This approach minimizes unexpected failures, reduces maintenance costs, and extends equipment lifespan [33].

For example, in hydrogen electrolysis plants, AI algorithms analyze fluctuations in voltage, temperature, and current density to predict electrode wear and catalyst degradation. In hydrogen storage facilities, AI models monitor pressure stability and leakage risks, alerting operators to potential failures before they escalate. Additionally, in hydrogen fuel cell vehicles (FCEVs), AI-based diagnostics assess fuel cell stack efficiency, coolant levels, and power output, ensuring optimal performance for extended durations. The predictive capabilities of AI-driven maintenance systems enhance safety, especially in high-pressure hydrogen environments where leaks or failures can have severe consequences. AI-powered leak detection systems, using infrared sensors and real-time data analytics, can pinpoint hydrogen leaks at an early stage, reducing the risk of explosions or environmental hazards. As shown in Table 8.1, AI is being applied across various hydrogen technologies to improve performance, efficiency, and safety.

Table 8.1 Applications of AI in different hydrogen technologies.

Hydrogen technology	AI application	Benefits
Electrolysis	AI-driven process optimization	Increases efficiency, reduces energy consumption [34]
Fuel cells	AI-based membrane and catalyst design	Enhances durability and performance [35]
Hydrogen storage	Predictive modeling for materials and pressure conditions	Optimizes storage capacity and safety [36]
Hydrogen transportation	AI-powered network optimization	Reduces costs and improves distribution efficiency [37]
Leak detection	Real-time monitoring with AI sensors	Enhances safety and minimizes losses [38]
Predictive maintenance	AI-based diagnostics and failure prediction	Extends equipment lifespan and reduces downtime [39]

8.3 Next-gen Hydrogen Technologies

8.3.1 Quantum Materials for Hydrogen Applications

The field of quantum materials is revolutionizing hydrogen technology by enhancing hydrogen production, storage, and system modeling through quantum dots, high-efficiency quantum catalysts, and quantum computing. These materials exhibit unique electronic and optical properties that make them highly efficient in energy conversion and chemical reactions, offering innovative pathways for improving electrolysis efficiency, optimizing storage mechanisms, and enabling advanced computational modeling.

Quantum dots (QDs) are nanoscale semiconductor particles that exhibit quantum confinement effects, leading to size-dependent electronic and optical properties. In hydrogen production, QDs play a significant role in photocatalysis and photoelectrochemical (PEC) water splitting, where sunlight is directly used to drive the decomposition of water into hydrogen and oxygen. One of the key advantages of QDs in hydrogen production is their broad spectral absorption, allowing them to capture and utilize a wider range of sunlight compared to traditional photocatalysts. For instance, cadmium-based QDs have demonstrated high solar-to-hydrogen conversion efficiencies, making them promising materials for PEC water splitting. Moreover, QDs can be engineered to enhance charge separation and reduce recombination losses, improving overall hydrogen yield [40].

Another application of QDs in hydrogen technology is their integration into hybrid photocatalyst systems, where they serve as co-catalysts alongside traditional materials such as titanium dioxide (TiO_2) or MOFs. These hybrid systems leverage the high surface-to-volume ratio of QDs, enabling more efficient electron transfer and boosting photocatalytic performance [41]. As depicted in Figure 8.1, bio-inspired quantum catalysts are being explored to mimic natural enzymatic processes, further enhancing hydrogen production efficiency.

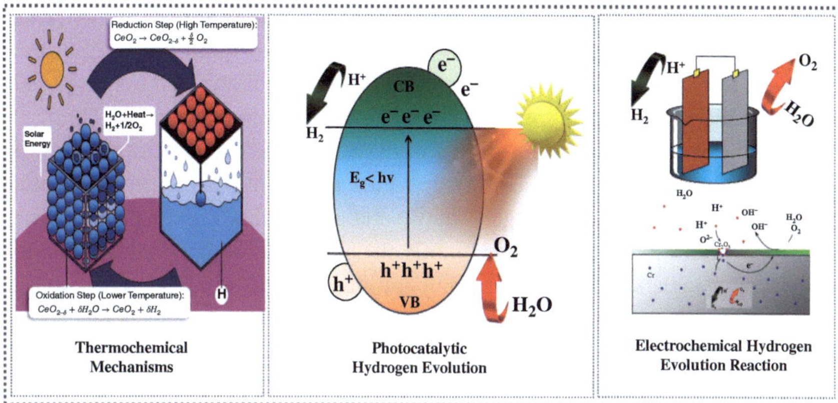

Figure 8.1 Conceptual schematic of bio-inspired catalysts for hydrogen production. [46]/MDPI/CC BY 4.0.

The efficiency of hydrogen electrolysis – the process of splitting water into hydrogen and oxygen using electricity – depends heavily on the catalysts used for the hydrogen evolution reaction (HER) and oxygen evolution reaction (OER). Quantum materials, particularly quantum catalysts, have emerged as next-generation alternatives to conventional platinum-group metal catalysts, offering high activity, stability, and cost-effectiveness. One promising class of quantum catalysts is single-atom catalysts (SACs), where individual metal atoms are dispersed on a conductive support, maximizing their catalytic efficiency. SACs based on transition metals like iron (Fe), nickel (Ni), and molybdenum (Mo) have shown exceptional HER activity, rivaling that of platinum while being significantly cheaper [42].

Additionally, quantum-confined metal oxides and sulfides such as MoS_2 QDs exhibit excellent catalytic properties due to their abundant edge sites, high charge mobility, and tunable band structures. These materials facilitate faster reaction kinetics, leading to higher hydrogen production rates at lower energy inputs. Beyond electrolysis, quantum catalysts are also being explored for hydrogen storage applications. Traditional hydrogen storage materials, such as metal hydrides and complex hydrides, often suffer from slow hydrogen absorption/desorption kinetics and high operating temperatures. Quantum materials, particularly quantum-tuned nanostructures, can modulate hydrogen adsorption energies, making storage and release processes more efficient [43].

The integration of quantum computing into hydrogen energy research is poised to revolutionize computational modeling, material discovery, and system optimization. Hydrogen technologies involve highly complex chemical reactions, phase transitions, and multi-variable energy systems that are challenging to model using classical computational methods. Quantum computing, with its unparalleled computational power, enables the simulation of quantum-scale interactions with unprecedented accuracy [44].

One of the most promising applications of quantum computing in hydrogen technology is in catalyst discovery. Traditional computational methods, such as DFT, require significant processing time to model the electronic structures of complex materials. Quantum algorithms, particularly variational quantum eigensolvers, can accelerate these calculations, allowing researchers to screen thousands of potential catalysts in a fraction of the time. In hydrogen storage, quantum computing is being used to analyze molecular interactions in storage materials, helping scientists design high-capacity, thermodynamically stable hydrogen carriers. Quantum simulations can model hydrogen diffusion in MOFs, metal hydrides, and porous carbons, leading to more efficient storage solutions. Moreover, quantum computing plays a crucial role in hydrogen supply chain optimization. Hydrogen distribution networks involve multiple dynamic variables, including production rates, transportation logistics, and demand fluctuations. Quantum algorithms, such as quantum annealing, can optimize these complex networks, ensuring cost-effective and efficient hydrogen distribution at a global scale [45].

Quantum materials are at the forefront of next-generation hydrogen technologies, offering high-efficiency solutions for hydrogen production, storage, and system modeling. QDs are revolutionizing photocatalysis for hydrogen generation, quantum catalysts are enabling faster and more cost-effective electrolysis, and quantum computing is driving breakthroughs in material discovery and system optimization. As depicted in Figure 8.1, bio-inspired catalysts incorporating quantum principles are paving the way for highly efficient and sustainable hydrogen production. With continuous advancements in quantum physics, materials science, and computational techniques, the synergy between quantum materials and hydrogen energy will play a pivotal role in achieving a clean, sustainable, and economically viable hydrogen economy.

8.3.2 Bio-inspired Catalysts and Hydrogen Production

As the global energy demand shifts towards sustainable alternatives, bio-inspired catalysts have emerged as promising solutions for hydrogen production. These catalysts mimic natural enzymatic processes found in biological systems, offering high-efficiency, selectivity, and environmental compatibility. Unlike traditional platinum-based catalysts, bio-inspired catalysts rely on earth-abundant materials, reducing costs while maintaining excellent catalytic activity. The key approaches in bio-inspired hydrogen production include enzyme-mimicking catalysts, microbial and enzymatic hydrogen generation, and synthetic biology-driven advancements [47].

Water splitting is a crucial process for hydrogen production, requiring efficient catalysts to drive the HER and OER. Nature provides a blueprint for efficient catalysis through hydrogenases and oxygen-evolving enzymes, which operate under mild conditions with high turnover rates. Bio-inspired catalysts aim to replicate these enzymatic functions using inorganic, organometallic, and hybrid materials. One of the most studied enzyme-mimicking catalysts is [Fe-Fe] and [Ni-Fe] hydrogenase-inspired catalysts, which mimic the active sites of natural hydrogenases responsible for biological hydrogen production. These synthetic catalysts incorporate iron, nickel, and sulfur-based coordination complexes, offering high catalytic efficiency and stability while eliminating the need for precious metals like platinum. Additionally, MOFs and covalent organic frameworks have been engineered with biomimetic active sites, further enhancing catalytic performance. Another promising bio-inspired approach involves manganese-based catalysts for OER, inspired by the oxygen-evolving complex in photosystem II. These catalysts leverage multi-electron transfer mechanisms, allowing for efficient water oxidation under ambient conditions. The use of bio-inspired ligands, such as porphyrins and macrocyclic compounds, further enhances their stability and electron transport properties, making them viable alternatives for large-scale hydrogen production [48].

In addition to mimicking natural enzymes, researchers are leveraging microbial systems and enzymatic pathways for direct hydrogen production. Biological hydrogen production occurs via fermentative, photofermentative, and photobiological processes, utilizing microorganisms such as bacteria, cyanobacteria, and algae. Dark fermentation is a microbial process where hydrogen is produced as a byproduct of organic matter breakdown under anaerobic conditions. Certain bacteria, such as Clostridium and Enterobacter species, possess hydrogenase and nitrogenase enzymes that facilitate hydrogen evolution from carbohydrates, organic acids, and waste biomass. This approach provides a sustainable method for hydrogen production using agricultural and industrial waste as feedstock [49].

Photofermentation, on the other hand, relies on photosynthetic bacteria, such as Rhodobacter and Rhodopseudomonas species, to convert light energy into hydrogen. These bacteria utilize bacteriochlorophyll pigments to drive electron transfer reactions, producing hydrogen in the presence of organic substrates. While photofermentation achieves higher hydrogen yields compared to dark fermentation, challenges such as low photon conversion efficiency and oxygen sensitivity limit its large-scale application. Photobiological hydrogen production, driven by microalgae and cyanobacteria, harnesses solar energy to split water via photosynthetic pathways. Species like Chlamydomonas reinhardtii possess oxygen-tolerant hydrogenases, enabling direct hydrogen production under low-oxygen conditions. However, limitations such as hydrogenase inhibition by oxygen and low quantum efficiency necessitate further genetic and metabolic engineering interventions to improve productivity [50].

The integration of synthetic biology and genetic engineering has opened new frontiers in bio-inspired hydrogen production, enabling tailored modifications to enzymes, metabolic pathways, and microbial strains for optimized performance. Genetic engineering focuses on enhancing hydrogenase stability, reducing oxygen sensitivity, and increasing metabolic flux towards hydrogen production. For instance, researchers have successfully engineered oxygen-tolerant hydrogenases by introducing site-specific mutations, improving their catalytic efficiency in aerobic environments. Additionally, gene editing techniques such as CRISPR-Cas9 have been employed to redirect metabolic pathways in bacteria and algae, increasing hydrogen output while minimizing byproduct formation.

Another breakthrough in bio-inspired hydrogen production involves the creation of artificial microbial consortia, where multiple engineered species work synergistically to enhance hydrogen yields. For example, co-culturing dark fermentative bacteria with photosynthetic microorganisms enables efficient substrate utilization, continuous hydrogen production, and improved light absorption. Recent advancements in nanobiotechnology have further enhanced the efficiency of bio-inspired

Table 8.2 Comparison of quantum materials and bio-inspired catalysts in hydrogen applications.

Parameter	Quantum materials	Bio-inspired catalysts
Composition	QDs, perovskites, 2D materials	Enzyme-mimicking metal complexes, hydrogenases
Hydrogen production mechanism	Photocatalysis, electrocatalysis, quantum-enhanced charge transfer	Biological and enzymatic pathways, biomimetic catalysis
Efficiency	High photon absorption and charge separation	High selectivity and enzymatic efficiency
Stability	Requires surface modifications for long-term stability	Sensitive to environmental conditions (oxygen, pH)
Scalability	Limited by synthesis complexity and cost	Scalable with metabolic engineering approaches
Operating conditions	High temperatures and controlled environments	Ambient conditions, biological systems
Material abundance	Can require rare elements (e.g., transition metals)	Utilizes earth-abundant materials
Environmental impact	Some materials require toxic precursors	Eco-friendly and biodegradable
Potential applications	Electrolysis, photocatalysis, hydrogen sensors	Biohydrogen production, enzymatic fuel cells
Current limitations	Expensive synthesis, long-term degradation	Sensitivity to oxygen, slow kinetics in some cases

catalysts by integrating nanomaterials with enzymatic systems. Functionalizing hydrogenase enzymes with nanoparticles, carbon nanotubes, or QDs improves electron transfer rates and catalytic stability, resulting in higher hydrogen evolution efficiencies [51]. As depicted in Table 8.2, a comparative analysis of quantum materials and bio-inspired catalysts highlights their respective advantages and limitations in hydrogen applications.

8.4 Hydrogen's Role in Achieving Net-zero Carbon Emissions

8.4.1 Hydrogen in Decarbonizing Heavy Industries

Heavy industries such as steel, cement, and chemicals are among the largest contributors to global carbon emissions, accounting for a significant share of industrial greenhouse gas (GHG) output. These sectors rely heavily on fossil fuels for both energy and feedstock, making decarbonization particularly challenging. Green hydrogen, produced using renewable energy-powered electrolysis, has emerged as a promising solution to reduce emissions in these carbon-intensive industries. By replacing conventional fossil fuel-based processes with hydrogen-based

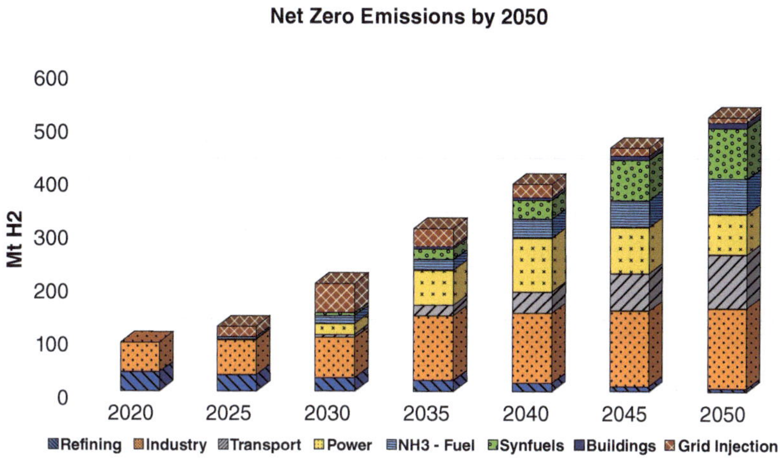

Figure 8.2 Hydrogen's contribution to net-zero carbons goals across industries. [61]/IEA/CC BY 4.0.

technologies, industries can transition toward more sustainable operations while maintaining production efficiency [52].

One of the primary areas where hydrogen can significantly reduce emissions is the steel industry. Traditional steel manufacturing relies on coal-fired blast furnaces that generate substantial amounts of carbon dioxide. Green hydrogen can be used as a reducing agent in the direct reduction of iron (DRI), replacing carbon-intensive coke and eliminating a major source of industrial emissions. This process, often referred to as hydrogen-based DRI, allows for the production of nearly zero-emission steel when paired with renewable energy sources. Major steel producers worldwide are investing in pilot projects to scale up this technology, with countries like Germany and Sweden leading the transition toward hydrogen-based steel production [53]. As shown in Figure 8.2, hydrogen can completely replace coke in the reduction process, significantly cutting emissions while maintaining steel quality. The cement industry is another critical sector where hydrogen can contribute to decarbonization. Cement production is responsible for nearly 8% of global CO_2 emissions, primarily due to the high-temperature kilns that burn fossil fuels and the inherent chemical reactions that release CO_2. Hydrogen can be utilized as a clean fuel in cement kilns, reducing reliance on coal and natural gas. Additionally, the integration of carbon capture and storage technologies alongside hydrogen use in cement manufacturing can further lower emissions, paving the way for a net-zero production process.

Similarly, the chemical industry, particularly ammonia and methanol production, stands to benefit from hydrogen-based processes. Currently, ammonia synthesis relies on the Haber-Bosch process, which uses natural gas as a feedstock, leading to substantial CO_2 emissions. By replacing natural gas-derived hydrogen with green hydrogen, ammonia production can become carbon-neutral. This is particularly important for sectors such as fertilizer manufacturing, where ammonia is a key ingredient. Hydrogen-based methanol production, another essential chemical process, can also significantly reduce carbon emissions, supporting the transition toward sustainable chemical synthesis [54]. Overall, hydrogen's role in decarbonizing heavy industries is critical for achieving net-zero carbon emissions. While challenges such as high costs, infrastructure development, and hydrogen availability remain, ongoing advancements in electrolyzer technology, policy support, and increasing investments in green hydrogen production are expected to accelerate this transition. With the right industrial strategies and government incentives, hydrogen has the potential to revolutionize hard-to-abate sectors, making them more sustainable and environmentally friendly in the long run.

8.4.2 Hydrogen as an Energy Carrier for Carbon-neutral Grids

As the world transitions toward renewable energy sources to combat climate change, the role of hydrogen as an energy carrier has gained significant attention. Hydrogen offers a versatile solution for grid balancing, large-scale energy storage, and backup power systems, addressing the intermittency challenges associated with renewable energy sources like wind and solar. By integrating hydrogen into the energy infrastructure, it is possible to build a more resilient, stable, and carbon-neutral grid that supports the global transition to net-zero emissions [55].

One of the primary applications of hydrogen in power grids is grid balancing and large-scale energy storage. Renewable energy generation from solar and wind is inherently intermittent, meaning that energy production fluctuates based on weather conditions and time of day. This variability can cause supply-demand mismatches, leading to either excess generation or shortfalls in electricity supply [56]. Hydrogen provides a long-duration energy storage solution by enabling surplus electricity from renewables to be used for electrolysis, splitting water into hydrogen and oxygen. The generated hydrogen can then be stored and later converted back into electricity using fuel cells or gas turbines during periods of low renewable output. This process ensures a steady and reliable power supply, reducing reliance on fossil fuel-based backup systems.

The integration of hydrogen with wind and solar energy systems presents a promising pathway to fully renewable power grids. Excess electricity from offshore and onshore wind farms or large-scale solar installations can be used to produce green hydrogen, which can then be stored and transported to areas with high energy demand. Countries with abundant renewable resources, such as Germany, Australia, and Japan, are investing heavily in hydrogen infrastructure to complement their wind and solar expansion [57]. Hydrogen pipelines and storage facilities can serve as strategic energy reserves, allowing for the seasonal storage of excess renewable energy and its distribution across regions where direct transmission via power lines is impractical or inefficient [58].

Beyond storage, hydrogen also plays a critical role in hydrogen-fueled backup power systems. Conventional backup power relies on diesel generators, which contribute to GHG emissions and air pollution. In contrast, hydrogen fuel cells offer a zero-emission alternative for backup power in critical infrastructure, data centers, hospitals, and remote locations. Hydrogen-powered microgrids can operate independently or supplement the main grid, enhancing energy security in case of power outages or extreme weather events. Moreover, hydrogen fuel cell generators are quieter, require less maintenance, and have longer operational lifespans compared to traditional fossil fuel-based generators [59]. The growing role of hydrogen in the energy sector is illustrated in Figure 8.2, which depicts hydrogen's contribution to net-zero carbon goals across industries. As advancements in electrolysis technology, fuel cells, and hydrogen infrastructure continue, the cost-effectiveness and efficiency of hydrogen-based energy solutions are expected to improve. Governments and energy companies worldwide are increasingly recognizing hydrogen as a key enabler of a decarbonized and resilient power grid, driving investments in hydrogen research, development, and deployment.

While challenges such as hydrogen transportation, storage efficiency, and infrastructure costs remain, ongoing technological innovations and policy support will accelerate the widespread adoption of hydrogen in energy systems. By leveraging hydrogen's capabilities as a clean and storable energy carrier, nations can achieve carbon-neutral grids, reduce dependency on fossil fuels, and ensure energy security in the face of climate change [60].

8.5 Hydrogen-powered Smart Cities and Off-grid Applications

Table 8.3 outlines various hydrogen-powered solutions for smart cities and off-grid applications, showcasing their potential in reducing emissions, promoting energy independence, and enhancing resilience. Hydrogen infrastructure can enable clean public transportation, decentralized energy systems (microgrids), and hydrogen-powered buildings that optimize energy consumption. Furthermore, hydrogen offers a sustainable alternative to diesel generators in off-grid and remote locations, while also supporting rural electrification and providing emergency power during disasters.

Table 8.3 Hydrogen-powered smart cities and off-grid applications.

Section	Application	Description	Benefits
Hydrogen infrastructure in smart cities	**Hydrogen-powered public transportation**	Use of hydrogen fuel cells for buses, trams, and trains within urban areas.	Zero-emission, efficient, and sustainable transportation solutions [62].
	Hydrogen fueling stations	Infrastructure for fueling hydrogen-powered vehicles in urban and metropolitan areas.	Reduction in GHG emissions, improving urban air quality [63].

(Continued)

Table 8.3 (Continued)

Section	Application	Description	Benefits
	Hydrogen microgrids	Decentralized energy systems powered by hydrogen, ensuring resilience and autonomy for cities.	Increased energy independence, resilience to grid disruptions, and improved reliability [64].
	Smart hydrogen-powered buildings and heating, ventilation, and air conditioning systems	Hydrogen-based systems used for heating, cooling, and energy management in smart buildings.	Reduced carbon footprint, enhanced energy efficiency, and sustainable building designs [65].
Hydrogen for off-grid and remote applications	**Hydrogen as an alternative to diesel generators**	Hydrogen fuel cells as a clean power source replacing diesel generators in remote or off-grid locations.	Reduction in pollution, lower operational costs, and energy independence in remote areas [66].
	Hydrogen for rural electrification	Providing hydrogen-based electricity for rural and underserved regions to promote development.	Reliable, sustainable electricity for rural areas without reliance on the main grid [67].
	Hydrogen storage solutions for disaster resilience	Hydrogen-based backup systems for emergency power during natural disasters or outages.	Immediate power availability in emergency situations, improving disaster preparedness [68].

8.6 Conclusion and Future Outlook

In this chapter, we have explored the emerging innovations in hydrogen technologies and their transformative potential across various sectors. From AI-driven optimization and quantum materials for hydrogen production, to the integration of bio-inspired catalysts and hydrogen storage solutions, these advancements are creating a foundation for a more sustainable and carbon-neutral energy future. Hydrogen's role as a key enabler of the clean energy transition is increasingly clear, particularly in sectors such as heavy industries, energy grids, smart cities, and off-grid applications.

Despite these groundbreaking developments, several challenges remain that need to be addressed for the widespread adoption of advanced hydrogen technologies. One of the major hurdles is the scalability of hydrogen production at competitive costs, particularly for green hydrogen derived from renewable sources. The infrastructure requirements, such as hydrogen storage, transportation, and distribution systems, remain a significant barrier to the effective deployment of hydrogen at a large scale. Additionally, the technological maturity of hydrogen-based solutions, such as fuel cells and advanced storage methods, still requires further research and development.

Moreover, policy and industrial strategies will play a critical role in accelerating the adoption of hydrogen technologies. Governments must implement supportive policies, such as incentives for green hydrogen production, subsidies for infrastructure development, and regulatory frameworks that support innovation. Private sector involvement will be key to advancing the commercialization of hydrogen-based solutions, with companies investing in scalable technologies and creating public-private partnerships to drive large-scale deployment. Global collaboration across regions will be essential to harmonize standards, facilitate knowledge exchange, and create a unified hydrogen economy.

Looking ahead, the future of hydrogen is bright, with exciting innovations on the horizon. However, addressing the remaining challenges, fostering strong partnerships between industry, government, and academia, and focusing on affordable, scalable, and efficient solutions will be crucial in unlocking hydrogen's full potential and enabling a net-zero carbon economy. The next decade will be pivotal in determining hydrogen's role as a cornerstone of the global clean energy transition.

References

1 Ball M, Wietschel M. The future of hydrogen – Opportunities and challenges. *Int J Hydrogen Energy*. 2009 Jan 1; 34(2):615–27.

2 Jayabal R. Hydrogen energy storage with artificial intelligent-powered strategies for a sustainable future: A review. *J Mech Sci Technol* [Internet]. 2025 Mar 3 [cited 2025 Mar 17];39(3):1503–10. Available from: https://link.springer.com/article/10.1007/s12206-025-0240-3.

3 Fang W, Chen J, Feng Y, Li XZ, Michaelides A. The quantum nature of hydrogen. *Int Rev Phys Chem* [Internet]. 2019 Jan 2 [cited 2025 Mar 17];38(1):35–61. Available from: https://www.tandfonline.com/doi/abs/10.1080/0144235X.2019.1558623.

4 Vyas Y, Chundawat P, Dharmendra D, Punjabi PB, Ameta C. Review on hydrogen production photocatalytically using carbon quantum dots: Future fuel. *Int J Hydrogen Energy*. 2021 Oct 29; 46(75):37208–41.

5 Paudel HP, Syamlal M, Crawford SE, Lee YL, Shugayev RA, Lu P. Quantum computing and simulations for energy applications: Review and perspective. *ACS Engineering Au* [Internet]. 2022 Jun 15 [cited 2025 Mar 17];2(3):151–96. Available from: https://pubs.acs.org/doi/full/10.1021/acsengineeringau.1c00033.

6 Ho MH, Chen S, Rousseau R, Dupuis M, Bullock RM, Raugei S. Bio-inspired molecular catalysts for hydrogen oxidation and hydrogen production. *ACS Symposium Series* [Internet]. 2013 [cited 2025 Mar 17];1133:89–111. Available from: https://pubs.acs.org/doi/abs/10.1021/bk-2013-1133.ch006.

7 Hassan Q, Sameen AZ, Olapade O, Alghoul M, Salman HM, Jaszczur M. Hydrogen fuel as an important element of the energy storage needs for future smart cities. *Int J Hydrogen Energy*. 2023 Sep 12; 48(78):30247–62.

8 Viteri JP, Viteri S, Alvarez-Vasco C, Henao F. A systematic review on green hydrogen for off-grid communities—Technologies, advantages, and limitations. *Int J Hydrogen Energy*. 2023 Jun22; 48(52):19751–71.

9 Parra D, Valverde L, Pino FJ, Patel MK. A review on the role, cost and value of hydrogen energy systems for deep decarbonisation. *Renew Sustain Energy Rev*. 2019 Mar1; 101:279–94.

10 Xie Z, Jin Q, Su G, Lu W. A review of hydrogen storage and transportation: Progresses and challenges. *Energies* [Internet]. 2024 Aug 16 [cited 2025 Mar 17];17(16):4070. Available from: https://www.mdpi.com/1996-1073/17/16/4070/htm.

11 Schönauer AL, Glanz S. Hydrogen in future energy systems: Social acceptance of the technology and its large-scale infrastructure. *Int J Hydrogen Energy*. 2022 Mar 19; 47(24):12251–63.

12 Abdin Z. Empowering the hydrogen economy: The transformative potential of blockchain technology. *Renew Sustain Energy Rev*. 2024 Aug 1;200:114572.

13 Acharya A. Scaling-up green hydrogen development with effective policy interventions. *Article J Develop*. 2022;15(5): 2022.

14 Lane B, Reed J, Shaffer B, Samuelsen S. Forecasting renewable hydrogen production technology shares under cost uncertainty. *Int J Hydrogen Energy*. 2021 Aug 5; 46(54):27293–306.

15 Iliffe ME. Unlocking bulk hydrogen production, storage and transport with learnings from real-life UK trials. *Proceedings of the Annual Offshore Technology Conference* [Internet]. 2023 Apr 24 [cited 2025 Mar 17];2023-May. Available from: https://dx.doi.org/10.4043/32445-MS.

16 Mullanu S, Chua C, Molnar A, Yavari A. Artificial intelligence for hydrogen-enabled integrated energy systems: A systematic review. *Int J Hydrogen Energy*. 2024 Aug 10; 141: 283–303. Available from: https://doi.org/10.1016/j.ijhydene.2024.08.013.

17 Le TT, Sharma P, Bora BJ, Tran VD, Truong TH, Le HC. Fueling the future: A comprehensive review of hydrogen energy systems and their challenges. *Int J Hydrogen Energy*. 2024 Feb 7; 54:791–816.

18 Dreher A, Bexten T, Sieker T, Lehna M, Schütt J, Scholz C. AI agents envisioning the future: Forecast-based operation of renewable energy storage systems using hydrogen with deep reinforcement learning. *Energy Convers Manag*. 2022 Apr 15; 258:115401.

19 Yang Y, Zheng Y, Liu S, Shan M, Guo J, Yang R. Deep learning prediction of photocatalytic water splitting for hydrogen production under natural light based on experiments. *Energy Convers Manag*. 2024 Feb 1;301:118007.

20 Ramkumar G, Tamilselvi M, D Sundarsingh Jebaseelan S, Mohanavel V, Kamyab H, Anitha G. Enhanced machine learning for nanomaterial identification of photo thermal hydrogen production. *Int J Hydrogen Energy*. 2024 Jan 2; 52:696–708.

21 Delali D, Wayo K, Goliatt L, Ganji D. AI and quantum computing in binary photocatalytic hydrogen production. 2024 Dec 31 [cited 2025 Mar 17]; Available from: https://arxiv.org/abs/2501.00575v1.

22 Vo Thanh H, Dai Z, Du Z, Yin H, Yan B, Soltanian MR. Artificial intelligence-based prediction of hydrogen adsorption in various kerogen types: Implications for underground hydrogen storage and cleaner production. *Int J Hydrogen Energy*. 2024 Feb 29; 57:1000–9.

23 Yang R, Mohamed A, Kim K. Optimal design and flow-field pattern selection of proton exchange membrane electrolyzers using artificial intelligence. *Energy*. 2023 Feb 1;264:126135.

24 Oladosu TL, Pasupuleti J, Kiong TS, Koh SPJ, Yusaf T. Energy management strategies, control systems, and artificial intelligence-based algorithms development for hydrogen fuel cell-powered vehicles: A review. *Int J Hydrogen Energy*. 2024 Apr3; 61:1380–404.

25 Rakić E, Grilc M, Likozar B. Liquid organic hydrogen carrier hydrogenation–dehydrogenation: From ab initio catalysis to reaction micro-kinetics modelling. *Chem Eng J*. 2023 Sep 15;472:144836.

26 Wang G, Luo Z, Desta HG, Chen M, Dong Y, Lin B. AI-driven development of high-performance solid-state hydrogen storage. *Energy Rev*. 2025 Mar 1; 4(1):100106.

27 Katterbauer K, Qasim A, Marsala A, Yousef A. A data driven artificial intelligence framework for hydrogen production optimization in waterflooded hydrocarbon reservoir. Society of petroleum engineers—*Abu Dhabi International Petroleum Exhibition and Conference*, ADIP 2021 [Internet]. 2021 Dec 9 [cited 2025 Mar 17]; Available from: https://dx.doi.org/10.2118/207847-MS.

28 Feng Z, Eiubovi I, Shao Y, Fan Z, Tan R. Review of digital twin technology applications in hydrogen energy. *Chain*. 2024 Jul 11; 1(1):54–74.

29 Sivri AH. A data-driven digital twin approach to optimize continuous production environment with deep reinforcement learning. 2023.

30 Cao Y, Dhahad HA, Zare SG, Farouk N, Anqi AE, Issakhov A. Potential application of metal-organic frameworks (MOFs) for hydrogen storage: Simulation by artificial intelligent techniques. *Int J Hydrogen Energy*. 2021 Oct 22; 46(73):36336–47.

31 Tian X, Zhou S, Hao H, Ruan H, Gaddam RR, Dutta RC. Machine learning and density functional theory for catalyst and process design in hydrogen production. *Chain* [Internet]. 2024 Jun [cited 2025 Mar 17];1(2):150–66. Available from: https://ieeexplore.ieee.org/document/10838497/.

32 Ding R, Zhang S, Chen Y, Rui Z, Hua K, Wu Y. Application of machine learning in optimizing proton exchange membrane fuel cells: A review. *Energy AI*. 2022 Aug 1;9:100170.

33 Patil RR, Calay RK, Mustafa MY, Thakur S. Artificial intelligence-driven innovations in hydrogen safety. *Hydrogen* [Internet]. 2024 Jun 8 [cited 2025 Mar 17]; 5(2):312–26. Available from: https://www.mdpi.com/2673-4141/5/2/18/htm.

34 Vidas L, Castro R, Cagno E, Soltani M, Pouresmaeil E. Recent developments on hydrogen production technologies: State-of-the-art review with a focus on green-electrolysis. *Applied Sciences* [Internet]. 2021 Dec 1 [cited 2025 Mar 17];11(23):11363. Available from: https://www.mdpi.com/2076-3417/11/23/11363/htm.

35 Al-Othman A, Tawalbeh M, Martis R, Dhou S, Orhan M, Qasim M. Artificial intelligence and numerical models in hybrid renewable energy systems with fuel cells: Advances and prospects. *Energy Convers Manag*. 2022 Feb 1; 253:115154.

36 Derakhshani R, Lankof L, GhasemiNejad A, Zaresefat M. Artificial intelligence-driven assessment of salt caverns for underground hydrogen storage in Poland. *Sci Rep* [Internet]. 2024 Jun 20 [cited 2025 Mar 17];14(1):1–15. Available from: https://www.nature.com/articles/s41598-024-64020-9.

37 Gan W, Yan M, Yao W, Guo J, Fang J, Ai X. Multi-network coordinated hydrogen supply infrastructure planning for the integration of hydrogen vehicles and renewable energy. *IEEE Trans Ind Appl*. 2022;58(2): 2875–86.

38 Menon SK, Kumar A, Mondal S. Advancements in hydrogen gas leakage detection sensor technologies and safety measures. *Clean Energy* [Internet]. 2025 Jan 22 [cited 2025 Mar 17];9(1):263–77. Available from: https://dx.doi.org/10.1093/ce/zkae122.

39 Onwusinkwue S, Osasona F, Ahmad I, Ahmad I, Anyanwu AC, Dawodu SO. Artificial intelligence (AI) in renewable energy: A review of predictive maintenance and energy optimization. https://wjarr.co.in/sites/default/files/WJARR-2024-0347.pdf [Internet]. 2024 Jan 30 [cited 2025 Mar 17];21(1):2487–99. Available from: https://wjarr.co.in/wjarr-2024-0347.

40 Kandi D, Martha S, Parida KM. Quantum dots as enhancer in photocatalytic hydrogen evolution: A review. *Int J Hydrogen Energy*. 2017 Apr 13; 42(15):9467–81.

41 Tahir M, Tasleem S, Tahir B. Recent development in band engineering of binary semiconductor materials for solar driven photocatalytic hydrogen production. *Int J Hydrogen Energy*. 2020 Jun11; 45(32):15985–6038.

42 Martindale BCM, Hutton GAM, Caputo CA, Reisner E. Solar hydrogen production using carbon quantum dots and a molecular nickel catalyst. *J Am Chem Soc* [Internet]. 2015 May 13 [cited 2025 Mar 17];137(18):6018–25. Available from: https://pubs.acs.org/doi/full/10.1021/jacs.5b01650.

43 Sun J, Duan L, Wu Q, Yao W. Synthesis of MoS2 quantum dots cocatalysts and their efficient photocatalytic performance for hydrogen evolution. *Chem Eng J*. 2018 Jan 15; 332:449–55.

44 Miháliková I, Friák M, Pivoluska M, Plesch M, Saip M, Šob M. Best-practice aspects of quantum-computer calculations: A case study of the hydrogen molecule. *Molecules* [Internet]. 2022 Jan 18 [cited 2025 Mar 17];27(3):597. Available from: https://www.mdpi.com/1420-3049/27/3/597/htm.

45 Ricciardi Celsi M, Ricciardi Celsi L. Quantum computing as a game changer on the path towards a net-zero economy: A review of the main challenges in the energy domain. *Energies* [Internet]. 2024 Feb 22 [cited 2025 Mar 17];17(5):1039. Available from: https://www.mdpi.com/1996-1073/17/5/1039/htm.

46 Kumar N, Aepuru R, Lee SY, Park SJ. Advances in catalysts for hydrogen production: A comprehensive review of materials and mechanisms. *Nanomaterials* [Internet]. 2025 Feb 8 [cited 2025 Mar 17];15(4):256. Available from: https://www.mdpi.com/2079-4991/15/4/256/htm.

47 Ho MH, O'Hagan M, Dupuis M, Dubois DL, Bullock RM, Shaw WJ. Water-assisted proton delivery and removal in bio-inspired hydrogen production catalysts. *Dalton*

Trans [Internet]. 2015 Jun 10 [cited 2025 Mar 17];44(24):10969–79. Available from: https://pubs.rsc.org/en/content/articlehtml/2015/dt/c5dt00782h.

48 Aralekallu S, Sannegowda LK, Singh V. Developments in electrocatalysts for electrocatalytic hydrogen evolution reaction with reference to bio-inspired phthalocyanines. *Int J Hydrogen Energy*. 2023 May 22; 48(44):16569–92.

49 Hallenbeck PC. Microbial paths to renewable hydrogen production. *Biofuels* [Internet]. 2011 May [cited 2025 Mar 17];2(3):285–302. Available from: https://www.tandfonline.com/doi/abs/10.4155/bfs.11.6.

50 Dutta S. Review on solar hydrogen: Its prospects and limitations. *Energy Fuels* [Internet]. 2021 Aug 5 [cited 2025 Mar 17];35(15):11613–39. Available from: https://pubs.acs.org/doi/abs/10.1021/acs.energyfuels.1c00823.

51 Ergal İ, Bochmann G, Fuchs W, Rittmann SKM. Design and engineering of artificial microbial consortia for biohydrogen production. *Curr Opin Biotechnol*. 2022 Feb 1; 73:74–80.

52 Benavides K. *Exploring the role of hydrogen in decarbonizing heavy industry*; 2023.

53 Borup R, Krause T, Brouwer J. Hydrogen is essential for industry and transportation decarbonization. *Electrochem Soc Interface* [Internet]. 2021 Dec 1 [cited 2025 Mar 17];30(4):79–83. Available from: https://iopscience.iop.org/article/10.1149/2.F18214IF.

54 Bhattacharyya R, Singh KK, Grover RB, Bhanja K. Nuclear hydrogen production for industrial decarbonization: Creating the business case for the near term. *Int J Energy Res* [Internet]. 2022 Apr 1 [cited 2025 Mar 17];46(5):6929–43. Available from: https://onlinelibrary.wiley.com/doi/full/10.1002/er.7572.

55 Muradov NZ, Veziroğlu TN. "Green" path from fossil-based to hydrogen economy: An overview of carbon-neutral technologies. *Int J Hydrogen Energy*. 2008 Dec 1; 33(23):6804–39.

56 Koivunen T, Khosravi A, Syri S. The role of power–to–hydrogen in carbon neutral energy and industrial systems: Case Finland. *Energy*. 2023 Dec 1;284:128624.

57 Liu X, Liu G, Xue J, Wang X, Li Q. Hydrogen as a carrier of renewable energies toward carbon neutrality: State-of-the-art and challenging issues. *Int J Miner Metall Mater* [Internet]. 2022 May 1 [cited 2025 Mar 17];29(5):1073–89. Available from: https://link.springer.com/article/10.1007/s12613-022-2449-9.

58 Zhou Y. Low-carbon transition in smart city with sustainable airport energy ecosystems and hydrogen-based renewable-grid-storage-flexibility. *Energy Rev*. 2022 Sep 1; 1(1):100001.

59 Bandyopadhyay D, Sutar PS, Sonawane SB, Rairikar S, Thipse SS, Jadhav A. Hydrogen as a carbon neutral ICE fuel for future India. *SAE Tech Pap* [Internet]. 2024 Jan 16 [cited 2025 Mar 17]; Available from: https://www.sae.org/publications/technical-papers/content/2024-26-0177/.

60 Hui Y, Wang M, Guo S, Akhtar S, Bhattacharya S, Dai B. Comprehensive review of development and applications of hydrogen energy technologies in China for carbon neutrality: Technology advances and challenges. *Energy Convers Manag*. 2024 Sep 1;315:118776.

61 Hydrogen Production and Infrastructure Projects Database—Data product—IEA [Internet]. [cited 2025 Apr 7]. Available from: https://www.iea.org/data-and-statistics/data-product/hydrogen-production-and-infrastructure-projects-database.

62 Han J, Wang J, He Z, An Q, Song Y, Mujeeb A. Hydrogen-powered smart grid resilience. *Energy Convers Econ* [Internet]. 2023 Apr 1 [cited 2025 Mar 17];4(2): 89–104. Available from: https://onlinelibrary.wiley.com/doi/full/10.1049/enc2.12083.

63 Bouramdane AA. Understanding and optimizing hydrogen transport technologies for 21st century smart cities: Innovation for Morocco's green interconnections in hydrogen, ammonia, and kerosene production. *The Emerald Handbook of Smart Cities in the Gulf Region: Innovation, Development, Transformation, and Prosperity for Vision 2040* [Internet]. 2024 Nov 22 [cited 2025 Mar 17];405–80. Available from: https://www.emerald.com/insight/content/doi/10.1108/978-1-83608-292-720241030/full/html.

64 Lin RH, Zhao YY, Wu BD. Toward a hydrogen society: Hydrogen and smart grid integration. *Int J Hydrogen Energy*. 2020 Aug 7; 45(39):20164–75.

65 Hussain SN, Ghosh A. Highly insulated hydrogen powered vacuum gasochromic smart windows for energy-efficient buildings. *J Build Eng*. 2024 Nov 15;97:110855.

66 Wang Y, Kazemi M, Nojavan S, Jermsittiparsert K. RETRACTED: Robust design of off-grid solar-powered charging station for hydrogen and electric vehicles via robust optimization approach. *Int J Hydrogen Energy*. 2020 Jul 31; 45(38):18995–9006.

67 Kaushal J, Chowdhury SPD. Hydrogen-powered fuel cell integration in low voltage microgrid systems: Performance evaluation and power quality analysis. *Int J Ambient Energy* [Internet]. 2025 Dec 31 [cited 2025 Mar 17];46(1):2446524. Available from: https://www.tandfonline.com/doi/abs/10.1080/01430750.2024.2446524.

68 Zhang F, Thanapalan K, Procter A, Carr S, Maddy J, Premier G. Power management control for off-grid solar hydrogen production and utilisation system. *Int J Hydrogen Energy*. 2013 Apr 15; 38(11):4334–41.

9

Hydrogen Safety, Regulations, and Standardization

9.1 Introduction to Hydrogen Safety and Standardization

Hydrogen is widely regarded as a key component of the future energy landscape, offering a sustainable and clean alternative to fossil fuels. However, despite its advantages, hydrogen presents significant safety challenges that must be carefully managed through strict regulations and standardization efforts. Hydrogen is a highly flammable gas with a broad combustion range of 4% to 75% in air, meaning that even small leaks can create a potentially explosive mixture when combined with oxygen. Additionally, its low ignition energy of approximately 0.02 mJ makes it susceptible to ignition from electrostatic discharges or minimal heat sources. These risks are further complicated by hydrogen's small molecular size, which allows it to diffuse through many materials, increasing the likelihood of leaks from pipelines, storage tanks, and containment vessels. Such leaks are particularly dangerous because hydrogen burns with an almost invisible flame, making it difficult to detect and control fires in case of an accident [1]. As shown in Figure 9.1, hydrogen safety challenges exist at every stage of the value chain, from production and storage to transport and end-use applications, necessitating a comprehensive approach to risk assessment and mitigation.

Hydrogen production involves processes such as electrolysis, steam methane reforming (SMR), and thermochemical cycles, each of which carries unique safety concerns. High-pressure electrolysis systems operate at pressures exceeding 30–80 bar, posing a risk of rupture and sudden gas release, while SMR processes involve high temperatures and flammable gases, creating explosion hazards. Thermochemical production methods require handling of corrosive chemicals, increasing the likelihood of leaks and material degradation. Without strict safety protocols, hydrogen production sites could face industrial accidents, necessitating real-time monitoring, automated shut-off mechanisms, and explosion-proof containment facilities. Similarly, hydrogen storage poses additional risks depending on the method used. Compressed hydrogen storage at pressures of 350–700 bar introduces a high risk of tank ruptures due to mechanical failure or material fatigue. Over time, hydrogen embrittlement can weaken storage containers, making them

Next-Generation Hydrogen Economy: Production, Storage, and Fuel Cell Technologies, First Edition.
Peeyush Phogat, Shreya Sharma and Satyam Rawat.

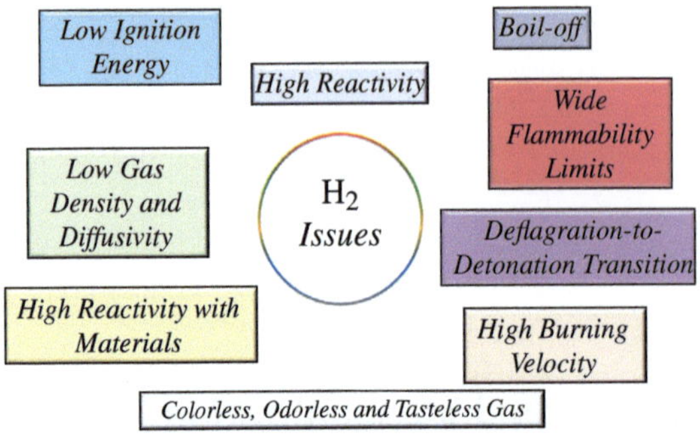

Figure 9.1 Hydrogen safety challenges across the value chain – This figure illustrates critical safety concerns at various stages of hydrogen deployment, highlighting risks in production, storage, and transportation, and emphasizing the need for global standardization efforts. [8]/MDPI/CC BY 4.0.

more prone to fractures. Cryogenic liquid hydrogen storage, which requires maintaining temperatures as low as −253 °C, is associated with risks such as rapid expansion in case of leaks, boil-off losses, and potential frostbite hazards. Chemical storage methods, such as ammonia and liquid organic hydrogen carriers (LOHCs), introduce concerns related to toxicity and controlled hydrogen release mechanisms. Advanced storage solutions incorporating reinforced composite tanks, thermal insulation, and automated venting systems have been developed to address these risks. Transportation of hydrogen further amplifies safety concerns, as it is often moved through pipelines, high-pressure tankers, and refueling stations. Hydrogen pipelines must be constructed using specialized materials resistant to embrittlement, as prolonged exposure can degrade traditional steel structures and lead to pipeline failures. Hydrogen fueling stations must operate at extremely high pressures, requiring robust dispenser mechanisms and automated leak detection systems to prevent accidental releases. Transporting cryogenic hydrogen in tankers presents additional challenges due to boil-off losses and the need for reinforced insulation to maintain stability at ultra-low temperatures. Innovations in hydrogen logistics, including real-time leak monitoring, artificial intelligence (AI)-based predictive maintenance, and reinforced containment technologies, are crucial to minimizing risks associated with large-scale hydrogen transport. The implementation of such advanced safety measures across hydrogen infrastructure is essential for ensuring reliability and public confidence in hydrogen as a clean energy source [2–4].

Due to these inherent safety risks, global standardization efforts are necessary to create a uniform framework for hydrogen handling, ensuring that safety practices remain consistent across different countries and industries. Currently, hydrogen safety regulations vary widely, creating challenges in infrastructure compatibility, transport policies, and emergency response planning [5]. For instance, hydrogen refueling stations in different regions operate at varying pressures – typically 350 bar for buses and heavy vehicles, while passenger vehicles require 700 bar. Without a standardized approach, interoperability between different hydrogen fuel stations remains a challenge, slowing down the transition to hydrogen-powered transportation. Similarly, material standards for pipelines and storage tanks differ between the

the United States, Europe, and Asia, leading to inconsistencies in safety practices. Additionally, industrial hydrogen applications, such as its use in ammonia production and steel manufacturing, require harmonized safety codes to prevent operational inefficiencies and potential hazards. Organizations such as the International Organization for Standardization (ISO), the US Department of Energy (DOE), the National Fire Protection Association (NFPA), and the European Commission have developed regulatory frameworks to address hydrogen safety concerns. Standards such as ISO 14687, which defines purity levels for hydrogen fuel, and ISO 19880-1, which sets guidelines for hydrogen refueling stations, play a vital role in ensuring the safe expansion of hydrogen infrastructure [6].

Despite these efforts, challenges remain in achieving a globally unified safety and regulatory framework for hydrogen. Many regions are still developing their hydrogen safety regulations, leading to inconsistencies in implementation. Some countries prioritize hydrogen refueling infrastructure, while others focus more on pipeline transport and industrial applications. This lack of alignment poses difficulties for companies looking to expand hydrogen projects internationally. Additionally, existing safety standards must continuously evolve to accommodate new hydrogen technologies, such as metal hydride storage, advanced electrolysis methods, and AI-driven leak detection systems. Establishing a global regulatory alliance for hydrogen safety, supported by cross-border collaboration and industry partnerships, will be essential to overcoming these hurdles and accelerating the widespread adoption of hydrogen energy [7]. As illustrated in Figure 9.1, ensuring a safe hydrogen value chain requires addressing risks at every stage – from production and storage to distribution and final usage – through a combination of regulatory oversight, technological advancements, and continuous safety monitoring. Hydrogen safety is a critical factor in its large scale deployment, requiring stringent regulations and global standardization to mitigate risks. Key hazards, such as high flammability, leakage risks, and material degradation, necessitate advanced safety measures, including reinforced storage tanks, real-time leak detection, and emergency shut-off systems. Global standardization efforts aim to unify hydrogen safety codes and ensure compatibility across different infrastructure components, but challenges remain in achieving full regulatory harmonization. By prioritizing safety innovations and international collaboration, the hydrogen industry can address these concerns, paving the way for a sustainable and risk-free hydrogen economy.

9.2 Hydrogen Safety Protocols in Production, Storage, and Transportation

9.2.1 Safety Considerations in Hydrogen Production

Hydrogen production processes involve high-temperature reactions, high-pressure systems, and volatile gases, making safety a critical concern in industrial-scale hydrogen generation. The two most common methods of hydrogen production – electrolysis and thermochemical processes – pose distinct safety risks, including high-pressure hazards, explosion potential, and fire risks. Without proper engineering controls and industrial safety protocols, these risks could lead to equipment failures, hazardous gas leaks, or catastrophic accidents. As summarized in Table 9.1, different hydrogen production methods come with varying levels of safety challenges based on their operating conditions, reaction mechanisms, and containment requirements.

One of the primary risks in hydrogen production is high-pressure electrolysis, which is widely used for green hydrogen generation through proton exchange membrane (PEM) electrolysis and alkaline electrolysis. These systems operate at pressures ranging from 30 to 80 bar, and in some cases, exceed 100 bar for high-efficiency industrial applications. The main danger associated with high-pressure electrolysis is sudden pressure vessel failure, which can lead to rapid hydrogen gas expansion, equipment rupture, and potential fire hazards. Additionally, electrolysis involves high-voltage electrical inputs, increasing the risk of electrical short circuits, overheating, and ignition in faulty systems.

Table 9.1 Comparative safety risks in different hydrogen production methods.

Hydrogen production method	Primary safety risks	Hazard potential	Mitigation strategies
Electrolysis (PEM & alkaline)	High-pressure gas leaks, electrical hazards, membrane degradation	Risk of hydrogen-oxygen crossover, potential for electrical short circuits, high-pressure explosions	Pressure relief valves, leak detection sensors, automated shut-off mechanisms, use of advanced membrane materials [13]
SMR	High-temperature operation, CO poisoning, flammable gas mixtures	Fire hazards due to methane and hydrogen presence, potential for carbon monoxide exposure	Flame-proof containment systems, exhaust gas treatment, continuous gas monitoring [14]
Thermochemical water splitting	Handling of corrosive and toxic chemicals, extreme operating temperatures	Chemical burns, toxic gas release, potential reactor overpressure	Corrosion-resistant materials, sealed chemical containment, emergency cooling systems [15]
Biological hydrogen production	Microbial contamination, slow hydrogen generation rates	Risk of pathogen exposure, flammability concerns with hydrogen off-gassing	Controlled bioreactor environments, sterilization protocols, hydrogen venting systems [16]
Chemical hydrogen production (ammonia cracking, LOHCs)	Toxic byproducts, catalyst degradation, heat management challenges	Release of hazardous ammonia or organic vapors, potential fire risks	Advanced catalyst formulations, gas containment enclosures, chemical neutralization systems [17]

Advanced safety mechanisms, including pressure relief valves, reinforced cell designs, and automated shutdown controls, help mitigate these risks. Another significant concern is oxygen and hydrogen cross-contamination, which can occur due to membrane failures in PEM electrolysis, leading to the formation of explosive gas mixtures [9].

Thermochemical hydrogen production, including SMR, biomass gasification, and metal oxide cycles, presents additional risks related to high-temperature operations and flammable gas mixtures. SMR, the most widely used method for large-scale hydrogen production, operates at temperatures exceeding 800–1,000 °C, creating an inherently hazardous environment. The process involves methane (CH_4), hydrogen (H_2), carbon monoxide (CO), and high-temperature steam (H_2O), all of which contribute to fire hazards. Gas leaks, improper thermal insulation, or incomplete combustion reactions can result in flash fires or carbon monoxide poisoning. Furthermore, thermochemical cycles, such as the Sulfur-Iodine cycle and metal oxide redox cycles, require handling of corrosive and toxic chemicals, which necessitate strict material containment and ventilation protocols to prevent hazardous exposure incidents [10]. As illustrated in Table 9.1, thermochemical production poses a higher fire risk compared to electrolysis due to the involvement of hydrocarbon feedstocks and exothermic reactions. The explosion risks in hydrogen generation plants are another major safety concern. Hydrogen is a highly flammable gas with a wide combustion range (4–75% in air), making leak detection and controlled containment critical. Industrial hydrogen plants must implement real-time gas monitoring, flame detection sensors, and emergency venting systems to minimize explosion hazards. Uncontrolled hydrogen accumulation in confined spaces can lead to detonations, particularly in steam reforming and gasification plants, where large volumes of flammable gases are processed. Proper ventilation, intrinsically safe electrical equipment, and explosion-proof storage tanks are necessary to reduce the likelihood of ignition [11].

Fire hazards and thermal runaway risks in hydrogen production facilities also present significant challenges. In PEM electrolyzers and high-pressure alkaline electrolysis, excessive electrical currents, electrode overheating, or cell membrane degradation can result in localized thermal runaway. This can lead to rapid increases in temperature and pressure, ultimately causing equipment failure or hydrogen ignition. Similarly, in biomass gasification and plasma-assisted hydrogen production, incomplete combustion reactions or insufficient cooling mechanisms can result in runaway thermal conditions, increasing the risk of fire and system damage. Implementing temperature monitoring sensors, emergency cooling systems, and controlled shutdown protocols can help mitigate these risks. Hydrogen production presents varied safety challenges, including high-pressure risks, explosion potential, and fire hazards, depending on the method used [12]. As shown in Table 9.1, electrolysis is primarily associated with high-pressure containment risks and electrical failures, whereas thermochemical methods involve extreme temperatures and flammable gas mixtures. Industrial hydrogen plants must integrate advanced leak detection, fire suppression, and explosion mitigation systems to ensure safe operation and large-scale deployment of hydrogen technologies.

9.2.2 Safety Protocols for Hydrogen Storage

Ensuring the safe storage of hydrogen is a critical challenge due to its high flammability, rapid diffusion rate, and tendency to cause material degradation over time. Hydrogen is typically stored in compressed gas tanks, cryogenic liquid containers, or chemically bonded carriers, each of which presents unique safety risks that must be mitigated through advanced leak detection, material innovations, and emergency protection systems [18]. As shown in Figure 9.2, hydrogen storage safety mechanisms include real-time leak monitoring, structural reinforcement of storage tanks, and automated venting systems designed to prevent hazardous failures.

One of the most significant concerns in hydrogen storage is leak detection and rapid containment. Due to hydrogen's small molecular size and high permeability, leaks are difficult to detect and can easily accumulate in confined spaces, leading to explosive conditions. Modern leak detection systems use electrochemical sensors, infrared spectroscopy, and ultrasonic gas detectors to identify even trace amounts of hydrogen leaks. These systems are integrated into high-pressure tanks, refueling stations, and pipeline networks to ensure early warning and rapid response. In addition, automated safety shut-off valves are used to isolate leaks and

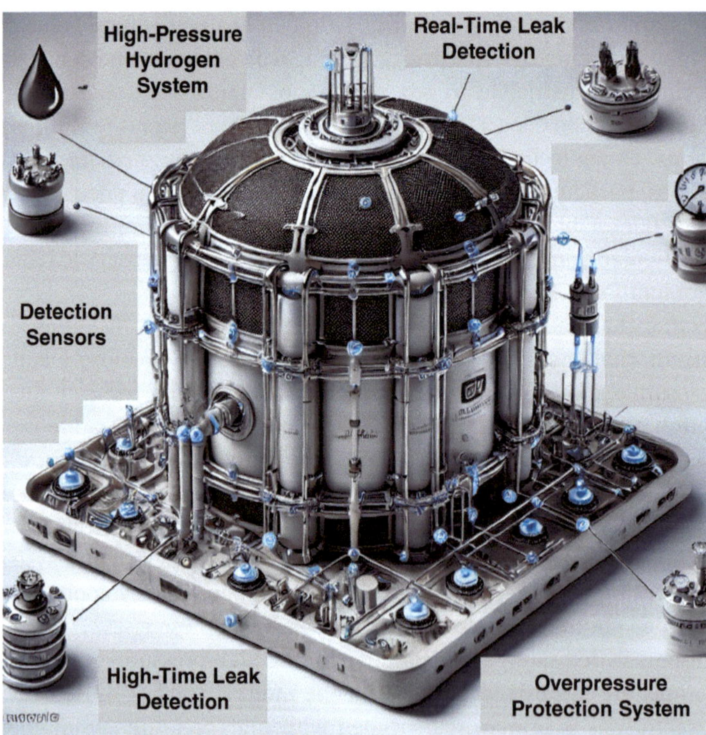

Figure 9.2 Hydrogen storage safety mechanisms: This figure illustrates the key safety technologies used in hydrogen storage, including high-pressure reinforced tanks, real-time leak detection sensors, embrittlement-resistant materials, and overpressure protection systems designed to enhance operational safety and prevent failures. Generated with AI using ChatGPT – OpenAI.

prevent uncontrolled gas escape in the event of a system failure. These valves are programmed to activate immediately if a pressure drop or abnormal gas flow is detected, significantly reducing the likelihood of fire or explosion [19].

Another major safety risk in hydrogen storage is hydrogen embrittlement, a phenomenon in which hydrogen atoms penetrate metal surfaces, leading to structural weakness, cracking, and potential tank rupture. This issue is particularly problematic in steel and metal alloy storage tanks, where long-term exposure to hydrogen can degrade mechanical strength. Material innovations, such as hydrogen-resistant composite tanks, protective nano-coatings, and non-metallic liners, have been developed to mitigate this risk. Carbon fiber-reinforced tanks and polymer-based storage materials are increasingly used to enhance structural durability and prevent embrittlement-related failures [20].

Emergency venting systems and overpressure protection mechanisms are essential safety components for hydrogen storage facilities. Given the high-pressure conditions required for hydrogen storage (typically 350–700 bar for compressed gas and vacuum-insulated cryogenic tanks for liquid hydrogen), there is a risk of overpressure buildup due to temperature fluctuations, gas expansion, or equipment malfunction. Pressure relief valves and burst discs are installed in storage tanks to automatically release excess hydrogen before pressure reaches critical levels. In cryogenic hydrogen storage, controlled venting systems prevent boil-off losses and minimize the risk of pressure explosions caused by trapped gaseous hydrogen [21]. As shown in Figure 9.2, the integration of advanced leak detection, structural reinforcement, and automated safety mechanisms is essential for the safe storage and handling of hydrogen. The continuous advancement of storage materials and monitoring technologies will further enhance hydrogen safety, ensuring its widespread adoption as a sustainable energy carrier.

9.2.3 Risk Mitigation in Hydrogen Transportation and Distribution

Hydrogen transportation and distribution present significant safety challenges due to hydrogen's high flammability, small molecular size, and tendency to cause material degradation over time. Whether transported through pipelines, high-pressure tankers, or hydrogen carriers, ensuring structural integrity, leak prevention, and explosion risk mitigation is essential for large-scale hydrogen deployment [22]. As summarized in Table 9.2, different hydrogen transportation methods come with unique risks and safety considerations, requiring the use of hydrogen-compatible materials, advanced monitoring technologies, and specialized transport infrastructure to ensure safe and efficient hydrogen delivery.

One of the primary concerns in hydrogen pipeline transportation is pipeline integrity and material compatibility. Unlike natural gas, hydrogen has a higher diffusion rate, making it prone to leakage through pipeline joints and microscopic cracks. Additionally, long-term exposure to hydrogen leads to hydrogen embrittlement, where hydrogen atoms penetrate metal surfaces, causing structural weakening and increased fracture risk. To mitigate this issue, hydrogen-compatible materials such as high-strength stainless steel, polymer-lined composite pipes, and hydrogen-resistant alloys are being developed. Additionally, real-time pressure

Table 9.2 Safety considerations for different hydrogen transportation methods.

Transport method	Primary safety risks	Challenges	Mitigation strategies
Hydrogen pipelines	Hydrogen embrittlement, leakage due to small molecular size, pressure buildup	Requires specialized steel or composite pipelines, potential for ruptures under high pressure	Use of hydrogen-compatible alloys, periodic integrity testing, automated leak detection [27]
Compressed hydrogen tankers (road transport)	High-pressure explosion risks, impact damage during transit	Hydrogen stored at 350–700 bar, requires reinforced carbon-fiber tanks	Crash-resistant tank design, safety release valves, temperature and pressure monitoring [28]
Hydrogen rail transport	Vibration-induced fatigue, risk of hydrogen leaks in enclosed railcars	Shock absorption and vibration-proof containment needed for stability	Reinforced storage compartments, vibration-damping mounting, real-time gas leak detection [29]
Cryogenic liquid hydrogen (maritime transport)	Boil-off losses, insulation failure, cryogenic spills	Hydrogen stored at −253 °C, requiring vacuum-insulated containers	Multi-layer insulation, controlled venting systems, automated boil-off recovery [30]
Ammonia & LOHC-based transport	Toxicity risks, chemical decomposition, accidental spills	Requires safe chemical handling, catalytic decomposition units for hydrogen extraction	Chemical-resistant tanks, real-time ammonia/LOHC leak sensors, closed-loop safety systems [31]

monitoring, automated leak detection, and inline pipeline inspection technologies are crucial for ensuring the safe and continuous transportation of hydrogen. Pipeline purging techniques using inert gases like nitrogen are also employed to minimize the risk of hydrogen-oxygen mixtures forming explosive concentrations [23].

Hydrogen fueling stations also require stringent safety measures to ensure secure handling during refueling. Hydrogen is typically stored at extremely high pressures (350–700 bar) in fueling dispensers, increasing the risks of hose failures, valve malfunctions, and accidental gas leaks. To address these challenges, modern hydrogen refueling stations incorporate reinforced high-pressure nozzles, automated shut-off mechanisms, and infrared leak detection systems to prevent gas escape. Real-time monitoring systems continuously measure pressure, temperature, and flow rates to detect irregularities and trigger emergency shutdowns if anomalies are detected [24]. Furthermore, blast-resistant refueling station designs with hydrogen venting systems and explosion-proof enclosures are used to minimize risks in case of accidental hydrogen release. Hydrogen transport via road, rail, and maritime shipping introduces additional risks, particularly related to gas containment, pressure regulation, and cryogenic handling. High-pressure hydrogen tankers used for road transport must comply with strict weight, pressure, and ventilation standards to prevent tank over pressurization or rupture. Hydrogen transport via rail requires specialized shock-absorbing containers and reinforced pipeline connections to withstand mechanical vibrations and sudden impacts. In maritime hydrogen transport, cryogenic liquid hydrogen carriers face challenges such as boil-off losses, insulation breakdown, and safe venting of excess hydrogen gas [25]. Table 9.2 highlights the specific safety challenges and mitigation strategies for various hydrogen transport methods. By integrating hydrogen-resistant materials, real-time monitoring, automated safety shut-offs, and reinforced containment technologies, risks associated with hydrogen transportation and distribution can be significantly minimized. Future advancements in hydrogen pipeline infrastructure, fuel cell-powered transport solutions, and cryogenic containment materials will further enhance the safety and scalability of hydrogen as a clean energy carrier [26].

9.3 Regulatory Landscape: International Hydrogen Safety Standards and Policies

9.3.1 Global Hydrogen Safety Standards

The widespread adoption of hydrogen as a clean energy carrier requires a comprehensive regulatory framework to ensure safe production, storage, transportation, and usage. Given hydrogen's high flammability, low ignition energy, and potential for leaks due to its small molecular size, global safety standards play a crucial role in mitigating risks and fostering public and industrial confidence in hydrogen technologies. Several international organizations and regulatory bodies have established hydrogen safety guidelines to address infrastructure development, equipment handling, and emergency response protocols. These standards are essential for ensuring interoperability, hazard mitigation, and regulatory compliance across different regions [32].

9.3.2 International Hydrogen Safety Organizations

Several key organizations oversee hydrogen safety regulations worldwide, each focusing on different aspects of infrastructure development, industrial operations, and transportation:

- International Organization for Standardization (ISO): The ISO 14687 standard defines purity requirements for hydrogen used in fuel cells, while ISO 19880-1 establishes safety guidelines for hydrogen refueling stations. The ISO 15916 standard focuses on basic safety considerations for hydrogen energy applications [33].
- Occupational Safety and Health Administration (OSHA, United States): OSHA regulates industrial hydrogen handling, with guidelines for workplace exposure limits, storage requirements, and equipment safety. The OSHA 29 CFR 1910.103 standard specifically covers hydrogen systems safety in industrial facilities [34].

- NFPA, United States: NFPA provides fire and explosion prevention codes related to hydrogen production and storage. The NFPA 2 Hydrogen Technologies Code outlines ventilation, leak detection, and emergency shutdown procedures for hydrogen facilities [35].
- European Commission (EC) and European Union (EU) Regulations: The EU Alternative Fuels Infrastructure Directive (AFID) promotes the development of safe hydrogen refueling networks across Europe, ensuring that hydrogen infrastructure meets stringent safety and efficiency standards [36].
- Society of Automotive Engineers (SAE): The SAE J2601 standard governs hydrogen refueling protocols for fuel cell vehicles, ensuring that hydrogen dispensers, pressures, and fueling times are standardized globally [33].

These organizations work collectively to harmonize hydrogen safety measures, addressing regional differences in regulatory frameworks while promoting safe hydrogen adoption across industries.

9.3.3 Hydrogen Safety Codes for Infrastructure Development and Risk Assessment

Hydrogen infrastructure development – including production plants, refueling stations, pipelines, and storage facilities – requires strict adherence to safety codes to prevent accidents, gas leaks, and fire hazards. Key safety codes and guidelines include:

- Hydrogen Storage and Pipeline Safety: Hydrogen is stored in high-pressure tanks, cryogenic liquid containers, and chemical carriers, each requiring specific safety measures. The ISO 16111 standard addresses metal hydride storage system safety, while ISO 17124 provides guidance on pipelines and pressurized hydrogen transport systems [37].
- Refueling Station Safety Standards: The ISO 19880-1 standard defines design requirements, emergency shutdown systems, and dispenser safety measures for hydrogen refueling stations. Additionally, NFPA 55 regulates hydrogen gas storage at fueling stations, ensuring proper ventilation and explosion prevention [38].
- Fire and Explosion Risk Mitigation: Standards such as NFPA 2, OSHA 29 CFR 1910.103, and IEC 60079 outline explosion-proof electrical components, safe venting mechanisms, and hydrogen gas dispersion modeling to prevent ignition hazards in industrial hydrogen plants [39].
- Hydrogen Fuel Cell System Safety: The ISO 26142 standard provides guidelines for hydrogen leak detection in fuel cell applications, while SAE J2578 addresses hydrogen vehicle safety requirements, ensuring that fuel cell electric vehicles (FCEVs) meet strict safety protocols [40].

The integration of these global safety standards ensures that hydrogen infrastructure meets uniform safety benchmarks, preventing technical inconsistencies and safety lapses. However, regional differences in safety regulations still exist, creating challenges in international hydrogen trade and cross-border technology deployment.

Efforts are ongoing to harmonize hydrogen safety regulations globally, fostering a standardized, risk-free hydrogen economy. Hydrogen safety standards play a critical role in ensuring the safe production, storage, transportation, and usage of hydrogen across industrial and commercial applications. Organizations such as ISO, OSHA, NFPA, EC, and SAE have developed comprehensive safety frameworks to mitigate hydrogen-related risks, ensuring compliance with fire protection, explosion prevention, and leak detection protocols. Despite these advancements, global standardization efforts must continue to address inconsistencies in regional regulations, ensuring seamless integration of hydrogen infrastructure worldwide. By adhering to strict safety guidelines and risk assessment measures, hydrogen can be securely implemented as a key pillar of the future clean energy economy [41].

9.3.4 Country-specific Hydrogen Safety Regulations

As hydrogen adoption accelerates globally, country-specific safety regulations are being developed to ensure secure production, storage, and transportation. While international organizations like ISO, OSHA, and NFPA provide broad guidelines, individual nations establish region-specific policies tailored to their infrastructure, industrial needs, and hydrogen economy growth strategies. As shown in Table 9.3, major hydrogen economies such as the United States, EU, Japan, China, and India have formulated unique safety standards and regulatory frameworks to govern hydrogen technologies.

Table 9.3 Comparison of hydrogen safety policies across major hydrogen economies.

Country/Region	Regulatory body	Key safety standards & policies	Focus areas
United States	DOE, OSHA, NFPA	NFPA 2 Hydrogen Technologies Code, OSHA 29 CFR 1910.103	Hydrogen storage, refueling safety, industrial handling
EU	EC	EU AFID	Hydrogen refueling station standards, pipeline transport safety
Japan	METI	Fuel cell vehicle safety, hydrogen infrastructure guidelines	Commercial hydrogen deployment, safety in fuel cells and power plants
China	NEA	Emerging hydrogen safety guidelines	Industrial hydrogen handling, refueling station safety
India	Ministry of New & Renewable Energy (MNRE)	National Green Hydrogen Mission (NGHM) Safety Regulations	Electrolysis safety, fuel cell deployment, hydrogen transport

In the United States, hydrogen safety is primarily regulated by the DOE, OSHA, and the NFPA. The NFPA 2 Hydrogen Technologies Code provides detailed safety requirements for hydrogen production, storage, and fueling stations, while OSHA 29 CFR 1910.103 ensures workplace hydrogen safety in industrial settings. Additionally, the DOE Hydrogen and Fuel Cells Program establishes safety guidelines for research, pilot projects, and commercial-scale hydrogen deployment [42]. The EU follows a centralized regulatory framework under the EU AFID, which mandates safety standards for hydrogen refueling stations, transportation, and integration with renewable energy sources. This directive ensures harmonized hydrogen infrastructure development across EU member states while prioritizing safety and environmental sustainability. Japan, a global leader in hydrogen fuel cell technology, regulates hydrogen safety through the Ministry of Economy, Trade, and Industry (METI). METI has developed strict guidelines for fuel cell vehicle safety, refueling infrastructure, and hydrogen power plants, ensuring safe commercialization of hydrogen technologies. Japan's Basic Hydrogen Strategy emphasizes risk management and safety innovations to promote public confidence in hydrogen adoption [43]. In China and India, hydrogen safety regulations are emerging, with policies focusing on infrastructure development and industrial safety standards. China has introduced hydrogen handling and storage safety policies under its National Energy Administration (NEA), while India's National Green Hydrogen Mission is working on safety guidelines for electrolysis, fuel cells, and hydrogen distribution networks. These nations are rapidly developing hydrogen deployment strategies, incorporating global best practices to ensure safe hydrogen adoption [44].

As shown in Table 9.3, while hydrogen safety policies differ across countries, efforts are underway to standardize regulations globally, ensuring safe, efficient, and scalable hydrogen deployment worldwide.

9.4 Challenges in Standardization for Hydrogen Infrastructure

9.4.1 Interoperability of Hydrogen Refueling Stations and Pipelines

The interoperability of hydrogen refueling stations and pipelines is a crucial factor in enabling a seamless hydrogen infrastructure for widespread adoption. Ensuring compatibility between refueling stations, fuel cell vehicles, and hydrogen pipelines requires standardized refueling pressures, dispenser designs, and material selection to prevent leakage, embrittlement, and operational inefficiencies. Addressing these challenges is key to developing a globally integrated hydrogen economy where refueling stations and transport networks can support diverse applications, including FCEVs, buses, heavy-duty trucks, and industrial hydrogen users [45].

One of the primary challenges in hydrogen refueling station interoperability is the difference in refueling pressures for various types of hydrogen-powered vehicles. Passenger vehicles typically require 700 bar (H70) refueling, while buses and heavy-duty trucks use 350 bar (H35) refueling due to their larger storage capacities and lower volumetric constraints. The lack of universal refueling dispensers that can accommodate both pressure standards poses a major compatibility issue for multi-use hydrogen stations. To address this, international organizations such as ISO (ISO 19880-1) and SAE (SAE J2601) have developed standardized refueling protocols to ensure that nozzle configurations, pressure regulators, and dispenser flow rates are harmonized across different vehicle types. Multi-pressure hydrogen refueling stations, equipped with dual-dispenser technology, are being deployed in some regions to provide both 350 bar and 700 bar refueling options at a single facility, improving accessibility for various vehicle categories [46].

Beyond refueling stations, the interoperability of hydrogen pipelines presents additional challenges, primarily related to material selection and leakage prevention. Unlike natural gas pipelines, hydrogen pipelines require specialized materials that can withstand hydrogen embrittlement, a process in which hydrogen atoms penetrate metals, causing cracks, structural degradation, and increased fracture risk. Conventional steel pipelines are highly susceptible to embrittlement, necessitating the use of hydrogen-resistant alloys, polymer-lined steel, or reinforced composite materials to ensure long-term durability and leak-proof transportation. Additionally, hydrogen's small molecular size allows it to diffuse through seals, gaskets, and pipeline joints, increasing the potential for undetected leaks [47]. To mitigate this, advanced sealing technologies, real-time leak detection sensors, and pipeline monitoring systems are being integrated into hydrogen transport networks. Standardizing hydrogen pipeline design is also critical for global infrastructure interoperability. Differences in pipeline pressure ratings, connection standards, and transmission protocols between countries and regions create inconsistencies that hinder international hydrogen trade and cross-border distribution. Efforts are being made by global regulatory bodies to establish harmonized pipeline specifications that ensure safe, efficient, and scalable hydrogen transport across various industries [48].

9.4.2 Certification of Hydrogen Storage and Transport Technologies

The safe and efficient storage and transportation of hydrogen require harmonized global certification standards to ensure interoperability, reliability, and compliance with safety protocols. Given hydrogen's high flammability, low ignition energy, and material compatibility issues, standardized safety testing procedures are essential to mitigate risks in high-pressure storage tanks, cryogenic hydrogen systems, and hydrogen transport pipelines. However, variations in regional certification frameworks pose challenges for global hydrogen trade and cross-border infrastructure

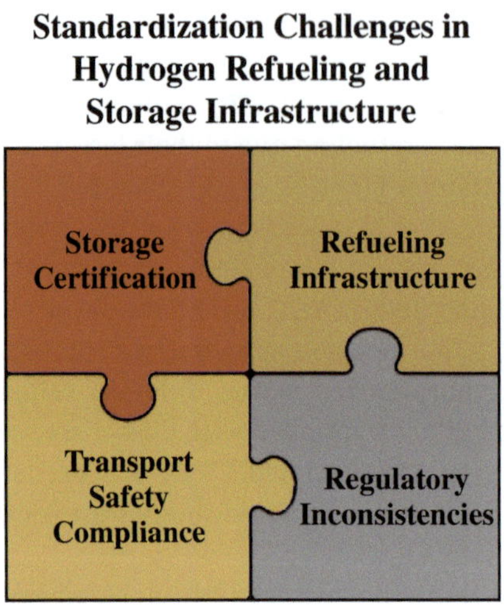

Figure 9.3 Standardization challenges in hydrogen refueling and storage infrastructure: This figure illustrates the key regulatory inconsistencies affecting hydrogen storage certification, refueling infrastructure, and transport safety compliance worldwide. Generated with AI using ChatGPT – OpenAI.

deployment. As shown in Figure 9.3, inconsistencies in hydrogen refueling, storage, and transportation standards create barriers to achieving a unified hydrogen economy [49].

One of the key certification challenges is ensuring compatibility between hydrogen production, storage, and end-use applications. Hydrogen storage tanks, whether compressed gas, cryogenic liquid, or chemical carriers, must comply with different pressure, temperature, and material standards depending on the region. For example, ISO 16111 provides international standards for hydrogen storage in metal hydrides, while Department of Transportation (DOT) regulations in the United States set specific requirements for hydrogen pressure vessels. Similarly, EU and Japanese regulations mandate distinct testing procedures for hydrogen refueling stations and storage infrastructure, leading to technical mismatches between different markets [50]. As summarized in Table 9.4, key differences in hydrogen safety testing and certification exist across major hydrogen economies, making it difficult to achieve seamless integration of hydrogen storage and transport technologies. To address these challenges, global efforts are underway to harmonize safety testing and certification processes, ensuring that hydrogen tanks, pipelines, and transport methods meet consistent safety and performance standards. The establishment of a unified global certification framework will be crucial in facilitating widespread hydrogen deployment while enhancing safety, efficiency, and scalability across the entire hydrogen value chain.

Table 9.4 Key differences in hydrogen safety testing and certification across regions.

Region	Regulatory body	Key certification standards	Safety testing requirements	Focus areas
United States	DOT, OSHA, NFPA	49 CFR 178 (Hydrogen Cylinder Certification), NFPA 2 Hydrogen Code	Hydrogen storage tanks must withstand pressure cycling, fire resistance tests, and impact tests	High-pressure tanks, industrial hydrogen handling, fuel cell vehicle safety [51]
EU	EC, ISO	ISO 16111 (Metal Hydride Storage), ISO 19880-1 (Hydrogen refueling stations)	Stringent leak testing, thermal expansion tests, and environmental durability assessments	Standardized hydrogen fueling stations, cross-border hydrogen transport [52]
Japan	METI, JPEC (Japan Petroleum Energy Center)	JPEC-S 0003 (Hydrogen Refueling Safety), METI Hydrogen Guidelines	Tanks must be tested for hydrogen embrittlement, structural integrity, and earthquake resistance	Fuel cell vehicle safety, earthquake-resistant hydrogen infrastructure [53]
China	NEA, SAC	GB/T 34537 (Hydrogen Storage Safety), GB/T 37768 (Hydrogen Refueling Guidelines)	Pressure cycling tests, material fatigue tests, and burst tests for storage systems	Industrial hydrogen production, pipeline safety, emerging hydrogen economy [54]
India	BIS, MNRE	NGHM Safety Standards (under development)	Hydrogen cylinder certification, fire safety assessments, and risk analysis for hydrogen storage	Electrolysis-based hydrogen production, transportation safety, fuel cell deployment [55]

9.5 Hydrogen's Role in Public Safety and Environmental Impact Assessment

As hydrogen emerges as a key player in the clean energy transition, ensuring public safety and minimizing environmental risks are critical to its widespread adoption. Hydrogen's unique properties – high energy density, zero carbon emissions at the point of use, and versatility across industries – make it an attractive energy carrier. However, concerns regarding its safety, infrastructure development, and environmental impact must be addressed through public education, stringent safety protocols, and comprehensive environmental assessments. The role of hydrogen in public safety and environmental impact extends beyond industrial applications, influencing policy decisions, infrastructure investments, and public perception [56].

9.5.1 Public Awareness and Safety Training for Hydrogen Adoption

A major challenge in hydrogen deployment is ensuring that industries, policymakers, emergency responders, and the general public are well-informed about hydrogen safety. Unlike conventional fossil fuels, hydrogen has different handling, storage, and combustion characteristics, which require specialized safety training and risk management strategies. One of the most critical areas of hydrogen safety awareness is training first responders such as firefighters, police officers, and emergency medical personnel to handle hydrogen-related incidents. Since hydrogen is odorless, colorless, and burns with an invisible flame, conventional firefighting techniques may not be effective in detecting and suppressing hydrogen fires. Emergency responders need specialized infrared detection equipment, ventilation strategies for hydrogen leaks, and training on handling high-pressure hydrogen storage systems. Several countries, including Japan, the United States, and Germany, have already introduced hydrogen safety training programs for first responders, ensuring that potential hazards at fueling stations, storage sites, and industrial hydrogen plants are well managed [57, 58].

Public perception is another key factor influencing hydrogen adoption. While hydrogen-powered vehicles and energy systems offer clean and efficient alternatives, public skepticism remains due to safety concerns associated with hydrogen leaks, refueling station explosions, and transportation risks. Effective risk communication strategies are essential to educate consumers on the safety measures and regulations in place to ensure hydrogen's safe usage. Public outreach campaigns, demonstration projects, and transparent communication on safety protocols can help increase trust and confidence in hydrogen-powered technologies. Hydrogen-powered public transportation systems, including fuel cell buses, hydrogen trains, and aviation projects, require specific public engagement initiatives to educate commuters and local communities on the environmental and safety benefits of hydrogen adoption. Governments and industries are increasingly focusing on public-private partnerships to develop hydrogen refueling infrastructure, ensuring that public concerns and safety considerations are addressed before large-scale deployment [59, 8].

9.5.2 Environmental Risks and Benefits of Hydrogen as a Clean Energy Source

While hydrogen is widely recognized as a clean energy carrier, its environmental impact depends on the method of production. Hydrogen can be classified into three primary types based on carbon emissions and sustainability:

a. Green Hydrogen – Produced through electrolysis powered by renewable energy sources (solar, wind, or hydroelectric). This method generates zero carbon emissions, making it the most environmentally sustainable option [60].

b. Blue Hydrogen – Derived from natural gas with carbon capture and storage. Although it significantly reduces CO_2 emissions compared to conventional fossil fuels, it still has some environmental footprint due to methane emissions and energy consumption [61].

c. Gray Hydrogen – Produced from SMR without carbon capture, releasing high levels of CO_2 into the atmosphere. It is the most commonly used hydrogen production method today but is unsustainable in the long term [62].

A direct carbon footprint comparison between green and blue hydrogen highlights the significant environmental benefits of green hydrogen. For example, producing 1 kg of gray hydrogen emits approximately 9–12 kg of CO_2, whereas blue hydrogen reduces this by nearly 50%. In contrast, green hydrogen emits no direct CO_2, making it the most viable option for achieving net-zero emissions. Another critical environmental consideration is water usage in hydrogen production [63]. Electrolysis-based green hydrogen requires large volumes of purified water to generate hydrogen and oxygen. On average, 9 liters of water are required to produce 1 kg of hydrogen. While this may not be a major concern in water-abundant regions, in water-scarce areas, reliance on electrolysis could pose challenges for long-term sustainability. Research is ongoing into seawater electrolysis and wastewater-based hydrogen production to minimize freshwater consumption and improve the sustainability of hydrogen production. Additionally, large-scale hydrogen deployment requires an assessment of hydrogen's impact on atmospheric chemistry. Hydrogen leaks into the atmosphere can indirectly contribute to climate change by increasing the lifespan of methane, a potent greenhouse gas. Although hydrogen itself does not trap heat like CO_2 or methane, leaked hydrogen can react with atmospheric oxidants, reducing their ability to break down methane. This is why pipeline integrity, leak detection systems, and strict regulatory measures are crucial to prevent uncontrolled hydrogen emissions [64].

9.6 Summary and Future Directions

As hydrogen becomes a cornerstone of the global clean energy transition, ensuring comprehensive safety measures and regulatory frameworks is critical for its widespread adoption. The challenges associated with hydrogen production, storage, transportation, and end-use applications require rigorous safety protocols, advanced risk mitigation strategies, and harmonized regulatory standards to facilitate secure and efficient hydrogen deployment worldwide.

A key aspect of hydrogen safety is its unique physical and chemical properties, which demand specialized handling and containment measures. Hydrogen production facilities, including electrolysis plants, SMR units, and thermochemical reactors, must adhere to strict safety guidelines to prevent gas leaks, fire hazards, and explosion risks. In storage, pressurized hydrogen tanks, cryogenic systems, and chemical carriers require reinforced materials, leak detection sensors, and emergency venting mechanisms to minimize risks. Hydrogen transportation, whether through pipelines, road tankers, rail networks, or maritime shipping, also presents unique safety challenges, necessitating hydrogen-compatible infrastructure and real-time monitoring to prevent embrittlement, pressure buildup, and accidental leaks.

Global safety standards and regulations play a crucial role in establishing uniform best practices across different regions. Organizations such as ISO, OSHA, NFPA, METI, and the EC have developed safety guidelines for hydrogen infrastructure, ensuring that storage tanks, refueling stations, and transportation networks meet international safety benchmarks. However, disparities in national policies and certification procedures create barriers to hydrogen standardization. As hydrogen trade expands, there is an urgent need for a globally unified regulatory framework that ensures interoperability between hydrogen infrastructure, refueling standards, and safety compliance measures across regions.

As hydrogen deployment scales up, future advancements in real-time hydrogen leak detection and explosion prevention technologies will significantly enhance operational safety. New sensor-based monitoring systems, AI-driven predictive maintenance, and automated emergency shutdown mechanisms will play a crucial role in preventing hydrogen-related accidents. Additionally, next-generation materials, such as hydrogen-resistant alloys, nanostructured coatings, and composite storage tanks, will further reduce hydrogen embrittlement risks and improve infrastructure durability.

Policy recommendations for harmonized global hydrogen regulations include:

- Establishing Unified Hydrogen Safety Codes – Governments and regulatory bodies should work towards a common international hydrogen standard that aligns production, storage, transport, and usage regulations to ensure seamless infrastructure development.
- Developing Integrated Hydrogen Safety Training Programs – Mandatory education and certification programs for first responders, industry professionals, and policymakers will enhance public confidence and operational safety in hydrogen systems.
- Strengthening International Collaboration – Cross-border initiatives between hydrogen-leading economies (e.g., United States, EU, Japan, China, and India) should focus on harmonizing refueling protocols, transportation safety standards, and risk assessment methodologies to facilitate global hydrogen trade and deployment.
- Encouraging Innovation in Hydrogen Safety Technologies – Governments should fund research into advanced leak detection systems, explosion-resistant materials, and automated safety solutions to further reduce hydrogen-related risks and improve infrastructure resilience.

In conclusion, while hydrogen offers a transformative pathway toward a clean energy future, achieving its safe and standardized deployment requires coordinated international efforts. The continuous improvement of safety protocols, real-time monitoring systems, and regulatory harmonization will be essential to making hydrogen a widely accepted, commercially viable, and environmentally sustainable energy source.

References

1 Yang Y, Xu H, Lin L, Bao W, Zhang B, Ai B. Development of standards for hydrogen safety. In: *E3S Web of Conferences* [Internet]. 2020 Oct 15 [cited 2025 Apr 1];194:02013. Available from: https://www.e3s-conferences.org/articles/e3sconf/abs/2020/54/e3sconf_icaeer2020_02013/e3sconf_icaeer2020_02013.html.

2 Chau K, Djire A, Khan F. Review and analysis of the hydrogen production technologies from a safety perspective. *Int J Hydrogen Energy*. 2022 Apr 5;47(29): 13990–4007.

3 Aprea JL. Quality specification and safety in hydrogen production, commercialization and utilization. *Int J Hydrogen Energy*. 2014 May 27;39(16): 8604–8.

4 Guandalini G, Campanari S, Valenti G. Comparative assessment and safety issues in state-of-the-art hydrogen production technologies. *Int J Hydrogen Energy*. 2016 Nov 9;41(42):18901–20.

5 Groth KM, Al-Douri A. Hydrogen safety, risk, and reliability analysis. *Hydrogen economy: Processes, supply chain, life cycle analysis and energy transition for sustainability*. 2023 Jan 1;487–510. Available from: https://doi.org/10.1016/B978-0-323-99514-6.00012-1.

6 Gasteiger HA, Baker DR, Carter RN, Gu W, Liu Y, Wagner FT, et al. Electrocatalysis and catalyst degradation challenges in proton exchange membrane fuel cells. *Hydrogen Energy*. 2010;3–16.

7 Cristello J, Dang Z, Hugo R, Park SS. Artificial intelligence based leak detection in blended hydrogen and natural gas pipelines. *Int J Hydrogen Energy*. 2024 Nov 19;91:744–64.

8 Calabrese M, Portarapillo M, Di Nardo A, Venezia V, Turco M, Luciani G, et al. Hydrogen safety challenges: A comprehensive review on production, storage, transport, utilization, and CFD-based consequence and risk assessment. *Energies* [Internet]. 2024 Mar 12 [cited 2025 Mar 25];17(6):1350. Available from: https://www.mdpi.com/1996-1073/17/6/1350/htm.

9 Grigoriev SA, Porembskiy VI, Korobtsev S V., Fateev VN, Auprêtre F, Millet P. High-pressure PEM water electrolysis and corresponding safety issues. *Int J Hydrogen Energy*. 2011 Feb 1;36(3):2721–8.

10 Wang Z, Secnik E, Naterer GF. Hazard analysis and safety assurance for the integration of nuclear reactors and thermochemical hydrogen plants. *Process Saf Environ Prot*. 2015 Jul 1;96:82–97.

11 Ade N, Alsuhaibani A, El-Halwagi MM, Goyette H, Wilhite B. Integrating safety and economics in designing a steam methane reforming process. *Int J Hydrogen Energy*. 2022 Jan 29;47(9):6404–14.

12 Mohsen R, Olewski T, Badreldin A, Abdel-Wahab A, Véchot L. Explosive mixture formation in PEM electrolyzers: A safety review using fault tree analysis. *Renew Sustain Energy Rev*. 2025 Apr 1;211:115225.

13 Wang J, Yang J, Feng Y, Hua J, Chen Z, Liao M, et al. Comparative experimental study of alkaline and proton exchange membrane water electrolysis for green hydrogen production. *Appl Energy*. 2025 Feb 1;379:124936.

14 Kadam RS, Sutar PR, Yadav GD. Energy and environmental assessment of industrial-scale hydrogen production: Comparison of steam methane reforming, electrolysis, and Cu–Cl cycles. *Int J Hydrogen Energy*. 2025 Mar 6;106:994–1005.

15 Yekta MA, Ghandehariun S. Thermal management of sodium-oxygen-hydrogen (Na-O-H) thermochemical water splitting for sustainable hydrogen production. *Int J Hydrogen Energy*. 2025 Mar 10;107:296–311.

16 Ahmaruzzaman M, Mishra SR, Gadore V, Sharma DK. Hydrogen generation by photolysis of water vis-à-vis other conventional and advanced non-conventional methods of hydrogen production—A review. *J Inorg Organomet Polym Mater* [Internet]. 2024 Aug 3 [cited 2025 Apr 1];35(1):22–45. Available from: https://link.springer.com/article/10.1007/s10904-024-03272-4.

17 Krishnan V, Hussain K. Hydrogen production from ammonia cracking. *Sustainable and green catalytic processes for renewable fuel production with net-zero emissions*. 2025 Jan 1;261–86. Available from: https://doi.org/10.1016/B978-0-443-21899-6.00011-2.

18 Sun H, Wang Z, Meng Q, White S. Advancements in hydrogen storage technologies: Enhancing efficiency, safety, and economic viability for sustainable energy transition. *Int J Hydrogen Energy*. 2025 Mar 4;105:10–22.

19 Hansen OR. Hydrogen infrastructure—Efficient risk assessment and design optimization approach to ensure safe and practical solutions. *Process Saf Environ Prot*. 2020 Nov 1;143:164–76.

20 Meda US, Bhat N, Pandey A, Subramanya KN, Lourdu Antony Raj MA. Challenges associated with hydrogen storage systems due to the hydrogen embrittlement of high strength steels. *Int J Hydrogen Energy*. 2023 Jun 1;48(47):17894–913.

21 Hu Q, Zhang X, Hao H. A review of hydrogen-air cloud explosions: The fundamentals, overpressure prediction methods, and influencing factors. *Int J Hydrogen Energy*. 2023 Apr 29;48(36):13705–30.

22 Sun D, Guo D, Xie D. Using multicriteria decision making to evaluate the risk of hydrogen energy storage and transportation in cities. *Sustainability* [Internet]. 2023 Jan 6 [cited 2025 Apr 1];15(2):1088. Available from: https://www.mdpi.com/2071-1050/15/2/1088/htm.

23 Weber M, Perrin J. Hydrogen transport and distribution. *Hydrogen Technol* [Internet]. 2008 Jul 17 [cited 2025 Apr 1];129–49. Available from: https://link.springer.com/chapter/10.1007/978-3-540-69925-5_4.

24 Khani H, El-Taweel NA, Farag HEZ. Supervisory scheduling of storage-based hydrogen fueling stations for transportation sector and distributed operating reserve in electricity markets. *IEEE Trans Industr Inform*. 2020 Mar 1;16(3):1529–38.

25 Chen D, Wu C, Li J. A fast and reliable model for predicting hydrogen-methane-air blast loading in unconfined spaces for blast-resistant design. *Int J Hydrogen Energy*. 2025 Jan 6;97:1316–26.
26 Feng W, Ruiz C. Risk management of energy communities with hydrogen production and storage technologies. *Appl Energy*. 2023 Oct 15;348:121494.
27 Panfilov M. Underground and pipeline hydrogen storage. *Compendium of hydrogen energy: Hydrogen storage, distribution and infrastructure*. 2016 Jan 1;2:91–115. Available from: https://doi.org/10.1016/B978-1-78242-362-1.00004-3.
28 Ustolin F, Lamb JJ, Burheim OS, Pollet BG. Energy and safety of hydrogen storage. *Hydrog Biomass Bioenergy*. 2020 Jan 1;133–53.
29 Xu Z, Zhao N, Hillmansen S, Roberts C, Yan Y. Techno-economic analysis of hydrogen storage technologies for railway engineering: A review. *Energies* [Internet]. 2022 Sep 5 [cited 2025 Apr 1];15(17):6467. Available from: https://www.mdpi.com/1996-1073/15/17/6467/htm.
30 Ustolin F, Campari A, Taccani R. An extensive review of liquid hydrogen in transportation with focus on the maritime sector. *J Marine Sci Eng* [Internet]. 2022 Sep 1 [cited 2025 Apr 1];10(9):1222. Available from: https://www.mdpi.com/2077-1312/10/9/1222/htm.
31 Teichmann D, Arlt W, Schlücker E, Wasserscheid P. Transport and storage of hydrogen via liquid organic hydrogen carrier (LOHC) systems. *Hydrog Sci Eng Mater Processes Syst Technol* [Internet]. 2016 Apr 29 [cited 2025 Apr 1];2:811–30. Available from: https://onlinelibrary.wiley.com/doi/full/10.1002/9783527674268.ch33.
32 San Marchi C, Hecht ES, Ekoto IW, Groth KM, LaFleur C, Somerday BP, et al. Overview of the DOE hydrogen safety, codes and standards program, part 3: Advances in research and development to enhance the scientific basis for hydrogen regulations, codes and standards. *Int J Hydrogen Energy*. 2017 Mar 16;42(11):7263–74.
33 Pique S, Weinberger B, De-Dianous V, Debray B. Comparative study of regulations, codes and standards and practices on hydrogen fuelling stations. *Int J Hydrogen Energy*. 2017 Mar 16;42(11):7429–39.
34 Skiba R. Competency standards for emerging hydrogen related activities. *Open J Saf Sci Technol* [Internet]. 2020 [cited 2025 Apr 1];10(02):42–52. Available from: https://shs.hal.science/halshs-02890289.
35 Rivkin C, Blake C, Burgess R, Buttner WJ, Post MB. A national set of hydrogen codes and standards for the United States. *Int J Hydrogen Energy*. 2011 Feb 1;36(3):2736–41.
36 Apak S, Atay E, Tuncer G. Renewable hydrogen energy regulations, codes and standards: Challenges faced by an EU candidate country. *Int J Hydrogen Energy*. 2012 Apr 1;37(7):5481–97.
37 Yang Y, Xu H, Lu Q, Bao W, Lin L, Ai B, et al. Development of standards for hydrogen storage and transportation. In: *E3S Web of Conferences* [Internet]. 2020 Oct 15 [cited 2025 Apr 1];194:02018. Available from: https://www.e3s-conferences.org/articles/e3sconf/abs/2020/54/e3sconf_icaeer2020_02018/e3sconf_icaeer2020_02018.html.

38 Genovese M, Cigolotti V, Jannelli E, Fragiacomo P. Current standards and configurations for the permitting and operation of hydrogen refueling stations. *Int J Hydrogen Energy*. 2023 Jun 16;48(51):19357–71.

39 Salehi F, Abbassi R, Asadnia M, Chan B, Chen L. Overview of safety practices in sustainable hydrogen economy – An Australian perspective. *Int J Hydrogen Energy*. 2022 Sep 22;47(81):34689–703.

40 Neef HJ. International overview of hydrogen and fuel cell research. *Energy*. 2009 Mar 1;34(3):327–33. Available from: https://doi.org/10.1016/j.energy.2008.08.014.

41 Guduru R, Patel R, Singh R, Vij RK. Government policies, guidelines, initiatives, and supports for underground hydrogen storage. *Subsurface hydrogen energy storage: Current status, prospects, and challenges*. 2025 Jan 1;321–57.

42 Čekerevac Z. Hydrogen hazards, risks, and protection: An in-depth review. *MEST J* [cited 2025 Apr 1]; Available from: https://orcid.org/0000-0003-2972-2472.

43 Baumgart M, Lavrijssen S. Exploring regulatory strategies for accelerating the development of sustainable hydrogen markets in the European Union. *J Energy Natural Resour Law* [Internet]. 2024 Apr 2 [cited 2025 Apr 1];42(2):137–66. Available from: https://www.tandfonline.com/doi/abs/10.1080/02646811.2023.2257528.

44 Rajagopal C, Velpandian M. Critical review of hydrogen safety assessment tools: Indian perspective. *Trans Indian National Academy Eng* 2024 9: 4 [Internet]. 2024 Jul 29 [cited 2025 Apr 1];9(4):737–50. Available from: https://link.springer.com/article/10.1007/s41403-024-00490-7.

45 Raghu D. Roadmap for repurposing natural gas infrastructure for hydrogen storage and transport—A review of the opportunities and challenges. 2024 Apr 29 [cited 2025 Apr 1]; Available from: https://dx.doi.org/10.4043/35217-MS.

46 Buschmann T, Gallep J. Mobile cross-modal infrastructure for the supply of electricity and hydrogen. *Lecture Notes Intell Transp Infrastruct* [Internet]. 2025 [cited 2025 Apr 1];Part F154:599–611. Available from: https://link.springer.com/chapter/10.1007/978-3-031-82818-8_45.

47 Hu D, Gao P, Cheng Z, Shen Y, He R, Yi F, et al. Comprehensive review of hydrogen leakage in relation to fuel cell vehicles and hydrogen refueling stations: Status, challenges, and future prospects. *Energy and Fuels* [Internet]. 2024 Mar 21 [cited 2025 Apr 1];38(6):4803–35. Available from: https://pubs.acs.org/doi/full/10.1021/acs.energyfuels.3c04557.

48 Joshi A, Sattari F, Lefsrud L, Tufail M, Khan MA. Mitigating uncertainty: A risk informed approach for deploying hydrogen refueling stations. *Int J Hydrogen Energy*. 2024 Jul 12;74:136–50. Available from: https://doi.org/10.1016/j.ijhydene.2024.06.085.

49 Arrieta MP, Villa JL. Challenges for the implementation of hydrogen's storage and transportation in Colombia. *2024 IEEE Technology and Engineering Management Society, TEMSCON LATAM* 2024. 2024. Available from: https://doi.org/10.1109/TEMSCONLATAM61834.2024.10717740.

50 Wang Z, Li M, Zhao F, Ji Y, Han F. Status and prospects in technical standards of hydrogen-powered ships for advancing maritime zero-carbon transformation. *Int J Hydrogen Energy*. 2024 Apr 10;62:925–46.

51 Chandler J, Paterson J. Selling sunshine: Emerging challenges in the certification of hydrogen. *J Energy Natural Resour Law* [Internet]. 2025 Jan 2 [cited 2025

Apr 1]; Available from: https://www.tandfonline.com/doi/abs/10.1080/02646811.2024.2344363.

52 Alarcon C, Alarcon S, Hoffer A, Pavez B. Hydrogen 5.0: Interdisciplinary development of a proof-of-concept smart system for green hydrogen leak detection. *Processes* [Internet]. 2025 Feb 24 [cited 2025 Apr 1];13(3):639. Available from: https://www.mdpi.com/2227-9717/13/3/639/htm.

53 Sakamoto J, Sato R, Nakayama J, Kasai N, Shibutani T, Miyake A. Leakage-type-based analysis of accidents involving hydrogen fueling stations in Japan and USA. *Int J Hydrogen Energy*. 2016 Dec 14;41(46):21564–70.

54 Li J, Li J, Guan C, Zhang S, Li X, Liu X. The development of hydrogen energy equipment safety certification in China. American society of mechanical engineers, pressure vessels and piping division (Publication) PVP [Internet]. 2024 Nov 14 [cited 2025 Apr 1];1. Available from: https://dx.doi.org/10.1115/PVP2024-123163.

55 Jayakumar A, Madheswaran DK, Kannan AM, Sureshvaran U, Sathish J. Can hydrogen be the sustainable fuel for mobility in India in the global context? *Int J Hydrogen Energy*. 2022 Sep 15;47(79):33571–96.

56 Min M, Yoon C, Yoo N, Kim J, Yoon Y, Jung S. Hydrogen risk assessment studies: A review toward environmental sustainability. *Energies* [Internet]. 2025 Jan 7 [cited 2025 Apr 1];18(2):229. Available from: https://www.mdpi.com/1996-1073/18/2/229/htm.

57 Huan N, Yamamoto T, Sato H, Tzioutzios D, Yin H, Sala R. Does accident awareness affect people's risk perception of hydrogen infrastructure and information-seeking behaviour? *Appl Energy*. 2024 Jun 15;364:123141.

58 Kim DH, Park HK, Lee MC. Barriers and strengthening strategies on the Korea safety laws for the application of hydrogen and ammonia as power generation fuels based on the expert survey. *Int J Hydrogen Energy*. 2024 Jun 12;70:568–78.

59 Bulat HH, Kaymakçı ÖT, Ilhan H. Human factors in hydrogen storage: An analysis of safety implications. *Int J Hydrogen Energy*. 2025 Mar 10;107:548–58.

60 Kourougianni F, Arsalis A, Olympios AV, Yiasoumas G, Konstantinou C, Papanastasiou P, et al. A comprehensive review of green hydrogen energy systems. *Renew Energy*. 2024 Sep 1;231:120911.

61 Shen H, Crespo del Granado P, Jorge RS, Löffler K. Environmental and climate impacts of a large-scale deployment of green hydrogen in Europe. *Energy Clim Change*. 2024 Dec 1;5:100133.

62 Saha P, Akash FA, Shovon SM, Monir MU, Ahmed MT, Khan MFH, et al. Grey, blue, and green hydrogen: A comprehensive review of production methods and prospects for zero-emission energy. *Int J Green Energy* [Internet]. 2024 May 2 [cited 2025 Apr 1];21(6):1383–97. Available from: https://www.tandfonline.com/doi/abs/10.1080/15435075.2023.2244583.

63 El-Adawy M, Dalha IB, Ismael MA, Al-Absi ZA, Nemitallah MA. Review of sustainable hydrogen energy processes: Production, storage, transportation, and color-coded classifications. *Energy Fuels* [Internet]. 2024 Dec 5 [cited 2025 Apr 1]; Available from: https://pubs.acs.org/doi/full/10.1021/acs.energyfuels.4c04317.

64 Schlehuber S, Gutsch M, Kronemeyer N, Frieden F, von Delft S. Assessment of environmental impacts and costs for hydrogen-powered cargo shipping in Europe. *Commun Eng* 2025 4: 1 [Internet]. 2025 Mar 17 [cited 2025 Apr 1];4(1):1–10. Avail able from: https://www.nature.com/articles/s44172-025-00388-4.

10

Industrial Scale-up and Commercialization of Hydrogen Technologies

10.1 Introduction to Industrial Hydrogen Scale Up

Hydrogen is emerging as a key energy carrier in the transition to a low-carbon economy, offering a viable solution for decarbonizing hard-to-abate industrial sectors such as steel, cement, chemicals, power generation, and transportation. As industries seek cleaner alternatives to fossil fuels, hydrogen presents an opportunity to replace carbon-intensive energy sources, significantly reducing greenhouse gas emissions. However, achieving industrial-scale hydrogen adoption comes with several challenges, including cost, scalability, and infrastructure readiness. Addressing these barriers is crucial for hydrogen to become a mainstream industrial energy source rather than remaining confined to demonstration projects and pilot-scale applications [1].

The global hydrogen industry has seen a surge in investments and policy commitments in recent years. Countries such as Germany, Japan, South Korea, and Australia have launched national hydrogen strategies, outlining ambitious plans for integrating hydrogen into industrial supply chains [2]. Similarly, the European Union's Hydrogen Strategy aims to install at least 40 GW of electrolyzer capacity by 2030, demonstrating a strong push toward large-scale hydrogen adoption. Despite this momentum, current hydrogen production is still largely dominated by fossil fuel-based methods, particularly gray hydrogen, which is produced from natural gas without carbon capture. Transitioning to low-carbon hydrogen alternatives, such as blue hydrogen (with carbon capture) and green hydrogen (produced via electrolysis using renewable energy), is critical to making hydrogen a truly sustainable solution [3].

A key challenge in industrial-scale hydrogen production is scalability, as existing technologies are not yet deployed at the level required to support large-scale industrial demand. Hydrogen production methods such as electrolysis, steam methane reforming (SMR) with carbon capture [4], and thermochemical processes need to be significantly scaled up to meet the energy requirements of heavy industries. For example, in the steel industry, hydrogen is being explored as a replacement for coke

Next-Generation Hydrogen Economy: Production, Storage, and Fuel Cell Technologies, First Edition.
Peeyush Phogat, Shreya Sharma and Satyam Rawat.

in the direct reduction of iron (DRI) process [5], enabling low-carbon steel production. However, scaling up hydrogen-based steelmaking requires massive investments in electrolyzer capacity, renewable energy generation, and hydrogen transportation infrastructure. Another major consideration in industrial hydrogen adoption is cost-effectiveness. Today, green hydrogen costs around $4–6 per kilogram, significantly higher than gray hydrogen ($1.5–2 per kilogram) [6]. This cost disparity limits green hydrogen's competitiveness, making it challenging for industries to justify the switch from cheaper fossil fuel-based alternatives. Electrolyzer efficiency improvements, economies of scale, and government subsidies will be crucial in driving down the cost of green hydrogen and making it financially viable for large-scale applications.

Beyond production challenges, hydrogen infrastructure readiness is a key factor in determining how efficiently hydrogen can be integrated into industrial processes. Unlike natural gas, which has an established global pipeline network, hydrogen infrastructure is still in its early stages of development. Hydrogen requires specialized pipelines, high-pressure storage solutions, and refueling stations, all of which need significant investments to support large-scale adoption. Currently, hydrogen transport is primarily done via compressed gas or liquid hydrogen tankers, both of which have high logistical costs and energy losses [7]. The development of hydrogen carrier technologies, such as ammonia and liquid organic hydrogen carriers (LOHCs), offers promising solutions for more efficient long-distance hydrogen transportation. Hydrogen is expected to play a transformative role in reducing industrial carbon emissions across multiple sectors [8]. The steel industry, responsible for nearly 7–9% of global CO_2 emissions, is already adopting hydrogen-based direct reduction processes to replace coal and coke. In the cement industry, hydrogen is being tested as an alternative high-temperature fuel to reduce reliance on fossil fuels [9]. Similarly, the chemical sector, which heavily depends on hydrogen for ammonia and methanol production, is shifting toward green hydrogen-based processes to cut down its carbon footprint. The power generation sector is also exploring hydrogen co-firing in gas turbines, while the transport industry is investing in fuel cell technologies for heavy-duty applications [10]. Figure 10.1 illustrates hydrogen's role in decarbonizing key industrial sectors, highlighting its potential applications in steel, cement, chemicals, power, and transport.

To fully realize hydrogen's industrial potential, policymakers and industry leaders must address several critical challenges. First, hydrogen production capacity must be rapidly scaled up, particularly in regions where renewable energy resources can support cost-competitive green hydrogen production [11]. Second, supply chain bottlenecks related to storage, distribution, and refueling infrastructure need to be resolved. Large-scale hydrogen hubs, where production, storage, and industrial end-use applications are co-located, could offer a solution to logistical challenges [12]. Third, governments must provide financial incentives such as tax credits, carbon pricing mechanisms, and direct subsidies to encourage industries to transition to low-carbon hydrogen [2]. The future of industrial hydrogen adoption will depend on how effectively technological advancements, policy frameworks, and private sector investments come together to overcome economic and logistical barriers. Companies in steel, cement, and chemical manufacturing are already piloting

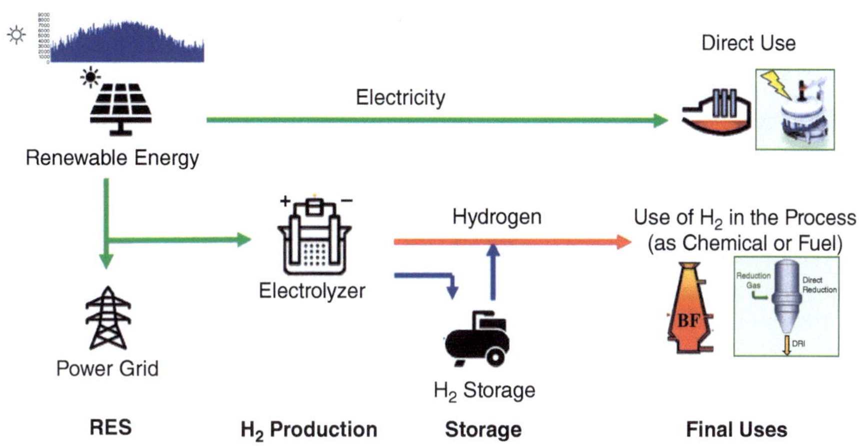

Figure 10.1 Hydrogen's role in industrial decarbonization across key sectors (steel, cement, chemicals, power, and transport). (Placed here to illustrate how hydrogen is being integrated into various industries to reduce carbon emissions.) [14]/MDPI/CC BY 4.0.

hydrogen-based solutions, but scaling these projects to full industrial capacity will require multi-billion-dollar investments and global coordination. By addressing the challenges of cost, scalability, and infrastructure readiness, hydrogen can become a cornerstone of the industrial energy transition, driving the world toward a cleaner and more sustainable future [13].

10.2 Challenges in Large-scale Hydrogen Production

10.2.1 Economic and Technical Barriers

The industrial-scale production of hydrogen presents significant economic and technical challenges that must be addressed to ensure its viability as a mainstream energy carrier. While hydrogen has the potential to decarbonize key industrial sectors such as steel, cement, chemicals, and power generation, its widespread adoption is hindered by efficiency concerns, high production costs, and infrastructure limitations. The three primary hydrogen production methods – electrolysis, SMR with carbon capture and storage (CCS), and thermochemical processes – each have unique challenges that impact their economic feasibility and scalability [15].

One of the most pressing concerns is the efficiency of hydrogen production technologies. Electrolysis, which is considered the cleanest method for producing green hydrogen, typically achieves 50–70% efficiency, meaning a substantial portion of the input energy is lost during conversion. This inefficiency raises concerns, particularly when renewable electricity is used, as it requires larger energy inputs to produce the same amount of hydrogen compared to fossil fuel-based methods. SMR with CCS. Thermochemical hydrogen production, which includes high-temperature chemical cycles such as the sulfur-iodine cycle, can achieve higher efficiencies, but these processes remain largely experimental and require significant heat inputs from nuclear or concentrated solar power, adding to operational complexity [16]. Another major challenge in industrial-scale hydrogen production is the high capital expenditure (CAPEX) and operational expenditure (OPEX) associated with building and maintaining hydrogen plants. Electrolyzer costs remain prohibitively high, with alkaline and PEM electrolyzers priced at $800–1,200 per kW and solid oxide electrolyzers (SOECs) exceeding $2,000 per kW. This significantly increases the upfront investment required for large-scale hydrogen plants. In comparison, SMR with CCS has a lower initial investment cost but requires additional infrastructure for CO_2 capture, transportation, and long-term storage, which adds to operational expenses [17]. Furthermore, thermochemical cycles and biomass gasification methods face challenges related to high-temperature reactor materials, process stability, and integration with existing industrial processes, further increasing costs.

The cost of hydrogen production is also influenced by regional energy prices and resource availability. In areas where renewable electricity is abundant, such as Australia, the Middle East, and parts of the United States, green hydrogen can be produced at a lower cost due to access to cheap solar and wind power. However, in regions where electricity prices remain high, the cost of green hydrogen remains uncompetitive with blue or gray hydrogen. The global levelized cost of hydrogen varies significantly, with gray hydrogen produced via SMR costing approximately $1.5–2.0 per kg, while blue hydrogen ranges from $2.0–3.0 per kg, and green hydrogen remains the most expensive at $4.0–6.0 per kg. These economic constraints make it difficult for industries to justify the transition to low-carbon hydrogen, particularly when cheaper fossil fuel-based alternatives are still available [18]. Another challenge in large-scale hydrogen production is infrastructure development and integration. Establishing a comprehensive hydrogen supply chain, including production plants, storage facilities, and distribution networks, requires substantial investment and coordination between governments and private sector players. The lack of existing hydrogen pipelines and storage infrastructure poses a significant barrier, as transporting hydrogen via trucks or ships remains costly and inefficient. Additionally, integrating hydrogen into industrial processes such as ammonia synthesis, steelmaking, and power generation requires modifications to existing equipment, further increasing transition costs [19]. Table 10.1 provides a comparative analysis of various hydrogen production methods at an industrial scale, focusing on CAPEX, OPEX, efficiency, and associated challenges. Gray hydrogen, derived from natural gas via SMR, offers low costs but emits significant CO_2. Blue hydrogen adds carbon capture, raising costs and storage complexities. Green hydrogen, produced via electrolysis, is clean but constrained by high electricity needs and costly electrolyzers. Thermochemical methods, though efficient, face feasibility issues due to extreme temperatures. Biological hydrogen, while sustainable, suffers from low efficiency and scalability. Each method presents trade-offs in cost, efficiency, and environmental impact.

10.2.2 Infrastructure Limitations and Energy Requirements

The large-scale adoption of hydrogen as an industrial energy carrier is heavily dependent on infrastructure readiness and energy availability. Despite the growing interest in hydrogen production, significant limitations exist in transportation,

Table 10.1 Cost comparison of different hydrogen production methods at an industrial scale.

Hydrogen production method	CAPEX ($/kW installed)	OPEX ($/kg H_2)	Efficiency (%)	Challenges
Gray hydrogen (SMR without CCS)	$500–1,000	$1.5–2.0	65–75	High CO_2 emissions, dependence on natural gas [20]
Blue hydrogen (SMR with CCS)	$1,000–1,500	$2.0–3.0	60–75	Costly CO_2 capture, storage challenges [21]
Green hydrogen (electrolysis)	$800–2,000	$4.0–6.0	50–70	High electricity demand, expensive electrolyzers [22]
Thermochemical hydrogen (sulfur-iodine cycle, etc.)	$2,500–4,000	$3.5–5.0	70–85	High process temperatures, experimental stage [23]
Biological hydrogen (microbial electrolysis, algae-based)	$2,000–3,500	$5.0–8.0	30–60	Low production rates, scalability challenges [24]

storage, and distribution networks, making industrial-scale deployment challenging. Hydrogen production processes, particularly electrolysis and thermochemical methods, are highly energy-intensive, requiring substantial electricity or heat input. Without a well-integrated renewable energy supply and a robust hydrogen distribution system, achieving widespread hydrogen commercialization remains a significant hurdle [25].

One of the biggest challenges is hydrogen's energy requirements. Electrolysis, which is the cleanest method for producing green hydrogen, demands vast amounts of electricity, typically sourced from renewable energy to maintain carbon neutrality. However, most electricity grids today are not equipped to handle the additional load required for large-scale hydrogen production. Expanding renewable energy capacity and improving grid resilience are critical to supporting a stable supply of electricity for electrolysis-based hydrogen plants. Additionally, many hydrogen hubs are located far from high-renewable-resource regions, requiring substantial grid expansions or on-site renewable installations to ensure cost-effective hydrogen generation [26].

Beyond production, hydrogen transportation and storage infrastructure also present significant challenges. Unlike natural gas, which has an extensive global pipeline network, hydrogen requires dedicated transport solutions due to its low energy density and high flammability. Currently, hydrogen is transported via high-pressure compressed gas cylinders, cryogenic liquid tanks, or chemically bound carriers such as ammonia or LOHCs. Each of these methods has technical and economic limitations, making long-distance hydrogen transport costly and inefficient. Building dedicated hydrogen pipelines, retrofitting existing gas infrastructure, and developing more efficient hydrogen carriers will be essential to reducing transportation costs

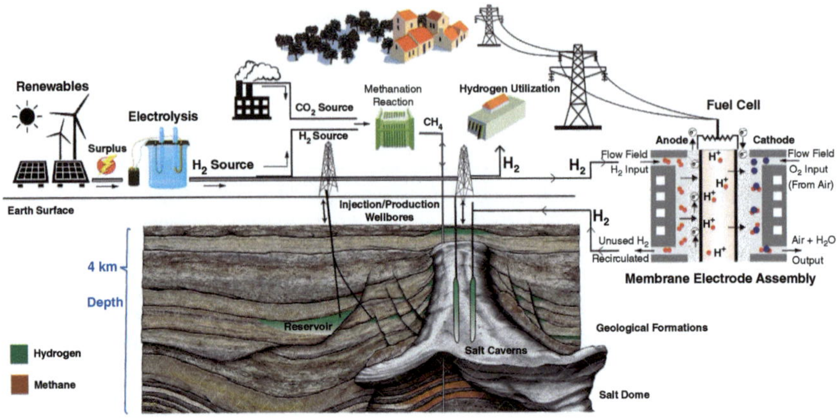

Figure 10.2 Schematic of large-scale hydrogen production and distribution infrastructure. (Placed here to illustrate hydrogen supply chains.) [28]/Springer Nature/CC BY 4.0.

and ensuring continuous hydrogen supply to industrial consumers. Another critical challenge is the need for large-scale hydrogen storage solutions. Hydrogen production does not always align with industrial demand, necessitating buffer storage to balance supply and consumption. However, current storage technologies, including compressed gas, cryogenic liquid, and metal hydrides, face capacity and efficiency constraints. Large-scale hydrogen storage requires significant advancements in underground caverns, high-density materials, and hydrogen liquefaction technologies to ensure safe and cost-effective storage solutions [27]. Figure 10.2 illustrates the schematic of large-scale hydrogen production and distribution infrastructure, depicting key elements such as electrolyzers, storage systems, transport networks, and industrial end-users.

10.3 Market Trends and Business Models for Hydrogen Commercialization

10.3.1 Investment Strategies and Funding Models

The successful commercialization of hydrogen technologies at an industrial scale relies heavily on investment strategies and sustainable financing models. Hydrogen projects require substantial capital investments for production facilities, transportation networks, and storage infrastructure, making it essential to secure long-term financial backing. Given the high upfront costs associated with electrolysis plants, carbon capture systems, and hydrogen distribution networks, governments and private investors are exploring various funding mechanisms to accelerate hydrogen adoption. One of the most effective financing models for large-scale hydrogen projects is public-private partnerships (PPPs) [29]. These partnerships involve collaboration between governments, energy companies, and private investors, ensuring that risks and capital expenditures are shared. Governments play a crucial role in providing financial incentives, tax credits, and research grants to lower the economic burden on private enterprises. For example, the European Union's Important Projects of Common European Interest (IPCEI) program has allocated billions of euros to fund green hydrogen initiatives across Europe. Similarly, the US Department of Energy's Hydrogen Earthshot Initiative is investing in hydrogen research and infrastructure to reduce hydrogen production costs. In Asia, Japan and South Korea have launched national hydrogen strategies that involve government-backed subsidies and industry collaborations to establish hydrogen supply chains [30].

Beyond government support, hydrogen is also gaining traction as a tradable commodity in global energy markets. The development of hydrogen trading hubs and price indices is paving the way for hydrogen to be bought and sold like oil and natural gas. Countries such as Germany, Australia, and the UAE are investing in hydrogen export infrastructure, positioning themselves as key players in the global hydrogen trade. Hydrogen price forecasting is becoming increasingly important as industries seek long-term purchasing agreements (PPAs) and investment security. Organizations like S&P Global and BloombergNEF are tracking hydrogen prices, enabling investors to analyze market trends and pricing dynamics [31]. Table 10.2

Table 10.2 Global investment trends and hydrogen project financing models.

Investment strategy	Key players	Description	Examples
PPPs	Governments, private companies, energy firms	Shared investment in large-scale hydrogen projects, with government subsidies and private sector expertise	EU IPCEI Hydrogen Fund, Japan's hydrogen strategy [32]
Government subsidies & tax credits	National governments, energy agencies	Direct financial support for hydrogen production and infrastructure development	US DOE Hydrogen Earthshot, Germany's H2Global Initiative [33]
Green bonds & sustainable finance	Investment banks, climate funds	Issuance of green bonds to finance hydrogen projects aligned with sustainability goals	World Bank Green Hydrogen Fund, European Green Bonds [34]
Hydrogen trading markets	Energy exchanges, commodity traders	Development of hydrogen as a tradable commodity with price forecasting and long-term contracts	Hydrogen exchange in Rotterdam, Australia-Japan hydrogen supply chain [35]
Private investment & venture capital	Investment firms, startups	Funding for hydrogen startups, innovative technologies, and new market entrants	Breakthrough energy ventures, HydrogenOne Capital [2]

provides an overview of global investment trends and financing models for hydrogen projects, highlighting various funding strategies, key players, and financial support mechanisms.

10.3.2 Supply Chain and Logistics Challenges

The successful commercialization of hydrogen as a global energy carrier depends on the scaling up of distribution networks, efficient storage solutions, and cost-effective shipping methods. While hydrogen production technologies are advancing rapidly, the supply chain infrastructure required to transport hydrogen from production sites to end-users remains a significant challenge. Unlike conventional fuels such as natural gas or gasoline, hydrogen has low volumetric energy density, making transport, storage, and delivery more complex and expensive [36, 37].

One of the most pressing issues in hydrogen logistics is the development of dedicated hydrogen pipelines. Currently, most hydrogen is transported using high-pressure gas cylinders, liquid hydrogen tankers, or chemical carriers such as ammonia and LOHCs. While these methods enable short- to medium-distance transport, they are not cost-effective for large-scale, long-distance hydrogen distribution. Developing hydrogen pipelines similar to natural gas networks would provide a cost-efficient and continuous supply of hydrogen to industrial hubs. However, existing natural gas pipelines cannot be directly used for pure hydrogen transport due to issues such as hydrogen embrittlement, leakage risks, and pressure constraints. To overcome this, countries like Germany and the Netherlands are retrofitting natural gas pipelines with hydrogen-compatible materials, while others, including Australia and Japan, are investing in new hydrogen pipeline projects [38].

Another challenge is hydrogen storage at large industrial scales. Since hydrogen has a very low boiling point (−252.87 °C), cryogenic storage is required for liquid hydrogen, which is energy-intensive and costly. In comparison, compressed hydrogen storage in underground salt caverns and high-pressure tanks is more efficient but requires geographical suitability and extensive safety measures. Research is also being conducted on solid-state hydrogen storage technologies, such as metal hydrides and porous materials (e.g., MOFs and carbon nanostructures), to enhance storage density and reduce energy losses. Shipping hydrogen across international markets presents additional logistical barriers. Liquid hydrogen shipping is energy-intensive, requiring liquefaction at cryogenic temperatures followed by re-gasification upon arrival, which results in high energy losses (~30–40%). A promising alternative is the conversion of hydrogen into ammonia, which can be transported using existing ammonia infrastructure and decomposed back into hydrogen at the destination [39]. Countries such as Japan, South Korea, and Germany are already testing ammonia-based hydrogen supply chains as part of their hydrogen import strategies [40]. Figure 10.3 illustrates the hydrogen supply chain from production to end-user applications, visualizing logistical challenges and potential solutions across production, storage, transportation, and industrial utilization.

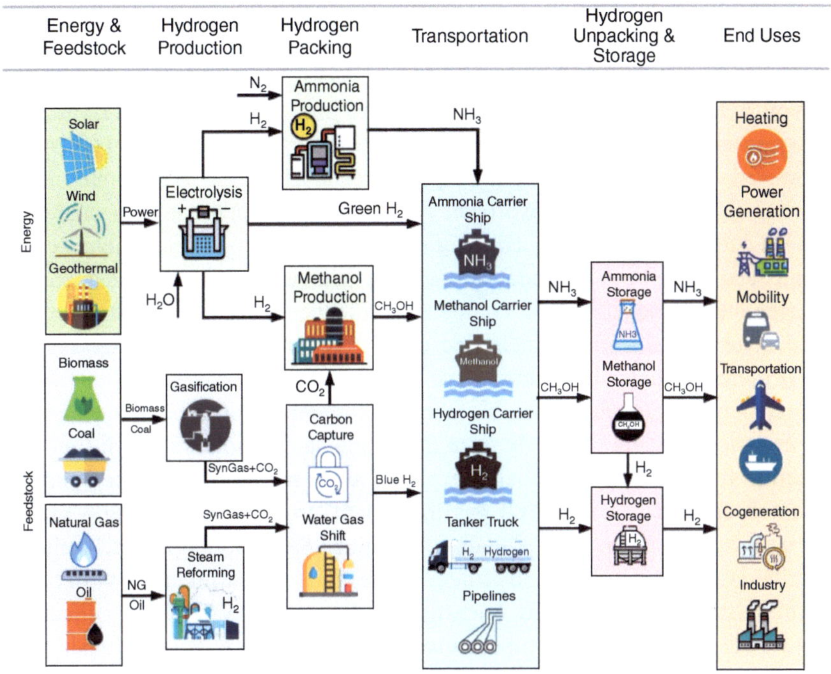

Figure 10.3 Hydrogen supply chain from production to end-user applications. [41]/MDPI/ CC BY 4.0.

10.4 Case Studies of Successful Industrial Hydrogen Projects

10.4.1 Hydrogen in Steel and Cement Industries

The steel and cement industries are among the most carbon-intensive sectors, collectively accounting for nearly 15% of global CO_2 emissions. Given their dependence on high-temperature processes and fossil fuel inputs, these industries face significant challenges in decarbonization efforts. Hydrogen has emerged as a promising alternative to reduce reliance on coal and natural gas, offering a low-carbon pathway for steel and cement production. By integrating green hydrogen into these industries, companies can significantly cut emissions and transition toward sustainable manufacturing practices.

The steel industry is currently one of the largest industrial consumers of hydrogen, primarily using it in hydrogen-based DRI. Traditionally, iron ore is processed in blast furnaces using coke (a coal derivative), which generates substantial CO_2 emissions. The hydrogen-based DRI method replaces coke with pure hydrogen, allowing iron ore to be reduced into metallic iron without carbon emissions. Instead of CO_2, the only byproduct of this reaction is water vapor, making it a cleaner alternative to conventional steelmaking. The transition to hydrogen-based steel production is already underway in several regions. European steel manufacturers such as SSAB, ArcelorMittal, and Thyssenkrupp have initiated pilot projects to produce fossil-free steel using green hydrogen. For instance, the HYBRIT project in Sweden aims to develop a completely hydrogen-powered steelmaking process, with plans to phase out coal-based blast furnaces by 2030. Similarly, Germany's Salzgitter AG and ArcelorMittal are investing in hydrogen-based DRI plants to scale up the production of low-carbon steel [42].

However, despite its potential, hydrogen-based steelmaking faces economic and technical challenges. Green hydrogen remains more expensive than coal-based methods due to high electricity costs for electrolysis. Scaling up electrolyzer capacity and securing affordable renewable energy sources will be critical to making hydrogen steel production cost-competitive. Furthermore, the industry requires infrastructure modifications, including new hydrogen pipelines, storage systems, and DRI reactors, to facilitate large-scale adoption.

The cement industry is responsible for approximately 8% of global CO_2 emissions, with emissions arising primarily from limestone calcination and fossil fuel combustion in kilns. Hydrogen is being explored as an alternative high-temperature fuel to replace coal and natural gas in cement production, offering a cleaner energy source for kilns and clinker production. Several companies, including HeidelbergCement, Holcim, and Lafarge, have started experimenting with hydrogen combustion in cement kilns. Early trials indicate that hydrogen can effectively generate the necessary high temperatures, although its lower energy density compared to coal presents efficiency challenges. Additionally, research is underway to use hydrogen plasma technology to reduce emissions from limestone processing, further enhancing the sustainability of cement manufacturing. The biggest challenge in integrating hydrogen into cement production lies in fuel availability and cost. Unlike steel, where hydrogen plays a direct role in chemical reduction, its role in cement production is primarily as a heat source, meaning cost considerations will determine how widely it is adopted. Scaling up renewable hydrogen production and establishing supply chains will be essential for hydrogen to become a viable alternative in cement manufacturing [43].

By adopting hydrogen-based processes, the steel and cement industries can significantly reduce their carbon footprints, contributing to global net-zero targets. However, for this transition to be economically viable, technological improvements, cost reductions, and supportive policy frameworks will be necessary to facilitate large-scale hydrogen adoption in heavy industries.

10.4.2 Hydrogen in Ammonia and Power Generation

Hydrogen is increasingly being recognized as a crucial feedstock for industrial applications, with ammonia production and power generation emerging as key areas where hydrogen can significantly contribute to decarbonization efforts. Green hydrogen, produced via electrolysis powered by renewable energy, offers a sustainable alternative to traditional fossil-fuel-based ammonia synthesis and can also be used in hydrogen co-firing for electricity generation. These applications not only reduce carbon emissions but also enhance energy security and fuel diversification.

Ammonia (NH_3) is one of the largest industrial applications of hydrogen, primarily used in fertilizer production, chemical synthesis, and emerging fuel technologies. Conventionally, ammonia is produced using the Haber-Bosch process, which relies on gray hydrogen derived from natural gas via SMR. This process is highly carbon-intensive, emitting approximately two metric tons of CO_2 per ton of ammonia produced. The transition to green ammonia, produced using electrolytic hydrogen instead of fossil-fuel-derived hydrogen, is seen as a critical step toward decarbonizing the chemical industry. Several large-scale green ammonia projects are already under development. For instance, Saudi Arabia's NEOM project is constructing a gigawatt-scale green ammonia plant powered entirely by solar and wind energy. Similarly, Australia, Japan, and Europe are investing in green ammonia production hubs to reduce dependence on fossil-fuel-based fertilizers and enable carbon-neutral ammonia exports. One of the major advantages of ammonia is its ability to serve as an effective hydrogen carrier, enabling the storage and transportation of hydrogen in a more energy-dense liquid form. Since ammonia is already widely traded and transported globally using established infrastructure, it presents a viable option for hydrogen supply chains [44].

Beyond fertilizers, green ammonia is gaining attention as a potential fuel for power generation and maritime applications. Several research initiatives and pilot projects are investigating direct ammonia combustion in gas turbines and marine engines. Companies like MAN Energy Solutions and Wärtsilä are developing ammonia-powered marine engines, while Japan's JERA and IHI Corporation are conducting trials on ammonia co-firing in thermal power plants to reduce reliance on coal. Despite its promise, challenges such as NO_x emissions during ammonia combustion and the need for advanced catalytic converters must be addressed before ammonia can become a mainstream low-carbon fuel alternative [45].

Power generation is another sector where hydrogen can play a significant role in reducing carbon emissions, particularly through hydrogen co-firing in gas turbines and combined-cycle power plants. Many existing power plants, especially those relying on natural gas, can be retrofitted to operate on a blend of hydrogen and methane, gradually transitioning to 100% hydrogen combustion as supply chains mature. A notable case study is the Hydrogen to Gas (H2G) project in the Netherlands, where hydrogen is being blended with natural gas in a Siemens SGT-600 gas turbine. The project demonstrated that up to 30% hydrogen blending could be achieved without significant modifications to existing infrastructure. Similarly, Japan's Fukushima Hydrogen Energy Research Field (FH2R) is exploring hydrogen co-firing in gas turbines as part of its broader strategy to create a hydrogen-powered energy system [46].

Hydrogen co-firing offers several benefits, including:

- Immediate emission reductions: Even at 20–30% hydrogen blending ratios, CO_2 emissions can be significantly reduced compared to pure natural gas combustion.
- Grid stability and dispatchability: Hydrogen-fueled gas turbines provide a reliable backup power source for intermittent renewable energy sources like solar and wind.
- Fuel flexibility: Many modern gas turbines can be retrofitted to accommodate hydrogen blends, reducing the need for entirely new infrastructure investments.

Despite these advantages, several challenges remain. Hydrogen has different combustion properties than natural gas, including a higher flame speed and lower ignition energy, requiring modifications to burner designs and flame stabilization techniques. Additionally, hydrogen storage and pipeline infrastructure must be expanded to support large-scale adoption. Projects such as GE's hydrogen-ready turbines and Mitsubishi Power's 100% hydrogen-capable gas turbines are actively addressing these challenges, demonstrating the feasibility of hydrogen-based power generation. The integration of green ammonia and hydrogen into power generation represents a transformative step toward decarbonization, offering sustainable solutions for both the chemical and energy industries. While technical and economic hurdles remain, continued innovation, policy support, and large-scale investments are expected to drive widespread adoption in the coming decades [47].

10.5 Future Roadmap for Hydrogen Adoption

The large-scale adoption of hydrogen as a clean energy solution requires a well-coordinated approach between industry leaders, governments, and international organizations. While technological advancements in hydrogen production, storage, and transportation are accelerating, strategic partnerships, cross-border collaborations, and supportive policy frameworks will be essential to drive commercialization and cost reductions.

Multinational corporations, energy firms, and industrial manufacturers are playing a crucial role in scaling up hydrogen commercialization. Companies such as Siemens, Air Liquide, Linde, Shell, BP, and Toyota are investing heavily in green hydrogen projects, electrolyzer manufacturing, and hydrogen-powered mobility solutions. These industry leaders are not only funding pilot projects but also forming global alliances to facilitate knowledge-sharing, infrastructure development, and policy advocacy. A key trend in hydrogen commercialization is the rise of cross-border hydrogen trade agreements. Countries with abundant renewable energy resources, such as Australia, Chile, the Middle East, and North Africa, are positioning themselves as hydrogen exporters, while energy-importing nations such as Japan, Germany, and South Korea are investing in hydrogen import infrastructure. Agreements such as the Japan-Australia Hydrogen Energy Supply Chain (HESC) and the Germany-Namibia green hydrogen partnership highlight the increasing globalization of hydrogen markets. The establishment of hydrogen trading hubs in Rotterdam, Singapore, and Abu Dhabi is expected to accelerate hydrogen commoditization, making it more accessible for industries worldwide. However, building an international hydrogen supply chain presents logistical challenges, including standardizing hydrogen transportation methods, ensuring cost-effective liquefaction and conversion technologies, and developing safety regulations for cross-border hydrogen shipments. Addressing these challenges will require intergovernmental coordination, financial investments in export terminals, and the deployment of hydrogen carriers such as ammonia and LOHCs.

Governments worldwide are playing a pivotal role in accelerating hydrogen deployment through subsidies, tax incentives, and carbon pricing mechanisms.

Many nations have introduced national hydrogen strategies that outline targets for production capacity, infrastructure development, and industrial applications. The European Union's Hydrogen Strategy aims to install 40 GW of electrolyzer capacity by 2030 and establish a hydrogen pipeline network across Europe. The United States' Inflation Reduction Act (IRA) and Hydrogen Production Tax Credit (45V) provide significant incentives for green hydrogen producers, making the cost of green hydrogen competitive with fossil fuels. Japan and South Korea have established hydrogen roadmap initiatives, promoting fuel cell vehicle adoption, hydrogen refueling stations, and industrial hydrogen applications. China and India are rapidly expanding their hydrogen pilot projects, with significant investments in electrolysis, hydrogen refueling stations, and fuel cell technologies. In addition to national efforts, global organizations such as the International Energy Agency (IEA), the United Nations (UN), and the World Economic Forum (WEF) are actively supporting hydrogen policy development. The IEA's Global Hydrogen Review provides market insights and policy recommendations, while the United Nations Industrial Development Organization (UNIDO) is promoting hydrogen adoption in emerging economies.

The future of hydrogen adoption will depend on industrial-scale deployment, international trade agreements, and strong policy backing. By fostering global collaboration, standardizing hydrogen infrastructure, and implementing financial incentives, governments and corporations can accelerate the transition to a hydrogen-powered future, ensuring a sustainable and carbon-free economy.

References

1 López CR, Pérez Garcés LF, Tepordei D, Moreno SP, Šljukić B, Santos DMF. Challenges in scaling low-carbon hydrogen production in Europe. *Solar-drlven green hydrogen generation and storage*. 2023 Jan 1;287–319. Available from: https://doi.org/10.1016/B978-0-323-99580-1.00022-4.

2 Harichandan S, Kar SK. Financing the hydrogen industry: Exploring demand and supply chain dynamics. *Environ Sci Pollut Res* [Internet]. 2023 Oct 9 [cited 2025 Mar 31];1:1–18. Available from: https://link.springer.com/article/10.1007/s11356-023-30262-9.

3 Quitzow R, Nunez A, Marian A. Positioning Germany in an international hydrogen economy: A policy review. *Energy Strategy Rev*. 2024 May 1;53:101361.

4 Cho HH, Strezov V, Evans TJ. Environmental impact assessment of hydrogen production via steam methane reforming based on emissions data. *Energy Rep*. 2022 Nov 1;8:13585–95.

5 Ji Y, Chi Z, Yuan S, Chen Y, Li Y, Jiang T, et al. Development and application of hydrogen-based direct reduction iron process. *Processes* [Internet]. 2024 Aug 28 [cited 2025 Mar 31];12(9):1829. Available from: https://www.mdpi.com/2227-9717/12/9/1829/htm.

6 Liu W, Wan Y, Xiong Y, Gao P. Green hydrogen standard in China: Standard and evaluation of low-carbon hydrogen, clean hydrogen, and renewable hydrogen. *Int J Hydrogen Energy*. 2022 Jul 8;47(58):24584–91.

7 Roy A, Pramanik S. A review of the hydrogen fuel path to emission reduction in the surface transport industry. *Int J Hydrogen Energy*. 2024 Jan 2;49:792–821.

8 Dickson R, Akhtar MS, Abbas A, Park ED, Liu J. Global transportation of green hydrogen via liquid carriers: Economic and environmental sustainability analysis, policy implications, and future directions. *Green Chemistry* [Internet]. 2022 Oct 31 [cited 2025 Mar 31];24(21):8484–93. Available from: https://pubs.rsc.org/en/content/articlehtml/2022/gc/d2gc02079c.

9 Zhang X, Jiao K, Zhang J, Guo Z. A review on low carbon emissions projects of steel industry in the World. *J Clean Prod*. 2021 Jul 15;306:127259.

10 Wang RR, Zhao YQ, Babich A, Senk D, Fan XY. Hydrogen direct reduction (H-DR) in steel industry—An overview of challenges and opportunities. *J Clean Prod*. 2021 Dec 20;329:129797.

11 Wappler M, Unguder D, Lu X, Ohlmeyer H, Teschke H, Lueke W. Building the green hydrogen market—Current state and outlook on green hydrogen demand and electrolyzer manufacturing. *Int J Hydrogen Energy*. 2022 Sep 15; 47(79):33551–70.

12 Kim C, Cho SH, Cho SM, Na Y, Kim S, Kim DK. Review of hydrogen infrastructure: The current status and roll-out strategy. *Int J Hydrogen Energy*. 2023 Jan 15;48(5):1701–16.

13 Wei S, Sacchi R, Tukker A, Suh S, Steubing B. Future environmental impacts of global hydrogen production. *Energy Environ Sci* [Internet]. 2024 Mar 19 [cited 2025 Mar 31];17(6):2157–72. Available from: https://pubs.rsc.org/en/content/articlehtml/2024/ee/d3ee03875k.

14 Franco A, Rocca M. Renewable electricity and green hydrogen integration for decarbonization of "Hard-to-Abate" industrial sectors. *Electricity* [Internet]. 2024 Jul 25 [cited 2025 Mar 31];5(3):471–90. Available from: https://www.mdpi.com/2673-4826/5/3/24/htm.

15 Olabi AG, bahri A saleh, Abdelghafar AA, Baroutaji A, Sayed ET, Alami AH, et al. Large-scale hydrogen production and storage technologies: Current status and future directions. *Int J Hydrogen Energy*. 2021 Jul 1;46(45):23498–528.

16 Younas M, Shafique S, Hafeez A, Javed F, Rehman F. An overview of hydrogen production: Current status, potential, and challenges. *Fuel*. 2022 May 15; 316:123317.

17 G. GS, Qian H. Developing and validating comprehensive indicators to evaluate the economic efficiency of hydrogen energy investments. *Operational research in engineering sciences: Theory and applications* [Internet]. 2024 Sep 30 [cited 2025 Mar 31];7(3):2620–1747. Available from: http://oresta.org/menu-script/index.php/oresta/article/view/801.

18 Zhang Z, Liu G, Lu X. Supply scale, carbon footprint, and levelized cost assessment of hydrogen production technologies during carbon neutrality transition in China. *Energy Strategy Rev*. 2024 Jul 1;54:101429.

19 Kountouris I, Bramstoft R, Madsen T, Gea-Bermúdez J, Münster M, Keles D. A unified European hydrogen infrastructure planning to support the rapid scale-up of hydrogen production. *Nat Commun* 2024 15: 1 [Internet]. 2024 Jun 29 [cited 2025 Mar 31];15(1):1–13. Available from: https://www.nature.com/articles/s41467-024-49867-w.

20 Rohilla T, Kumar M. Challenges with sustainable green hydrogen production: Role of materials, design, multi-scale modeling, and up-scaling. *Energy Environ Sustain* [Internet]. 2024 [cited 2025 Mar 31];Part F2879:455–90. Available from: https://link.springer.com/chapter/10.1007/978-981-97-1339-4_19.

21 Wu W, Zhai H, Holubnyak E. Technological evolution of large-scale blue hydrogen production toward the U.S. hydrogen energy earthshot. *Nat Commun* 2024 15:1 [Internet]. 2024 Jul 6 [cited 2025 Mar 31];15(1):1–14. Available from: https://www.nature.com/articles/s41467-024-50090-w.

22 Amini Horri B, Ozcan H. Green hydrogen production by water electrolysis: Current status and challenges. *Curr Opin Green Sustain Chem*. 2024 Jun 1; 47:100932.

23 Mehr AS, Phillips AD, Brandon MP, Pryce MT, Carton JG. Recent challenges and development of technical and technoeconomic aspects for hydrogen storage, insights at different scales; A state of art review. *Int J Hydrogen Energy*. 2024 Jun 12;70:786–815.

24 Lee HS, Vermaas WFJ, Rittmann BE. Biological hydrogen production: Prospects and challenges. *Trends Biotechnol* [Internet]. 2010 May 1 [cited 2025 Mar 31]; 28(5):262–71. Available from: https://www.cell.com/action/showFullText?pii=S0167779910000193.

25 Kuterbekov KA, Kabyshev A, Bekmyrza K, Kubenova M, Kabdrakhimova G, Ayalew AT. Innovative approaches to scaling up hydrogen production and storage for renewable energy integration. *Int J Low-Carbon Technol* [Internet]. 2024 Jan 18 [cited 2025 Mar 31];19:2234–48. Available from: https://dx.doi.org/10.1093/ijlct/ctae176.

26 Rolo I, Costa VAF, Brito FP. Hydrogen-Based energy systems: Current technology development status, opportunities and challenges. *Energies* [Internet]. 2023 Dec 28 [cited 2025 Mar 31];17(1):180. Available from: https://www.mdpi.com/1996-1073/17/1/180/htm.

27 Patel SKS, Gupta RK, Rohit MV, Lee JK. Recent developments in hydrogen production, storage, and transportation: Challenges, opportunities, and perspectives. *Fire* [Internet]. 2024 Jul 3 [cited 2025 Mar 31];7(7):233. Available from: https://www.mdpi.com/2571-6255/7/7/233/htm.

28 Osman AI, Mehta N, Elgarahy AM, Hefny M, Al-Hinai A, Al-Muhtaseb AH, et al. Hydrogen production, storage, utilisation and environmental impacts: A review. *Environ Chem Lett* 2021 20:1 [Internet]. 2021 Oct 6 [cited 2025 Mar 30];20(1): 153–88. Available from: https://link.springer.com/article/10.1007/s10311-021-01322-8.

29 Arensbach SD. Re-designing public–private partnerships: Case study—Green Hydrogen Hub, Denmark. *Journal of Design, Business and Society* [Internet]. 2023 Oct 9 [cited 2025 Apr 1];9(2):193–211. Available from: https://intellectdiscover.com/content/journals/10.1386/dbs_00053_1.

30 Vezzoni R. A “return of the state” in energy transitions? The making of a hydrogen economy in the European Union. *ZFW Adv Econ Geogr* [Internet]. 2024 Nov 1 [cited 2025 Apr 1];68(3–4):195–212. Available from: https://www.degruyter.com/document/doi/10.1515/zfw-2024-0050/html.

31 Zhao T, Liu Z, Jamasb T. A business model design for hydrogen refueling stations: A multi-level game approach. *Int J Hydrogen Energy*. 2024 Jan 2;52:577–88.

32 Upham P, Cruells Maristany MP. The global yet local nature of energy imaginaries: The cases of Dutch and Spanish hydrogen valleys. *Energy Res Soc Sci*. 2025 May 1;123:104028.

33 Curcio E. Techno-Economic analysis of hydrogen production: Costs, policies, and scalability in the transition to net-zero. 2025 Feb 17 [cited 2025 Apr 1]. Available from: https://arxiv.org/abs/2502.12211v1.

34 Zubairu N, Al Jabri L, Rejeb A. A review of hydrogen energy in renewable energy supply chain finance. *Discov Sustain* 2025 6:1 [Internet]. 2025 Mar 20 [cited 2025 Apr 1];6(1):1–21. Available from: https://link.springer.com/article/10.1007/s43621-025-01007-0.

35 Forsberg CW. Future hydrogen markets for large-scale hydrogen production systems. *Int J Hydrogen Energy*. 2007 Mar 1;32(4):431–9.

36 Klumpp M, Meisel F, Thies C, Buscher U. A multi-perspective framework to address manufacturing and transportation challenges in green hydrogen supply chains. *Procedia Comput Sci*. 2025 Jan 1;253:673–82.

37 Ratnakar RR, Gupta N, Zhang K, van Doorne C, Fesmire J, Dindoruk B, et al. Hydrogen supply chain and challenges in large-scale LH2 storage and transportation. *Int J Hydrogen Energy*. 2021 Jul 9;46(47):24149–68.

38 Agnolucci P, Akgul O, McDowall W, Papageorgiou LG. The importance of economies of scale, transport costs and demand patterns in optimising hydrogen fuelling infrastructure: An exploration with SHIPMod (Spatial hydrogen infrastructure planning model). *Int J Hydrogen Energy*. 2013 Aug 30;38(26):11189–201.

39 Zoback M, Smit D. Meeting the challenges of large-scale carbon storage and hydrogen production. *Proceedings of the National Academy of Sciences* [Internet]. 2023 Mar 14 [cited 2025 Apr 1];120(11):e2202397120. Available from: https://www.pnas.org/doi/abs/10.1073/pnas.2202397120.

40 Simon J, Ferriz AM, Correas LC. HyUnder – Hydrogen underground storage at large scale: Case study Spain. *Energy Procedia*. 2015 Jun 1;73:136–44.

41 Salimi M, Hosseinpour M, Borhani TN. The role of clean hydrogen value chain in a successful energy transition of Japan. *Energies* [Internet]. 2022 Aug 21 [cited 2025 Mar 29];15(16):6064. Available from: https://www.mdpi.com/1996-1073/15/16/6064/htm.

42 Liu W, Zuo H, Wang J, Xue Q, Ren B, Yang F. The production and application of hydrogen in steel industry. *Int J Hydrogen Energy*. 2021 Mar 8;46(17):10548–69.

43 El-Emam RS, Gabriel KS. Synergizing hydrogen and cement industries for Canada's climate plan—Case study. *Energy Sources Part A Recovery Util Environ Effects* [Internet]. 2021 Dec 2 [cited 2025 Apr 1];43(23):3151–65. Available from: https://www.tandfonline.com/doi/abs/10.1080/15567036.2021.1936699.

44 Cha J, Park Y, Brigljević B, Lee B, Lim D, Lee T. An efficient process for sustainable and scalable hydrogen production from green ammonia. *Renew Sustain Energy Rev*. 2021 Dec 1;152:111562.

45 Makhloufi C, Kezibri N. Large-scale decomposition of green ammonia for pure hydrogen production. *Int J Hydrogen Energy*. 2021 Oct 11;46(70):34777–87.

46 Gabriel KS, El-Emam RS, Zamfirescu C. Technoeconomics of large-scale clean hydrogen production—A review. *Int J Hydrogen Energy*. 2022 Aug 22; 47(72):30788–98.

47 Rostrup-Nielsen JR, Rostrup-Nielsen T. Large-scale hydrogen production. *CATTECH* [Internet]. 2002 Sep [cited 2025 Apr 1];6(4):150–9. Available from: https://link.springer.com/article/10.1023/A:1020163012266.

Index

Note: Page numbers in *italics* and **bold** refers to figures and tables respectively.

Next-Generation Hydrogen Economy: Production, Storage, and Fuel Cell Technologies, First Edition.
Peeyush Phogat, Shreya Sharma and Satyam Rawat.

d

e